18 Aug 2011

W. G. T.

Robust Control
of
Linear Dynamical Systems

To the memory of my father and mother

P.C. RAMAN

DAMAYANTI RAMAN

ROBUST CONTROL OF LINEAR DYNAMICAL SYSTEMS

P.C. Chandrasekharan
Anna University
Madras, India

ACADEMIC PRESS
Harcourt Brace & Company, Publishers
London Boston San Diego New York
Sydney Tokyo Toronto

ACADEMIC PRESS LIMITED
24–28 Oval Road
LONDON NW1 7DX

U.S. Edition Published by
ACADEMIC PRESS INC.
San Diego, CA 92101

This book is printed on acid free paper

A catalogue record for this book is available from the British Library

ISBN 0-12-167885-7

Typeset by P&R Typesetters, Salisbury
Printed in Great Britain by Hartnolls Ltd, Bodmin, Cornwall

Contents

Preface

Purpose and Scope

Robust control may be defined as the control of uncertain plants with unknown disturbance signals and unknown dynamics making use of fixed controllers. During the past decade much progress has been made in the area of robust control of linear time invariant dynamic systems. A sucession of key concepts such as $\mathcal{H}_\infty$ optimization and Kharitonov theory developed during this period has made robust control an area rich in theory and potential applications.

Unfortunately, most of the work done in this field is scattered in many research publications and is accessible only to a select group of experts. Frequently, the original ideas and the motivation for pursuing a particular path are lost in a maze of mathematical formalism. Therefore, there is a need to bring these ideas and techniques to the attention of a wider audience so that robust controller design may be better appreciated. This has provided the prime motivation for the author in writing this book.

There are many in the control community with a fairly good grounding in control theory but whose exposure to some of the concepts explained in this book may be described as minimal. Therefore, this book attempts to guide such persons through the labyrinths of robust control theory, right up to the frontiers of this area in a step by step fashion. The objective of elucidation and simplification which the author has set for himself has its own limitations. There are certain concepts in science and engineering which can best be expressed in mathematical language and any attempt at simplification can hardly do justice to the grandeur of these concepts. The 'unreasonable effectiveness of mathematics in natural sciences' to quote the physicist and Nobel laureate Eugene Wigner is nowhere more explicitly demonstrated than in control theory. Hence one has to strike a balance between the twin 'orthogonal' objectives of mathematical rigour and readability. To facilitate smooth flow of ideas untrammelled by frequent side diversions, the author has been forced to omit proofs of certain important results, during the course of the discussion. It is hoped that the reader will be willing to accept these results on trust, secure in the knowledge that they have been rigorously proved elsewhere.

Organization of chapters

This book deals with linear time invariant continuous time systems with fixed feedback controllers which also belong to the same category. The modelling error considered takes the form of unstructured uncertainty. This is the best possible way of taking into account high frequency modelling errors. Parametric errors could also be brought within its ambit. However, the author has made an exception in the penultimate chapter where Kharitonov theory which deals with structured parametric uncertainty is considered. The reason for doing so is explained in the introduction to that chapter.

The book consists of 12 Chapters and 3 Appendices. Chapter 1 gives a brief overview of robust control theory from a historical perspective. Chapter 2 is intended to convey to the reader the rudiments of robust control theory and thus to provide a backdrop against which subsequent developments could be viewed. Some of the fundamental concepts repeatedly made use of in subsequent chapters are introduced in Chapter 3. This chapter deals with various norms associated with both signals and systems. Important topics discussed include calculation of both $\mathcal{H}_2$ and $\mathcal{H}_\infty$ norms of a given transfer matrix. Further, the reader is introduced to the concepts of all pass systems, adjoint operators and inverse systems.

Matrix Fraction Description (MFD) is an important tool in control engineers tool box. An understanding of this method is a prerequisite in controller design. Chapter 4 is devoted to a discussion of MFD. $\mathcal{H}_\infty$ optimal control applied to systems is considerably simplified by the application of Youla parametrization technique. This topic is dealt with in Chapter 5. This chapter also discusses stable co-prime factorization of transfer matrices using state space techniques. Chapter 6 is concerned with two important issues in robust control – sensitivity minimization and robust stabilization. Combining MFD and Youla parametrization concepts it is shown how the feedback stabilization problem could be reduced to a model matching problem wherein the free parameter matrix Q enters the final results in an affine manner. In this context the 2-block and 4-block problems are discussed. Using spectral factorization both these problems are reduced to a 1-block problem. This chapter concludes with the derivation of sufficient conditions for ensuring robust stability of multivariable systems subjected to both additive and multiplicative model perturbutions.

The need to approximate a given rational function within prescribed tolerance limits by another rational function of smaller degree often arises in control theory. In this context Chapter 7 deals with the twin concepts of balanced realization and Hankel norm approximation. They are made use of to solve the $\mathcal{H}_\infty$ optimization problem in state space and the algorithm developed by Glover in this connection is fully explained. In Chapter 8, both $\mathcal{H}_2$ and $\mathcal{H}_\infty$ optimization problems are considered side by side. To start with, this chapter provides a quick survey of LQG methodology with particular reference to the properties of associated algebraic Riccati equations. This is followed by the elucidation of two different approaches (as compared to the state space approach) for solving $\mathcal{H}_\infty$ optimization problems. These relate to the function-theoretic approach of Nevanlinna and Pick and the operator theoretic approach formalized by Zames, Francis and Helton. While the state space method may be computationally more effective, it lacks the structural simplicity and physical insight provided by these two methods. This is justification for their inclusion in this book.

Chapter 9 is devoted to a discussion of plant-controller configuration in a generalized setting. In this context it is shown how a stabilizing controller, designed for a subsystem of the plant using Youla parametrization can be used to stabilize the entire plant. Once again, we come across the model matching problem with the Q parameter affinely related to the relevant transfer matrix. This chapter also demonstrates how to reduce a variety of optimization problems to the so called standard form which may then be solved using known techniques. Chapter 10 reports on some of the more recent developments in $\mathcal{H}_\infty$ optimization. The control problem studied here pertains to the design of stabilizing controllers that impose an upper bound on the $\mathcal{H}_\infty$ norm of a closed loop transfer matrix using state space method. The solution to this problem as reported by Doyle, Glover, Khargonekar and Francis is discussed in detail. It is shown how the output feedback problem separates itself into two parts – the full information problem and the $\mathcal{H}_\infty$ filter. This approach is reminiscent of the seperation principle successfully used in LQG control. However, there are significant differences and these are explained in this chapter.

Chapter 11 marks a departure from earlier chapters in the sense that for the first time parametric uncertainty over a bounded set is considered. The main discussion centres around Kharitonov theorem. Some of the post-Kharitonov era developments are also discussed. An important result in this context is the so called 'edge theorem'. In Chapter 12, in the light of current research activity, predictions are made about the directions in which robust control theory is likely to move in the coming years. The three appendices at the end relate to mathematical preliminaries, control theory fundamentals and singular value decomposition. The purpose of including these topics is to collect in one place information pertaining to diverse subjects which otherwise the reader would have to seek from scattered sources. The list of references given at the end merely catalogues the books and articles consulted by the author and should by no means be construed as a detailed bibliography on the subject. A common feature of all chapters is the notes given at the end followed by exercises on relevant topics. To conclude, robust control is an area which straddles across a number of disciplines like engineering, mathematics and physics and its topicality and relevance are bound to increase in the coming years.

In this book the focus is mainly on mathematical techniques rather than on computational methods. The importance of the latter cannot be underestimated, but it is just beyond the purview of this book. For the same reason computer software packages such as MATLAB are not discussed. Another omission relates to $\mathcal{H}_\infty$ synthesis theory for discrete time systems. Here again in the interest of keeping the size of the book within limits, this topic had to be omitted.

Readership

The book is intended to be used as a text book by graduate students belonging to diverse disciplines, besides practising engineers and researchers who have had an earlier exposure to control theory and who are interested in a rigorous approach to controller design under uncertainty conditions. The mathematical pre-requisites are modest and well within the reach of a graduate student in engineering and science. The subject matter discussed in the book could be grouped together to organize two one-semester courses on robust control – one at an elementary level for senior undergraduate students and the other at an advanced level for graduate students and doctoral candidates.

Acknowledgements

This book is an outgrowth of a series of lectures delivered by the author to graduate students and researchers at Anna University during his two-year tenure as UGC Emeritus fellow in 1990–92. The author would like to gratefully acknowledge the support given by the University Grants Commission (UGC) New Delhi in this connection. The author owes a debt of gratitude to Prof. M. Anandakrishnan, Vice-chancellor, Anna University for placing at his disposal all the facilities available at the Anna University. The intellectually stimulating and relaxed atmosphere at the university helped the author a great deal in writing this book. Among the persons who have helped the author, special thanks are due to Dr R.M. Umesh, former UGC Research Associate at Anna University, for painstakingly scanning through the original manuscript and making many valuable suggestions. The author wishes to thank Mr Bal Menon and his staff at Laser Words, Madras, for the excellent work done in connection with the electronic setting of the book from the original hand-written manuscript. It was a pleasure working with the thoroughly professional editorial and production staff of Academic Press, London. I thank all of them. Last but not least, the author wishes to thank his wife Jaya for her forbearance and encouragement during the long period when he was totally engrossed in writing the book.

P.C. Chandrasekharan
Madras, India
April 1996

1

Background and Motivation

1.1 Introduction

One of the central problems in control theory relates to the design of controllers for systems with uncertain dynamics and imprecisely known parameters. In the traditional treatment of this problem, probability distributions are assigned to the various uncertainties involved and it is recast into another problem wherein the expected value of a suitable cost functional is required to be minimized. For a long time such a probabilistic approach, supported by extensive mathematical theory, was the principal method used by designers while dealing with uncertain systems. However, this method has many disadvantages. Firstly it requires the *a priori* knowledge of the probability density distributions of the quantities involved, and in most cases such information is not available. Secondly, the end result only guaranteed the expected value of the performance functional, leaving open the question regarding deviations from the optimal value, sometimes over a wide range.

It was in this context that an alternative non-probabilistic approach to dealing with uncertain systems was proposed. This has found wide acceptance among the control community in recent years. In this approach uncertainties instead of being modelled as random variables with known probability density distributions are considered as totally unknown except for the fact that they belong to a given subset of an appropriately defined vector space. The design problem is reduced to one of finding a controller within an admissible class of controllers which minimizes the maximum value of a cost functional. This of course envisages a worst case scenario and therefore results in conservative designs. However, the merit of the method lies in the fact that the performance specifications – whatever they may be – are met with a fair degree of certainty on all occasions. This may not be a high price to pay, especially in situations which demand that the specifications are strictly adhered to without exception at all times. The philosophy underlying this approach is the essence of robust control theory.

1.2 What is Robust Control?

Robust control refers to the control of uncertain plants with unknown dynamics subjected to unknown disturbance signals. The problem is to design a fixed controller which guarantees acceptable performance norms in the presence of plant and input uncertainty. The performance specification may include properties such as stability and disturbance attenuation. Right from the beginning, control engineers were aware that any design of a controller based on a fixed plant transfer function was totally unrealistic. There was always a nagging doubt about the performance specifications if the transfer function on which the design was based deviated from the assumed value over a certain range. In the case of single input single output (SISO) systems this was roughly taken care of by concepts such as gain and phase margins. However, in the multiple input multiple output (MIMO) case matters became quite complicated and an easy extension of gain and phase margins was not found to be possible. This led to new approaches and new techniques to deal with the situation. In this book, we are primarily concerned with reporting these new developments. We limit our discussion to linear time invariant systems in the case of both the plant and the controller. Further, the controller configuration once it is designed remains fixed. The area of adaptive control which involves fine tuning of controllers and possible change of configuration, does not come within the purview of this book.

1.3 Historical Perspective

The compulsions and the motivations behind some of the recent developments in robust control theory are best understood when viewed from a historical perspective. Control theory is a comparatively new discipline and was recognised as such only during the early thirties of this century. To start with, it was the seminal contributions made by Nyquist (1932) and Bode (1945) in those early days which placed this discipline on firm theoretical foundations. Then in the early forties came the outstanding contribution of Norbert Wiener (1949) whose elegant solution of the problem of filtering and prediction of stationary *ensembles* constitutes a landmark in stochastic control theory. It is a measure of Wiener's percipience that he saw both the problem and the lines on which solutions may be sought. This period also witnessed the noteworthy contributions of Kolmogrov (1941) relating to the prediction of discrete time stationary processes. In all these developments the emphasis was on frequency domain techniques. Bode and Nyquist did understand and appreciate the concept of robustness which is embodied in their definitions of phase margin and gain margin.

A change in direction in control theory development took place in the fifties with the emergence of the state space approach. The principal architect in this field was R.E. Kalman. There was a flurry of activity pertaining to the solution of optimization problems in state space culminating in the dynamic programming approach of Richard Bellman and the enunciation of the maximum principle by L.S. Pontryagin. Simultaneously, Lyapunov stability theory was resurrected and used as a powerful tool in solving non-linear stability problems. The high point of these developments was reached during the sixties with the formulation and solution of what is known as the Linear Quadratic Gaussian (LQG) problem (See Athans 1971a). This development was primarily inspired by the definitive contributions of Kalman (1960, 1963, 1964, 1965). The LQG approach provided a mathematically elegant method for designing feedback controllers of systems working in a noisy environment. The method was immediately hailed as a powerful tool for solving the regulator design problem.

However, the euphoria generated by the LQG method did not last long. It was soon realized that one of its principal drawbacks was its inability to guarantee a robust solution. Control engineers found it difficult to incorporate robustness criteria in a quadratic integral performance index used in the LQG problem.

During the early seventies an attempt was made to generalize some of the useful concepts like phase margin and gain margin so that they may be made applicable to MIMO systems. The British school took the lead in this respect, spearheaded by some outstanding contributions from Rosenbrock (1970, 1974) and Macfarlane (1979).

This brings us to the 1980s which saw the burgeoning of a number of ideas of far reaching significance as far as robust control theory is concerned. It was indeed fortunate that this was also the period which registered striking advances in micro-processor design and computer technology. As a direct result of these advances, computational methods which would have proved daunting in an earlier era were found to be practical and within reach. Some of the important developments that occurred during the 1980s were:

1. Use of singular values as a measure of gain in transformations (Doyle and Stein 1981)
2. The factorial approach in controller synthesis (Vidyasagar 1985, Callier and Desoer 1982)

3. Parameterization of stabilizing controllers (Youla *et al.* 1976a and 1976b, Desoer *et al.* 1980)
4. H_∞ optimization (Zames 1981, Zames and Francis 1983, Helton 1985, Francis 1987)
5. Robust stabilization and sensitivity minimization (Kimura 1984, Vidyasager and Kimura 1986, Francis and Zames 1984, Chang and Pearson 1984)
6. Computational aspects of H_∞ optimization
   (a) Interpolation methods based on Nevanlinna–Pick interpolation theory (Delsarte *et al.* 1981)
   (b) Hankel Norm approach (Glover 1984)
   (c) Operator–theoretic approach (Francis, Helton and Zames 1984)
   (d) State space approach using separation principle (Doyle *et al.* 1989)
7. Kharitonov theory and related approaches (Barmish 1994)

With all these developments taking place in a span of approximately ten years, it is not surprising that robust control theory gained a lot of momentum. Currently it is one of the important sub-areas of research within the broad field of system theory. In subsequent chapters of this book all the above listed topics will be presented in some detail and an attempt will be made to show how they are linked together to provide a common conceptual framework for robust control theory.

1.4 Models and Modelling

We are primarily concerned with the design of controllers. However there cannot be any design without there first being a proper model to represent the plant. Hence the first step is to find a mathematical model to describe the behaviour of the plant. This task is at least as important as the task of designing the controller. Dynamic system modelling presupposes a thorough understanding of the working of the physical system which is to be modelled. In order to arrive at an authentic model, one must have a correct appreciation of the relative importance of the parameters involved. The diversity of dynamical systems with mutual interaction between electric, mechanical and thermodynamic forces makes it difficult to expound a general methodology for modelling.

The primary difficulty with modelling is that there exists no single model which can take care of all situations. Modelling specifications are intimately linked with application requirements. For example in power systems a 50-km long transmission line may be modelled as a lumped parameter system, whereas in telecommunication circuits, transmission lines a few meters long have to be treated as a distributed parameter system. The fact that the two transmission lines work at different frequencies – one at 50–60 Hertz and the other at a frequency of the order of Megahertz – makes all the difference as far as modelling is concerned.

We cannot avoid modelling imperfections. Generally the imperfections arise because of the following factors:

1. Imperfect knowledge of system parameters. This may include identification errors and parameter variation during operation.
2. Neglected non-linearities, time delays and diffusion processes.
3. Neglect of higher order dynamics either intentionally to reduce the complexity of the model or because of ignorance of the working of the plant.

These uncertainties may be broadly classified under two categories. They are:

i) Structured uncertainties such as parameter variations in a sub-set of parameter space.
ii) Unstructured uncertainties where nothing is known about the exact nature of the uncertainties except for the fact that they are bounded.

A good model of a physical system has to reckon with both these types of uncertainty. Of particular interest to us is how to take care of neglected higher order dynamics. These effects cannot be captured by modelling methods based purely on parameter variation. After arriving at a nominal model either through experiment on the plant or through the application of physical principles, the model has to be validated using simulation techniques. The next step is the design of a controller (based on the model) which will exhibit the required robustness properties. It must be borne in mind that the controller has to work in conjunction with the plant and not the model of the plant. Because of the mismatch between the plant and the model, the response of the plant controller configuration may not be as expected. However, if the controller design is based on robustness principle, minor errors in modelling will not matter and the closed loop system would still be able to maintain prescribed performance levels.

It is generally observed that man-made systems such as power systems and communication networks are much simpler than systems which occur in the physical and biological world. There is a mistaken belief that the more complex a model is the more accurately it will represent the real system. Complexity is certainly not a substitute for good modelling. In modelling, an awareness of the different timescales inherent in the system is most important. Eigen values differing in orders of magnitude from each other lead to ill-conditioned matrices and computational instability. Combining the microcosm with the macrocosm to yield a single universal model might sound an attractive philosophical proposition, but it certainly does not work in engineering.

Depending upon system complexity, mathematical models may be represented by sets of algebraic equations in simple cases or sets of coupled non-linear partial differential equations in highly complex systems. An example of the former is a network of linear resistors. A well known example of the latter is a long-term weather forecasting model.

As already stated, the concern here is with linear, constant coefficient ordinary differential equations which govern the system. It is then possible to represent input-output behavior of systems in the s-domain via transfer functions. In this framework complex systems, e.g. aerospace systems may be represented by a few matrix valued transfer functions. Indeed a lot of coded information is stored in the coefficients of these transfer function matrices and much effort – both experimental and analytical – is bestowed in determining their values. But it is truly amazing that without solving a single differential equation, we are able to make wide ranging predictions about the behavior of systems. Furthermore, we are able to make use of the information embedded in the coefficients to synthesisze controllers in order to make the feedback system behave in a pre-ordained manner.

1.5 Model Reduction

The need to find a simple model to replace a relatively complicated one often arises in engineering. In control theory the need for model reduction generally arises because of two factors. First, the order of the controller can be reduced if the order of the model on which its design is based is small. This points towards the necessity of having a reduced order model for the plant. Alternatively, the controller designed using the full order plant model may be reduced to a lower order, at the stage of implementation so that one may work with a simpler controller. Whatever approach one may take, ultimately the controller has to work in conjunction with the physical plant (not its model) and the ultimate closed loop plant controller configuration should be kept in mind while designing the controller.

A number of model reduction techniques have been tried over the years. Most of these techniques have one serious drawback. While they work well in specific tailor-made situations, the results are not satisfactory under more general conditions. This is because they do not specify an error norm which is valid under all circumstances. This lacuna is sought to be rectified in some of the more recent methods involving truncation of balanced realizations and optimal Hankel norm approximations. Both these methods are explained in Chapter 7.

1.6 Summary

We live in a world full of uncertainty. The physical systems that we come across are no exceptions to this rule. Therefore when we model such systems we should bear in mind that our model is only a nominal one and that variations about this model are most likely to occur. Under these circumstances our objective is to design a controller which will maintain prescribed performance levels under specified levels of uncertainty. This is precisely what a robust controller aims to do.

The appeal of robust control theory is universal and its applications need not be confined to specific fields of engineering. There are many disciplines in which the system concept plays an important role and in all these cases the issues raised in this chapter regarding robustness of performance are significant. Briefly this provides the motivation for making a detailed study of robust control theory as applied to linear dynamical systems.

Notes and Additional References

The following books published during the past decade, arranged in chronological order, specifically cover different aspects of robust control theory.

1. Vidyasagar M. (1985) *Control System Synthesis – A Factorization Approach*, M.I.T. Press, Cambridge MA.
2. Francis B.A. (1987) *A Course in H_∞ Control Theory*, Springer, New York.
3. Maciejowski J.M. (1989) *Multivariable Feedback Design*, Addison Wesley, New York.
4. Morari M. and Zafiriou E. (1989) *Robust Process Control*, Prentice Hall, New Jersey
5. Doyle J.C., Francis B.A. and Tannenbaum A.R. (1992) *Feedback Control Theory*, Macmillan, New York.
6. Stoorvogel A. (1992) *The H_∞ Control Problem*, Prentice Hall, UK.
7. Barmish B.R. (1994) *New Tools for Robustness of Linear Systems*, Macmillan, New York.
8. Green M. and Limebeer D.J.N. (1995) *Linear Robust Control*, Prentice Hall, New Jersey.

Book 1 gives a comprehensive treatment of the so called 'factorization approach' as applied to multivariable systems.

Book 2 is the first book to appear on H_∞ control and gives a rigorous treatment of the subject.

Book 3 clearly brings out the connection between classical frequency domain techniques of Bode and Nyquist and the more modern LQG and H_∞ techniques.

Book 4 discusses robust control as applied to process industry.

Book 5 is a concise and well organized summary of recent developments in robust control theory, restricted to SISO systems.

Book 6 is essentially a research monograph which reports the state of the art in H_∞ control with special reference to the author's own work in this area.

Book 7 is entirely devoted to Khartinov's approach to robustness and possible applications in linear robust control.

Book 8, the latest in the series gives a full treatment of linear robust control at an advanced level and covers for the first time in a single book both time varying and time invariant systems in a unified manner.

Besides the above mentioned text books, two IEEE publications give a collection of articles that have appeared in recent times having a bearing on robust control theory. The books are:

1. P. Dorato (Editor), *Robust control*, IEEE Press, Institution of Electrical and Electronic Engineers, New York, 1987.
2. P. Dorato and R.K. Yedavalli (Editors), *Recent Advances in Robust Control*, IEEE Press, Institution of Electrical and Electronic Engineers, New York, 1990.

2

Introduction to Robust Control

2.1 Introduction

This chapter is intended to serve as a curtain raiser for robust control theory which is presented in more depth in subsequent chapters. There is a belief among some practitioners of control theory that the sole purpose of feedback is to alter the dynamics of the plant. This is not correct, for if it were to be so, then an open loop controller could have served the purpose equally as well. The main purpose of feedback is to take into account plant uncertainties and ensure that performance specifications are adhered to within guaranteed limits. This particular aspect of feedback is explained in this chapter. In order to keep matters simple a SISO system is chosen to demonstrate the power and versatility of feedback. Having grasped the fundamentals in this simple case, the reader will be in a better position to appreciate the more complex nuances arising out of working with multivariable systems. To begin with, different performance specifications are explained. Subsequently the constraints that such specifications impose on loop gain are examined. It is shown how some of the performance requirements make conflicting demands on loop gain. Clearly, in such a situation there is scope for trade-offs between different specifications. This exercise is far from trivial even in the SISO case. It could be quite complicated in the MIMO context. This justifies the deployment of powerful new techniques to solve the problem in multivariable control. These methods are outlined in subsequent chapters.

2.2 Elements of Robust Control Theory

When we design controllers for specific plants, we necessarily have to use mathematical models to represent the plant. However any model can at best only be an approximation of the real system. This is even more so when we are dealing with linear time invariant models. Nature is essentially non-linear and the insistence upon linearity and time invariance makes our models inaccurate. The controller design is based on this inaccurate model. When such a controller is used in conjunction with the plant, the combination is likely to behave in a manner different from what the designer had originally intended. Take for example the inputs into the system. There is a range of inputs bounded in amplitude and rate of change for which the model can be taken as a reasonable approximation of the plant. Outside this range the responses diverge widely. This is to be expected because the neglected non-linearities and effects of higher order dynamics become important at high frequency ranges. The mismatch between the model and the plant may result in controllers which give rise to stability problems in the closed loop system. The role of robust control has to be viewed in this context. Briefly, robustness is that attribute of a system which enables it to maintain prescribed performance norms under specified levels of uncertainty. By specified performance norms we mean specification on the size of various signals of importance such as tracking errors and actuator signals. A design methodology which incorporates the concept of robustness is therefore of great practical importance.

2.3 Design Objectives and Specifications

The discussions in this chapter are confined to SISO systems. Both the plant model and the controller are assumed to be linear and time invariant. Only unstructured uncertainty is considered. The reason for this is that the unstructured uncertainty assumption leads to simple mathematically tractable results. Perhaps more important is the fact that this is the only way we can take care of unmodelled dynamics in a system.

2.3.1. Additive and Multiplicative Perturbations

Provided the plant is stable, its transfer function could be obtained by conducting frequency response experiments on it. Such a transfer function is called the nominal transfer function as it approximates the input–output data pertaining to the plant. There are various ways in which unstructured uncertainty could be represented. Two of the most common are:

1. Additive perturbation
2. Multiplicative perturbation

Let $\mathbf{p}(s)$ be the nominal transfer function of the plant and $\tilde{\mathbf{p}}(s)$ be the perturbed transfer function.

For additive perturbation we have

$$\tilde{\mathbf{p}}(s) = \mathbf{p}(s) + \boldsymbol{\delta}_{\mathbf{a}}(s) \tag{2.1}$$

where $\boldsymbol{\delta}_{\mathbf{a}}(s)$ is the additive perturbation with $|\boldsymbol{\delta}_{\mathbf{a}}(j\omega)| < \mathbf{l}_{\mathbf{a}}(\omega)$ for $\forall\omega \geq 0$.

Here $\mathbf{l}_{\mathbf{a}}(\cdot)$ denotes a positive scalar function which confines $\tilde{\mathbf{p}}(j\omega)$ to a neighbourhood of $\mathbf{p}(j\omega)$ with magnitude $\mathbf{l}_{\mathbf{a}}(\omega)$.

For multiplicative perturbation we have

$$\tilde{\mathbf{p}}(s) = [1 + \boldsymbol{\delta}_{\mathbf{m}}(s)]\mathbf{p}(s) \tag{2.2}$$

with $|\boldsymbol{\delta}_{\mathbf{m}}(j\omega)| < \mathbf{l}_{\mathbf{m}}(\omega)$ $\forall\omega \geq 0$, where $\mathbf{l}_{\mathbf{m}}(\cdot)$ is a positive scalar function which confines $\tilde{\mathbf{p}}(j\omega)$ to a normalized neighbourhood of $\mathbf{p}(j\omega)$ with magnitude $\mathbf{l}_{\mathbf{m}}(\omega)$.

We will further assume that in both cases $\tilde{\mathbf{p}}$ and $\mathbf{p}$ have the same **number** of unstable poles. However these unstable poles need not be identical. The above requirement on the number of unstable poles appears strange, but it is required to simplify discussions pertaining to stability.

We note from (2.2) that

$$\frac{\tilde{\mathbf{p}}(j\omega)}{\mathbf{p}(j\omega)} - 1 = \boldsymbol{\delta}_{\mathbf{m}}(j\omega)$$

and hence

$$\left|\frac{\tilde{\mathbf{p}}(j\omega) - \mathbf{p}(j\omega)}{\mathbf{p}(j\omega)}\right| = |\boldsymbol{\delta}_{\mathbf{m}}(j\omega)| < \mathbf{l}_{\mathbf{m}}(\omega) \tag{2.3}$$

The left hand side of (2.3) represents the normalized deviation and therefore is a measure of relative rather than absolute magnitude of deviation. This appears to be more realistic and hence multiplicative perturbations are preferred when the emphasis is on relative deviations. A typical plot of $\mathbf{l}_{\mathbf{m}}(\omega)$ as a function of ω is given in Figure 2.1. Note that $\mathbf{l}_{\mathbf{m}}(\omega) \ll 1$ at low frequencies but invariably moves towards unity and well above (modelling error $\gg$ 100%) as frequency increases.

2.3.2. Plant–Controller Configuration

A layout of a general feedback system is given in Figure 2.2. In the figure **G** represents the plant, **H** represents sensor dynamics, **F** the feedback compensator, **C** the cascade compensator and **P** the pre-filter. Plant disturbance is denoted by d, sensor noise by n, reference/command by r and output by y. In most applications **P** may be taken to be the unit matrix. **C** and **F** represent two degrees of freedom for the controller. However one degree of freedom would suffice in most cases. When the sensor has no

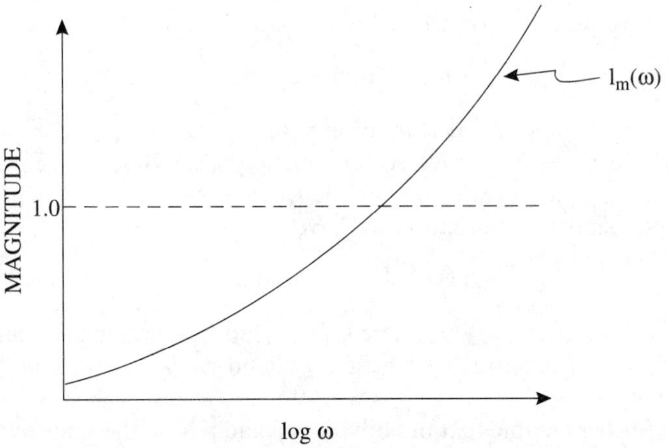

Figure 2.1. Uncertainty profile for multiplicative perturbation.

dynamics, **H** can be taken as the unit matrix. With these simplifications we arrive at a configuration shown in Figure 2.3.

From the figure it is readily seen that

$$y = \frac{\mathbf{pc}}{1+\mathbf{pc}}r - \frac{\mathbf{pc}}{1+\mathbf{pc}}n + \frac{1}{1+\mathbf{pc}}d \tag{2.4}$$

$$\begin{aligned} e &= r - y \\ &= \frac{1}{1+\mathbf{pc}}r + \frac{\mathbf{pc}}{1+\mathbf{pc}}n - \frac{1}{1+\mathbf{pc}}d \end{aligned} \tag{2.5}$$

The following definitions are standard in control theory and are given below for easy reference:

Loop transfer function: **pc**
Return difference: $(1 + \mathbf{pc})$
Sensitivity function (**s**): $\frac{1}{1+\mathbf{pc}}$
Complementary sensitivity function (**t**): $\frac{\mathbf{pc}}{1+\mathbf{pc}}$

We also have $(\mathbf{s} + \mathbf{t}) = 1$

Sensitivity function **s** derives it name from the fact that it represents the sensitivity of the closed loop transfer function **t** to an infinitesimal change in **p**. Thus we have

$$\mathbf{s} = \lim_{\Delta\mathbf{p}\to 0} \frac{\Delta\mathbf{t}}{\mathbf{t}} \bigg/ \frac{\Delta\mathbf{p}}{\mathbf{p}} = \frac{d\mathbf{t}}{d\mathbf{p}}\frac{\mathbf{p}}{\mathbf{t}} \tag{2.6}$$

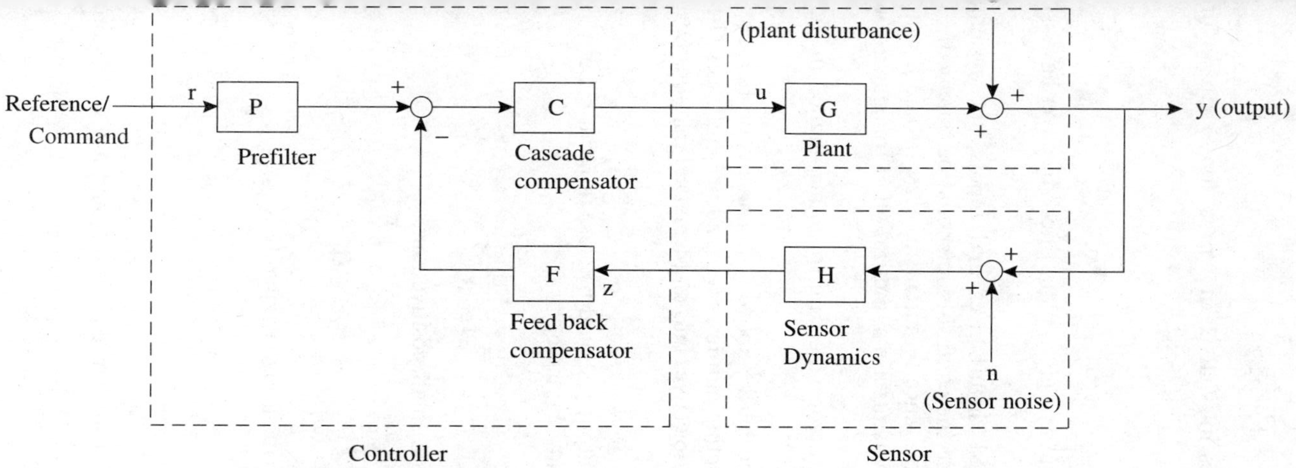

Figure 2.2. Feedback control system.

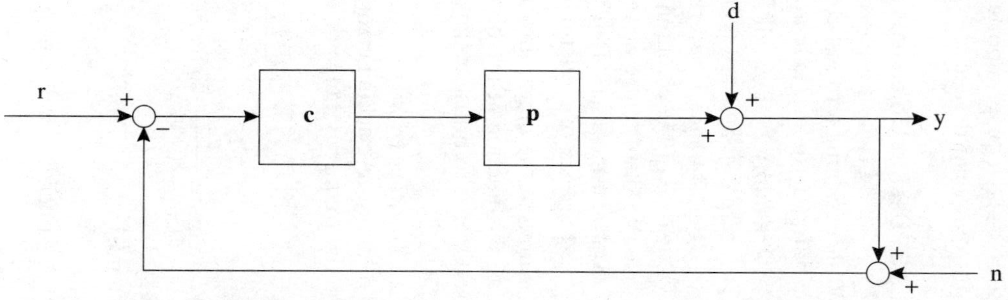

Figure 2.3. Simplified feedback system.

Direct evaluation of (2.6) yields the relation

$$\mathbf{s} = \frac{1}{1 + \mathbf{pc}}$$

Both **s** and **t** play important roles in what is known as 'loop shaping' in control system design.

2.3.3. Design Objectives

While designing a robust controller for a given plant three objectives have to be borne in mind. These are

1. **Stability** – For all bounded disturbances and inputs, the system response at every point inside the control loop should be bounded.
2. **Performance** – Several performance requirements may be laid down. Some of the common requirements are minimization of tracking error and disturbance attenuation.
3. **Robustness** – Stability and performance are to be maintained in the presence of model uncertainty.

Some of the constraints imposed by the above requirements on the loop transfer function **pc** are of a conflicting nature as we shall presently see. The next step is to translate the design objectives into mathematical relationships.

(a) Tracking error
To minimize tracking error, the output y should be close to input r. Let Ω_1, be the frequency range of the input. We have

$$\frac{r(j\omega) - y(j\omega)}{r(j\omega)} = \frac{e(j\omega)}{r(j\omega)} = \frac{1}{1 + \mathbf{pc}(j\omega)}$$

We require $\frac{e(j\omega)}{r(j\omega)} < \epsilon_1$ where ϵ_1 is a small positive number

Hence $$\frac{1}{1 + \mathbf{pc}(j\omega)} < \epsilon_1 \quad \forall \omega \in \Omega_1$$

A sufficient condition to satisfy the above inequality is

$$|\mathbf{pc}(j\omega)| > 1 + \frac{1}{\epsilon_1} \quad \forall \omega \in \Omega_1 \tag{2.7}$$

(b) Disturbance rejection
In order to minimize the effect of disturbance d on the output y we require

$$\frac{|y(j\omega)|}{|d(j\omega)|} < \epsilon_2 \quad \forall \omega \in \Omega_2$$

where ϵ_2 is a small positive number and Ω_2 is the frequency range of the disturbance signal. Hence

$$\frac{1}{1+\mathbf{pc}(j\omega)} < \epsilon_2 \quad \forall \omega \in \Omega_2$$

A sufficient condition for the above is

$$|\mathbf{pc}(j\omega)| < 1 + 1\epsilon_2 \quad \forall \omega \in \Omega_2 \tag{2.8}$$

(c) Insensitivity to noise
We require

$$\frac{|y(j\omega)|}{|n(j\omega)|} < \epsilon_3 \quad \forall \omega \in \Omega_3$$

where ϵ_3 is a small positive number and Ω_3 is the frequency range in which noise is active. We have

$$\frac{|y(j\omega)|}{|n(j\omega)|} = \frac{|\mathbf{pc}(j\omega)|}{|1+\mathbf{pc}(j\omega)|} < \epsilon_3 \quad \forall \omega \in \Omega_3$$

A conservative bound for $\mathbf{pc}(j\omega)$ is given by

$$|\mathbf{pc}(j\omega)| < \frac{\epsilon_3}{1+\epsilon_3} \qquad \forall \omega \in \Omega_3 \tag{2.9}$$

(d) Model uncertainty (low frequency)
This uncertainty is mostly due to imperfect knowledge of the parameters.
Let the nominal transfer function $\mathbf{p}(s)$ be perturbed to yield the transfer function $\tilde{\mathbf{p}}(s)$

$$\tilde{\mathbf{p}}(s) = \mathbf{p}(s) + \boldsymbol{\delta}\,\mathbf{p}(s)$$

Let the corresponding outputs be $y(s)$ and $\tilde{y}(s)$. We have

$$\tilde{y}(s) = y(s) + \delta y(s)$$

for good performance robustness under model perturbation we require

$$\frac{|\delta y(j\omega)|}{|y(j\omega)|} < \epsilon_4 \qquad \forall \omega \in \Omega_4$$

where ϵ_4 is a small positive number and Ω_4 the frequency range of interest. We have

$$y(j\omega) = \frac{\mathbf{pc}(j\omega)}{1+\mathbf{pc}(j\omega)}$$

and

$$\tilde{y}(j\omega) = \frac{\tilde{\mathbf{p}}\mathbf{c}(j\omega)}{1 + \tilde{\mathbf{p}}\mathbf{c}(j\omega)}$$

Hence

$$\delta y(j\omega) = \left[\frac{\tilde{\mathbf{p}}\mathbf{c}(j\omega)}{1 + \tilde{\mathbf{p}}\mathbf{c}(j\omega)} - \frac{\mathbf{pc}(j\omega)}{1 + \mathbf{pc}(j\omega)}\right] r(j\omega)$$
$$= \frac{\boldsymbol{\delta}\mathbf{pc}(j\omega)}{(1 + \tilde{\mathbf{p}}\mathbf{c}(j\omega))(1 + \mathbf{pc}(j\omega))} r(j\omega)$$

and

$$\frac{|\delta y(j\omega)|}{|y(j\omega)|} = \frac{|\boldsymbol{\delta}\mathbf{p}(j\omega)|}{|\mathbf{p}(j\omega)|} \cdot \frac{1}{|1 + \tilde{\mathbf{p}}\mathbf{c}(j\omega)|}$$

The requirement that $\dfrac{\delta y(j\omega)}{y(j\omega)} < \epsilon_4$ leads to

$$\frac{|\boldsymbol{\delta}\mathbf{p}(j\omega)|}{|\mathbf{p}(j\omega)|} \cdot \frac{1}{|1 + \tilde{\mathbf{p}}\mathbf{c}(j\omega)|} < \epsilon_4 \quad \forall \omega \in \Omega_4$$

Since $|\boldsymbol{\delta}\mathbf{p}(j\omega)| \ll |\mathbf{p}(j\omega)|$

$$|1 + \tilde{\mathbf{p}}\mathbf{c}(j\omega)| \approx |1 + \mathbf{pc}(j\omega)|$$

We therefore obtain a conservative bound

$$|\mathbf{pc}(j\omega)| > \frac{1}{\epsilon_4} \frac{|\boldsymbol{\delta}\mathbf{p}(j\omega)|}{|\mathbf{p}(j\omega)|} + 1 \quad \forall \omega \in \Omega_4 \tag{2.10}$$

(e) Model uncertainty (high frequency)
This type of uncertainty may be traced to the neglect of higher order dynamics. Such uncertainties cause stability problems. The controller has to ensure stability of the nominal plant and also the stability of a class of admissible plants in its neighbourhood. By admissible plants we mean that the nominal and perturbed plants possess the same number of closed right half plane poles.

Figure 2.4 shows the Nyquist plots for loop transfer functions in the case of nominal and perturbed plants. Assume that the closed loop system is stable for the nominal plant. According to Nyquist stability criteria, in order to ensure closed loop stability, the number of anticlockwise encirclements about the critical point $(-1, 0)$ by the $\mathbf{pc}(j\omega)$ locus must be equal to the total

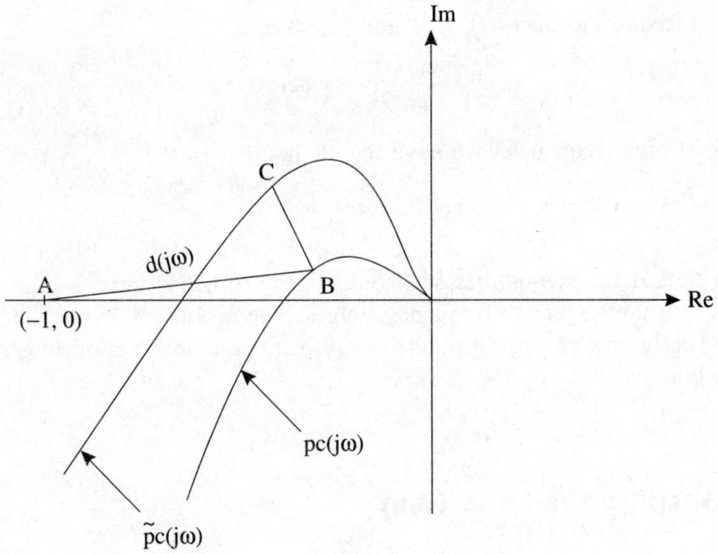

Figure 2.4. Nyquist plot.

number of poles of $\mathbf{p}$ and $\mathbf{c}$ in Re $s \geq 0$. As the perturbed plant has the same number of poles in Re $s \geq 0$ as the nominal plant, the closed loop stability of the perturbed plant is guaranteed if the number of encirclements of the $\tilde{\mathbf{p}}\mathbf{c}(j\omega)$ locus is the same as that of $\mathbf{pc}(j\omega)$ locus about $(-1, 0)$. This can happen only if the $\mathbf{pc}(j\omega)$ locus, which is continuously deformed due to perturbation and ultimately merges with the $\tilde{\mathbf{p}}\mathbf{c}(j\omega)$ locus, does not pass through the critical point even once. From a geometric viewpoint, a sufficient condition for this to happen is that

$$|\tilde{\mathbf{p}}\mathbf{c}(j\omega) - \mathbf{pc}(j\omega)| < \mathbf{d}(j\omega) \quad \forall 0 \leq \omega < \infty$$

Referring to Figure 2.4, for a particular value of ω, $|\tilde{\mathbf{p}}\mathbf{c}(j\omega) - \mathbf{pc}(j\omega)|$ is represented by BC and $\mathbf{d}(j\omega)$ is represented by AB. Note that B never merges with A as the nominal closed loop system is assumed stable. Being a sufficient condition the result derived by using this condition is invariably conservative. It is clear from Figure 2.4 that $\mathbf{d}(j\omega) = |1 + \mathbf{pc}(j\omega)|$. Hence

$$|\tilde{\mathbf{p}}\mathbf{c}(j\omega) - \mathbf{pc}(j\omega)| < |1 + \mathbf{pc}(j\omega)| \quad \forall 0 \leq \omega < \infty$$

Now consider a multiplicative perturbation where $\tilde{\mathbf{p}}(j\omega) = (1 + \delta_{\mathbf{m}}(j\omega))\mathbf{p}(j\omega)$. Substituting for $\tilde{\mathbf{p}}(j\omega)$ in the above inequality

$$\frac{|\mathbf{pc}(j\omega)|}{|1 + \mathbf{pc}(j\omega)|} < \frac{1}{\delta_{\mathbf{m}}(j\omega)}$$

At high frequencies $|\mathbf{pc}(j\omega)| \ll 1$ and we have

$$\frac{|\mathbf{pc}(j\omega)|}{|1+\mathbf{pc}(j\omega)|} \approx \mathbf{pc}(j\omega)$$

Hence at high frequencies we have the inequality

$$|\mathbf{pc}(j\omega)| < \frac{1}{|\delta_{\mathbf{m}}(j\omega)|} \tag{2.11}$$

We note that the inequalities spelt out in (2.7) to (2.11) impose bounds on $|\mathbf{pc}(j\omega)|$ in the respective frequency ranges. The problem is thus reduced to one of designing an appropriate loop transfer function conforming to the above bounds.

2.4 Shaping the Loop Gain

In the previous section we analysed the various performance objectives and obtained quantitative measures to describe them. On closer examination we note a dichotomy between the gain requirements of the loop transfer function at high and low frequencies. For example, minimization of tracking error entails a high value of $|\mathbf{pc}(j\omega)|$ in the frequency range Ω_1 as is clear from (2.7). The same trend is also apparent in the case of disturbance rejection which according to (2.8) demands a high value of $|\mathbf{pc}(j\omega)|$ in the Ω_2 frequency range. However, when it comes to insensitivity to sensor noise, (2.9) reveals that $|\mathbf{pc}(j\omega)|$ should be small in the frequency range Ω_3. Fortunately, Ω_3 corresponds to the high frequency range and hence there basically is no conflict in the gain requirements demanded in frequency ranges Ω_1, Ω_2 and Ω_3. Meeting the gain requirements due to model perturbation at low frequencies also does not pose a problem as seen from (2.10). However the real problem arises in accommodating the gain requirements to ensure stability for model perturbations in the high frequency range. According to (2.11) $|\mathbf{pc}(j\omega)| < 1/\mathbf{l_m}(\omega)$ where $\mathbf{l_m}(\omega)$ represents an upper bound of the uncertainty profile. This is a crucial requirement which is of fundamental importance as it affects stability.

A sketch of loop gain as frequency varies form 0 to ∞, with the performance requirements juxtaposed is given in Figure 2.5.

The primary task is to design the controller in such a way that the magnitude requirements of $\mathbf{pc}(j\omega)$ are met over the entire frequency spectrum. Since $\mathbf{p}$ is fixed and cannot be altered, this can only be done by suitably designing $\mathbf{c}$. If the plant has a good model at high frequencies then $\mathbf{l_m}(\omega)$ will be small and its reciprocal will be large. In that case the gain cross-

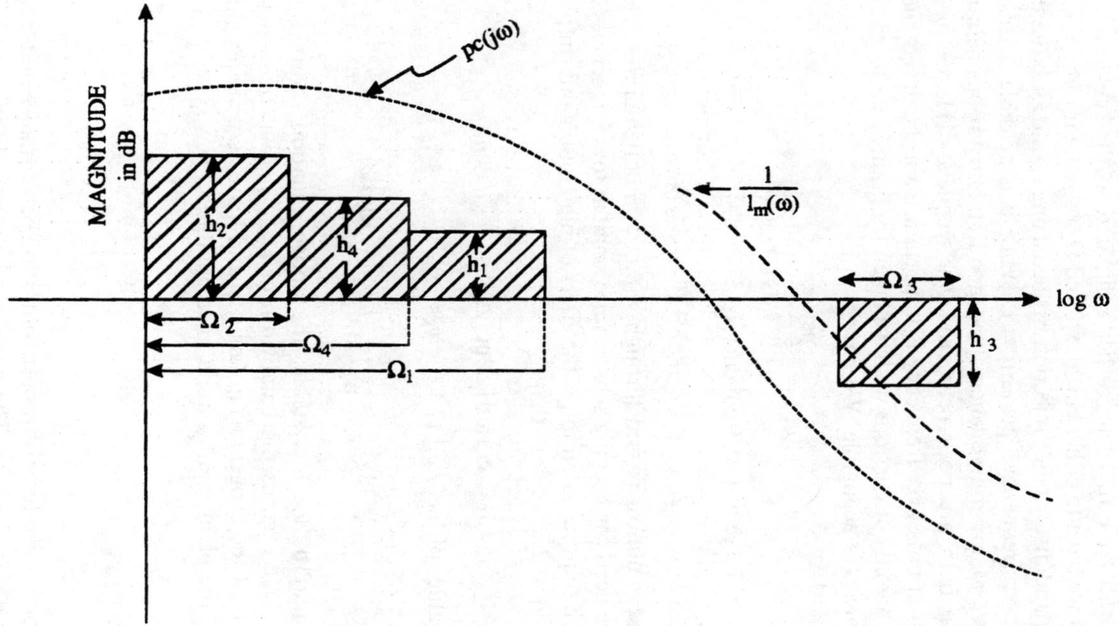

Figure 2.5. Bode magnitude plot for loop transfer function.

over frequency (the frequency at which the loop gain plot crosses the zero *db* line) can occur at a higher value. This allows for more stringent specifications regarding tracking and disturbance rejection. We may therefore either increase the frequency range Ω_1 and Ω_2 or decrease ϵ_1 and ϵ_2. Alternatively if we are working with a highly inaccurate model (especially in the high frequency range) the uncertainty profile will be high and the gain crossover frequency will have to occur earlier compared to the previous case. This means that we have to either curtail Ω_1 and Ω_2 or settle for larger inaccuracy in tracking and disturbance rejection by increasing ϵ_1 and ϵ_2. Note that at the gain crossover frequency, $|\mathbf{pc}(j\omega)| = 1$. While it is possible to ensure good performance at lower frequencies and good stability robustness at higher frequencies, both are poor near crossover frequency. Hence steepness of attenuation at crossover frequency is important from the design view point.

In the foregoing discussions we have taken for granted that sensor noise at low frequencies is negligible. We have

$$y(j\omega) = \frac{\mathbf{pc}(j\omega)}{1 + \mathbf{pc}(j\omega)} n(j\omega)$$

As $|\mathbf{pc}(j\omega)|$ is large at low frequencies

$$|y(j\omega)| \approx |n(j\omega)|$$

Hence noise is almost passed through without any attenuation. This points towards the need for keeping sensor noise small at low frequencies. Another factor which requires attention is the effect of high loop gains on plant input u. From Figure 2.3 we note that

$$u(j\omega) = \mathbf{c}(j\omega)[1 + \mathbf{pc}(j\omega)]^{-1}(r - d - n)(j\omega) \tag{2.12}$$

For high values of $|\mathbf{pc}(j\omega)|$ (2.12) gives

$$u(j\omega) \approx \frac{1}{\mathbf{p}(j\omega)}(r - d - n)(j\omega)$$

Assuming that $\mathbf{p}^{-1}(j\omega)$ is stable, the above equation shows that if the frequency range exceeds the bandwidth of $\mathbf{p}(j\omega)$, i.e. at frequencies where $|\mathbf{p}(j\omega)| \ll 1$ all the exogenous inputs r, d and n are actually amplified at u. This makes the plant input u quite unacceptable.

2.5 Summary

In this chapter, the effects of robust control on performance and stability are discussed for SISO systems. These notions cannot be directly carried over to MIMO systems without a proper definition of 'size'. It turns out that by replacing absolute values by singular values, the properties of robust control discussed in this chapter could be extended to the multivariable case. In summary the following facts have been established for SISO systems.

1. Tracking accuracy and disturbance attenuation are enhanced by designing for a high loop gain at low frequencies.
2. Noise attenuation at high frequencies demands that the system be designed for low loop gain at high frequencies.
3. High loop gain over an extended frequency range beyond plant bandwidth may lead to unacceptable values of plant input, and this is undesirable.
4. The gain crossover frequency and the rate at which gain is attenuated in the neighbourhood of this frequency are important. This depends upon how good the model is in the high frequency range and the stringency of performance specifications in the low frequency range.

Notes and Additional References

Doyle and Stein (1981) reviews the basic issues connected with feedback design in the presence of uncertainties and generalizes the properties established in this chapter to the MIMO case. The treatment regarding SISO case given in this chapter is adapted from Kazerooni (1988). For a thorough understanding of robust control theory as applicable to SISO systems the reader may refer to the book by Doyle *et al.* (1992).

Exercises

1. Assume that the feedback system shown in Figure 2.3 is internally stable and let $n = d = 0$ and define $e(t) = r(t) - y(t)$. Show that
   (a) If $r(t)$ is a step function, then $e(t) \to 0$ as $t \to \infty$ if and only if the sensitivity function $\mathbf{s}(s)$ has at least one zero at the origin.
   (b) If $r(t)$ is a ramp function, then $e(t) \to 0$ as $t \to \infty$ if and only if $\mathbf{s}(s)$ has at least two zeros at the origin.
2. In a closed loop feedback system the plant transfer function $\mathbf{p}(s) = \frac{1}{s}$ and the controller transfer function $\mathbf{c}(s) = \frac{1}{s}$. Show that the output will asymptotically track a ramp input.
3. Refer to Figure 2.3 where $r = n = 0$ and $d = 1(t) \sin \omega t$ where $1(t)$ is a unit step function. If the feedback system is internally stable show that $y(t) \to 0$ as $t \to \infty$ if and only if either $\mathbf{p}(s)$ or $\mathbf{c}(s)$ has a pole at $s = j\omega$.

4. In a feedback system, $\mathbf{p}(s) = \dfrac{4(s+2)}{(s+1)^2}$ and $\mathbf{c}(s) = 1$. If $\mathbf{s}(s)$ denotes the sensitivity function, obtain $\max_{\omega} |\mathbf{s}(j\omega)|$ for $0 \leq \omega \leq \infty$. Try another stabilizing controller (i.e a controller which internally stabilizes the system) and obtain the new value of $\max_{\omega} |\mathbf{s}(j\omega)|$, $0 \leq \omega \leq \infty$.
Is there a lower bound for this value as the controller is varied over all stabilizing controllers?
5. A unity feedback system has the plant transfer function $\mathbf{p}(s) = \dfrac{1}{10s+1}$ and the cascade compensator $\mathbf{c}(s) = k$. Find the smallest k which, while stabilizing the feedback system, ensures that $e(\infty) \leq 0.1$ when the system is subjected to a unit step input.
6. Consider a unity feedback system with a plant $\mathbf{p}$ and a controller $\mathbf{c}$. Let $\mathbf{p}$ be perturbed to $\mathbf{p} + \delta\,\mathbf{p}$ and let the input to output transfer function be $\mathbf{h}$ and $\tilde{\mathbf{h}}$ for the unperturbed and perturbed closed loop systems. Let the corresponding open loop transfer functions be $\mathbf{h}_0$ and $\tilde{\mathbf{h}}_0$ respectively. Further define $\delta\mathbf{h} = \tilde{\mathbf{h}} - \mathbf{h}$ and $\delta\mathbf{h}_0 = \tilde{\mathbf{h}}_0 - \mathbf{h}_0$. Show that $\delta\mathbf{h} = (1 + \tilde{\mathbf{p}}\mathbf{c})^{-1}\delta\mathbf{h}_0$.

3

Signals and Systems

3.1 Introduction

This chapter is devoted to a general study of signals and systems. To begin with, we explain methods for representing and classifying diverse types of signals. Sets of signals may be delineated by describing their attributes and also defining norms to quantitatively measure their strength. The most commonly used signals belong to $\mathcal{L}_2$ and $\mathcal{L}_\infty$ spaces. Their properties and definitions of norms form the subject matter of Section 3.2. Systems map signals from one signal space to another. Signal processing is thus closely linked with the properties of systems. Just as we have various norms for signals we do have different norms for systems also. These norms are a measure of signal gain as it passes through the system. For example, the gain may be reckoned from the energy content of input and output signals. This gives rise to a particular kind of norm for the system. The concept of the induced norm is in fact an extension of this idea. Section 3.3 discusses these matters in some detail. Sections 3.4 and 3.5 are devoted to two important topics – methods to calculate the $\mathcal{H}_2$ and $\mathcal{H}_\infty$ norms of systems. Though $\mathcal{H}_\infty$ is basically a frequency domain concept, it could be tied up with state-space concepts via algebraic methods. Section 3.6 explains how this could be done. The next three sections, 3.7, 3.8 and 3.9, are devoted to presenting three classes of systems. These are the all pass system, the adjoint system and the inverse system. Section 3.10 provides a summary. The contents of this chapter are basic and constitute a pre-requisite for appreciating the subject matter of the chapters that are to follow.

3.2 Signals and their Norms

A signal may be thought of as an entity which in some manner conveys information about a physical system. All of us are familiar with graphical display of signals using devices such as oscilloscopes and plotters. But signals could also be visualized in an abstract sense as points in appropriately defined vector spaces. Viewed from this angle we may consider signals as elements belonging to a set. The set membership could be described in terms of the set satisfying certain properties denoted by $\mathcal{P}$. Thus the set S may be described as

$$S = \{x : \mathcal{P}\}$$

which means the set of all x such that $\mathcal{P}$ is true. Examples of sets thus defined are given below. Signals whose energy content is bounded by some real number k may be described as

$$S_E(k) = \left\{ x : \int_{-\infty}^{\infty} x^2(t)\, \mathrm{d}t \leq k \right\}$$

Band limited signals may be characterized by

$$S_B(W) = \left\{ x : x(f) = \int_{-\infty}^{\infty} x(t) e^{-j2\pi ft}\, \mathrm{d}t = 0 \quad \forall |f| > W \right\}$$

3.2.1 Signal Spaces

We first explain some preliminary concepts. A subset S of a real number system $\mathbb{R}$ is said to be of *measure zero* if S contains a finite or countably infinite set of elements. A function $f : \mathbb{R} \rightarrow \mathbb{R}$ is said to be *measurable* if it is continuous everywhere except on a set of *measure zero*. A property is said to hold 'almost everywhere' when it holds everywhere except on a set of measure zero.

The label $\mathcal{L}$ refers to *Lebesgue measurable functions.* While this concept is important in theoretical studies, for our purpose we will treat all functions as continuous (except when stated otherwise) and all integrals as Rieman integrals.

The most commonly referred signal spaces are $\mathcal{L}_2$ and $\mathcal{L}_\infty$ spaces. These spaces characterize some well defined behavior of the signal over an infinite time interval. The space $\mathcal{L}_2(-\infty, \infty)$ is known as *Lebesgue 2-space*. It is defined as

$$\mathcal{L}_2(-\infty, \infty) = \left\{ f : \int_{-\infty}^{\infty} \|f(t)\|^2\, \mathrm{d}t < \infty \right\} \tag{3.1}$$

where

$$||f(t)|| = \sqrt{f'(t)f(t)} \tag{3.2}$$

we may refer to the 'size' of the signal by a non negative scalar measure called the *norm*. It is denoted by $|| \cdot ||_2$. In the above case, we have

$$||f||_2 = \left[\int_{-\infty}^{\infty} ||f(t)||^2 \mathrm{d}t \right]^{\frac{1}{2}} \tag{3.3}$$

The square of a signal may be associated with power. The analogy here is that of a voltage signal applied across a unit resistance. Extending the same analogy, the integral of the square of a signal over a specified time interval is associated with energy. Thus $\mathcal{L}_2(-\infty, \infty)$ represents signals with finite energy. $\mathcal{L}_2(-\infty, 0]$ and $\mathcal{L}_2[0, \infty)$ are subsets of $\mathcal{L}_2(-\infty, \infty)$ and they are signal spaces in their own right.

These signals vanish over the intervals $[0, \infty)$ and $(-\infty, 0]$ respectively. It can be shown that the signal spaces $\mathcal{L}_2(-\infty, \infty)$, $\mathcal{L}_2(-\infty, 0]$ and $\mathcal{L}_2[0, \infty)$ are complete normed spaces (Banach spaces) with their respective norms induced by inner products. Hence they are also classified under Hilbert spaces.

Let f, g belong to $\mathcal{L}_2(-\infty, \infty)$. Then the inner product $\langle f, g \rangle$ is defined as

$$\langle f, g \rangle = \int_{-\infty}^{\infty} g'(t)f(t)\, \mathrm{d}t \tag{3.4}$$

Inner product conceptualizes the notion of perpendicularity in abstract spaces. Thus $\langle f, g \rangle = 0$ implies that f is orthogonal to g. A subspace S_1 is orthogonal to a subspace S_2 if each and every vector of S_1 is orthogonal to each and every vector of S_2 and vice versa. If S_1 and S_2 together cover the entire space S we way write

$$S = S_1 \overset{\perp}{\oplus} S_2$$

Clearly, spaces $\mathcal{L}_2(-\infty, 0]$ and $\mathcal{L}_2[0, \infty)$ are orthogonal to each other and together they span the entire $\mathcal{L}_2(-\infty, \infty)$ space. Hence we have

$$\mathcal{L}_2(-\infty, \infty) = \mathcal{L}_2(-\infty, 0] \oplus \mathcal{L}_2[0, \infty)$$

The 2-norm induced by the inner product is defined as

$$||f||_2 = \langle f, f \rangle^{\frac{1}{2}} = \left[\int_{-\infty}^{\infty} f'(t)f(t)\, \mathrm{d}t \right]^{\frac{1}{2}}$$

The *Cauchy-Schwarz* inequality relates inner products with 2-norms. We thus have

$$\langle f, g\rangle \leq \|f\|_2 \cdot \|g\|_2$$

3.2.2 Signals in the Frequency Domain

The strength of a signal at different frequencies provides us with useful information. Signal description in the frequency domain is one way of providing this information. The 2-norm of a signal in the frequency domain is defined as

$$\|f\|_2 = \left[\frac{1}{2\pi}\int_{-\infty}^{\infty} f^*(j\omega)f(j\omega)\,\mathrm{d}\omega\right]^{\frac{1}{2}}$$

where $f(j\omega)$ is the frequency domain signal and ω is the angular frequency and $(\cdot)^*$ refers to the complex conjugate transpose.

Frequency domain signals are obtained as the Fourier transform of time domain signals with real coefficients in $\mathcal{L}_2(-\infty,\infty)$.

We have $f^*(j\omega) = \overline{f'(j\omega)} = f'(\overline{j\omega}) = f'(-j\omega)$ where the over-bar indicates complex conjugate. The frequency domain counter part of $\mathcal{L}_2(-\infty,\infty)$ space is denoted by $\mathcal{L}_2$. This notation serves to distinguish between the two spaces.

In $\mathcal{L}_2$ space inner product is defined as

$$\langle f, g\rangle = \frac{1}{2\pi}\int_{-\infty}^{\infty} g^*(j\omega)f(j\omega)\,\mathrm{d}\omega \tag{3.5}$$

where $f(j\omega) = \int_{-\infty}^{\infty} f(t)e^{-j\omega t}\,\mathrm{d}t$ is the Fourier transform of $f(t) \in \mathcal{L}_2(-\infty,\infty)$. It can be shown that the Fourier transform is a Hilbert space isomorphism from $\mathcal{L}_2(-\infty,\infty)$ on to $\mathcal{L}_2$. Because of this isomorphism, the inner products and the 2-norms are preserved in the respective spaces.

To distinguish a time signal f from its Fourier transform, we represent the latter by $\hat{f}$. Using this notation we have

$$\langle f, g\rangle = \langle \hat{f}, \hat{g}\rangle \quad \text{and} \quad \|f\|_2 = \|\hat{f}\|_2$$

3.2.3 Hardy Spaces

Hardy spaces are named after the well known British mathematician G.H. Hardy (1877–1947). These spaces are defined in the frequency domain.

The Hardy space $\mathcal{H}_2$ refers to the class of functions which are analytic and bounded in the open right half plane. By boundedness we mean that the norm defined in (3.6) is finite. We have

$$\|f\|_2 = \left\{ \sup_{\alpha>0} \frac{1}{2\pi} \int_{-\infty}^{\infty} f^*(\alpha + j\omega) f(\alpha + j\omega)\, \mathrm{d}\omega \right\}^{\frac{1}{2}} \tag{3.6}$$

We may therefore define $\mathcal{H}_2$ space as

$$\mathcal{H}_2 = \{ f : f(s) \text{ is analytic in Re } s > 0 \text{ and} \|f\|_2 < \infty \}$$

Now define the boundary function of $f(s)$ as

$$f_b(j\omega) = \lim_{\alpha \to 0_+} f(\alpha + j\omega)$$

It can be shown that such a boundary function $f_b(j\omega)$ exists and that it belongs to $\mathcal{L}_2$. Further

$$\|f_b\|_2 = \|f\|_2$$

The supremum in (3.6) always occurs on the boundary $\alpha = 0$ and this enables us to evaluate $\mathcal{H}_2$ norm making use of the boundary function. We may therefore write

$$\|f\|_2 = \left[\frac{1}{2\pi} \int_{-\infty}^{\infty} f_b^*(j\omega) f_b(j\omega)\, \mathrm{d}\omega \right]^{\frac{1}{2}} \tag{3.7}$$

Further, since $f_b(j\omega)$ exists for almost all ω, in (3.7) we may replace $f_b(j\omega)$ by $f(j\omega)$. According to Paley-Wiener theorem, the $\mathcal{H}_2$ space is isomorphic to $\mathcal{L}_2[0, \infty)$ space under the Laplace transform

$$\hat{f}(s) = \int_0^{\infty} f(t) e^{-st}\, \mathrm{d}t$$

$\mathcal{H}_2$ is a closed sub-space of $\mathcal{L}_2$. Its orthogonal complement in $\mathcal{L}_2$ is denoted by $\mathcal{H}_2^{\perp}$. We thus have

$$\mathcal{L}_2 = \mathcal{H}_2 \oplus \mathcal{H}_2^{\perp} \tag{3.8}$$

$\mathcal{H}_2^{\perp}$ refers to the space of all functions analytic and bounded in the open left half plane Note that $\mathcal{H}_2^{\perp}$ is isomorphic to $\mathcal{L}_2(-\infty, 0]$ under the Laplace transform. Because of the isomorphism between $\mathcal{L}_2[0, \infty)$ and $\mathcal{H}_2$ on the one hand and $\mathcal{L}_2(-\infty, 0]$ and $\mathcal{H}_2^{\perp}$ on the other, we will not make any distinction between signals belonging to the respective spaces and their counterparts, during the course of our discussion Thus, we may represent a signal w as belonging to $\mathcal{L}_2[0,\infty)$ and subsequently treat it as if it belongs to $\mathcal{H}_2$.

3.2.4 $\mathcal{L}_\infty$ Signals and Bounded Power Signals

$\mathcal{L}_\infty$-Signals

Among the class of signals widely used and understood by control engineers are those belonging to $\mathcal{L}_2$ and $\mathcal{L}_\infty$ spaces. Consider the following signal

$$u(t) = \begin{bmatrix} u_1(t) \\ u_2(t) \\ \vdots \\ u_m(t) \end{bmatrix} \in \mathbb{R}^m$$

Recall that $\mathcal{L}_2$ norm of the above signal is given by

$$||u||_2 = \left[\int_{-\infty}^{\infty} ||u(t)||^2 \, \mathrm{d}t \right]^{\frac{1}{2}} \tag{3.9}$$

where

$$||u(t)||^2 = \sum_{i=1}^{m} [u_i(t)]^2$$

We now define a norm known as $\mathcal{L}_\infty$ norm. The $\mathcal{L}_\infty$ norm of the signal $u(t)$ is given by

$$||u||_\infty = \text{ess } \sup_t ||u(t)|| \tag{3.10}$$

where

$$||u(t)|| = \max_{i\epsilon[1,m]} |u_i(t)|$$

The *essential supremum* mentioned in (3.10) refers to supremum almost everywhere except on a set of measure zero. However for all practical purposes, we may omit the word "essential" while defining $\mathcal{L}_\infty$ norm.

If $u(t)$ is a scalar, then

$$||u||_\infty = \sup_t |u(t)|$$

which translated into words means the least upper bound of the absolute value of $u(t)$.

Bounded power signals
The average power of a signal may be defined as the time average of its instantaneous power. For a signal $u(t)$, this is given by

$$\lim_{T\to\infty} \frac{1}{2T} \int_{-T}^{T} u'(t)u(t)\,dt$$

If the average power of a class of signals is bounded, we may refer to this class by $\mathcal{P}$ where

$$\mathcal{P} = \{u(t) : u(t) \in \mathcal{L}(-\infty, \infty) \quad \text{and} \quad ||u||_{\mathcal{P}} < \infty\} \tag{3.11}$$

$||u||_{\mathcal{P}}$ the norm for a bounded power signal may be defined as

$$||u||_{\mathcal{P}} = \left\{ \lim_{T\to\infty} \frac{1}{2T} \int_{-T}^{T} ||u(t)||^2\,dt \right\}^{1/2} \tag{3.12}$$

Note that $||u||_{\mathcal{P}}$ is a semi-norm because $||u||_{P}$ can be zero for $u \neq 0$

3.3 Systems and their Norms

Signals are subjected to a variety of transformations as they pass from input to output through different types of devices such as transducers, scanners, amplifiers, modulators and transmission channels. The basic role played by these devices is one of signal transformation. Mathematically speaking, we may consider signal transformation as a mapping from input space S_1 to output space S_2. If $\boldsymbol{G}$ represents such a transformation then

$$\boldsymbol{G} : S_1 \to S_2$$
$$w \to z = \boldsymbol{G}w$$

One way of classifying systems is to separate them into two categories – causal and non-causal. In *causal systems*, the output at time t depends only upon the input into the system up to and including the time t for every t considered. In *non-causal systems*, the output at the present time depends upon inputs which occur at a future time.

There is yet another way of classification – time-invariant and time-varying systems. In a *time invariant-system*, if the response of a system to an input $w(t)$ is $z(t)$ then its response when the input is $w(t - \tau)$ is $z(t - \tau)$. The response therefore does not depend upon the time at which the input is applied.

A third way of system classification is to categorize it as linear or non linear. A system $G : S_1 \rightarrow S_2$ is *linear* if the principle of super position holds good for the system.
Thus

$$G(\alpha_1 w_1 + \alpha_2 w_2) = \alpha_1(Gw_1) + \alpha_2(Gw_2)$$

for all w_1 and $w_2 \in S_1$ and α_1 and α_2 are scalars.

In this book we confine our attention to systems which are causal, time invariant and linear. An immediate fall-out from these assumptions is that the underlying mathematics is considerably simplified. For a linear time invariant system the output may be obtained from the input via the convolution integral. We have

$$z(t) = \int_{-\infty}^{\infty} G(t-\tau)w(\tau)\,d\tau \tag{3.13}$$

where $G(t)$ is the inverse Laplace transform of $G(s)$. Further, if the system is causal,

$$G(t-\tau) = 0 \quad \forall \tau > t$$

By taking the Laplace transform of (3.13) we have

$$z(s) = G(s)w(s)$$

$G(s)$ is called the transfer matrix of the system.

3.3.1 Transfer Matrices – Some Miscellaneous Properties

At this stage we digress a little to define certain terms and to present some properties associated with transfer matrices.

A matrix $A(s)$ is said to be *rational* if each of its elements is rational in the indeterminate s. We thus have

$$A_{ij} = \frac{p_{ij}(s)}{q_{ij}(s)}$$

where $p_{ij}(s)$ and $q_{ij}(s)$ are polynomials in s. If the coefficients of $p_{ij}(s)$ and $q_{ij}(s)$ are real, then $A(s)$ is real.

The following notations are used in the sequel. If A is an arbitrary matrix, then $A', \overline{A}, A^*$ and A^{-1} denote respectively the transpose, the complex conjugate, the adjoint (transpose conjugate) and the inverse of the matrix A.

Using the above notation, if $A(s)$ is real then $\overline{A}(s) = A(\overline{s})$. In particular $\overline{A}(\jmath\omega) = A(-\jmath\omega)$.

The non-negative integer $r(A)$ is the *normal rank* of $A(s)$ if (i) There exists at least one sub-minor of order r, which does not vanish identically and (ii) All minors of order greater than r vanish identically. Clearly, the normal rank of a rational matrix can decrease at most at a finite set of points in the s plane.

A non-square matrix does not possess an inverse in the ordinary sense. However it may possess either a right or left inverse. Thus, if A has dimension $m \times n$, A has a *right inverse* A^{-1} such that $AA^{-1} = I_m$ if and only if $m \leq n$ and $r(A) = m$.

A square matrix is elementary if and only if its determinant is a constant, independent of s. A matrix function $A(s)$ is analytic in a region of the s plane if all its entries are analytic in this region. s_0 is a pole of $A(s)$ if some elements of $A(s)$ have poles at $s = s_0$.

If s_0 is a pole of the rational matrix $A(s)$, each element of $A(s)$ may be expanded in partial fractions and after collecting all those terms having poles at s_0, there is obtained for $s_0 \neq 0$.

$$A(s) = (s - s_0)^{-k} A_k + (s - s_0)^{-k+1} A_{k-1} + \cdots + (s - s_0)^{-1} A_1 + A_0(s)$$

where $A_0(s_0)$ is finite, $A_k \neq 0$ and $A_i, 1 \leq i \leq k$ are constant matrices.

If $s_0 = \infty$ (this happens when the elements of $A(s)$ are improper rational functions) $(s - s_0)^{-i}$ is replaced by $s^i, 1 \leq i \leq k$. All of $A_0(s), A_1, \ldots, A_k$ are uniquely defined by their construction from $A(s)$. In the above partial fraction expansion, k is the order of pole of $A(s)$ at $s = s_0$.

A rational matrix $A(s)$ is said to be *para-conjugate hermitian* if $A^*(s) = A(-\bar{s})$. Hence on the imaginary axis where $s = j\omega$ we have, $A^*(j\omega) = A(j\omega)$ and $A(j\omega)$ is *hermitian* in the ordinary sense.

For real $A(s)$, we have $A^*(-\bar{s}) = A'(-s)$ and the para-conjugate hermetian condition simplifies to $A(s) = A'(-s)$. A real para-conjugate hermitian matrix is called *para-hermitian*. A rational $m \times n$ matrix $A(s)$ is said to be *para-conjugate unitary* if either

$$A^*(-\bar{s}) \cdot A(s) = I_n \quad \text{or} \quad A(s)A^*(-\bar{s}) = I_m$$

On the imaginary axis we have $s = j\omega$ and $A^*(-\bar{s}) = A^*(j\omega)$ and $A(j\omega)$ is *unitary* in the usual sense.

For real $A(s)$, the para-conjugate unitary condition simplifies to

$$A'(-s) \cdot A(s) = I_n \quad \text{or} \quad A(s)A'(-s) = I_m$$

A real para-conjugate unitary matrix is called *para-unitary*.

For convenience in notation we represent $A'(-s)$ by $A^{\sim}(s)$ in our subsequent discussions.

3.3.2 ∞-Norms and 2-Norms

We shall first consider the $\mathcal{L}_\infty$ norm of a system. The relevant mapping function in the time domain is

$$G : \mathcal{L}_2(-\infty, \infty) \to \mathcal{L}_2(-\infty, \infty)$$

In the frequency domain this may be expressed as

$$G : \mathcal{L}_2 \to \mathcal{L}_2$$

Let w be a signal belonging to $\mathcal{L}_2$ space which is transformed to $Gw \in \mathcal{L}_2$. From the definition of $\mathcal{L}_2$ norm as applied to signals, we have

$$||Gw||_2^2 = \frac{1}{2\pi} \int_{-\infty}^{\infty} ||G(j\omega)w(j\omega)||^2 \, d\omega$$

But $||G(j\omega)w(j\omega)|| \leq \overline{\sigma}[G(j\omega)]||w(j\omega)||$ (because of the property relating to maximum singular value)
Hence

$$||Gw||_2^2 \leq \frac{1}{2\pi} \int_{-\infty}^{\infty} \overline{\sigma}[G(j\omega)]^2 ||w(j\omega)||^2 \, d\omega$$

$$\leq \sup_\omega \overline{\sigma}[G(j\omega)]^2 \frac{1}{2\pi} \int_{-\infty}^{\infty} ||w(j\omega)||^2 \, d\omega$$

Thus

$$\frac{||Gw||_2}{||w||_2} \leq \sup_\omega \overline{\sigma}[G(j\omega)] \tag{3.14}$$

we denote $\sup_\omega \overline{\sigma}[G(j\omega)]$ by the symbol $||G||_\infty$
Hence

$$||Gw||_2 \leq ||G||_\infty ||w||_2 \tag{3.15}$$

for all $w \in \mathcal{L}_2$.

Clearly if $||G||_\infty$ is finite then an $\mathcal{L}_2$ input will give rise to an $\mathcal{L}_2$ output. The $\mathcal{L}_\infty$ space of transfer function matrices may therefore be defined as

$$\mathcal{L}_\infty = \{G : ||G||_\infty < \infty\} \tag{3.16}$$

and the $\mathcal{L}_\infty$ norm may be defined as

$$||G||_\infty = \sup_\omega \overline{\sigma}[G(j\omega)] \tag{3.17}$$

That $||\cdot||_\infty$ is a norm can be easily verified. Further, the ∞-norm has an important property known as the *sub-multiplicative property*. According to this property

$$||\boldsymbol{AB}||_\infty \leq ||\boldsymbol{A}||_\infty ||\boldsymbol{B}||_\infty \tag{3.18}$$

where A and B are matrix value transfer functions of compatible dimensions. The proof of this property is straightforward and is left as an exercise. It is relevant to point out here that the sub-multiplicative property is not satisfied by the 2-norm. It is this property that makes the ∞-norm ideally suited to tackle robust stability problems.

REMARKS

In SISO systems, the size of a signal at a particular frequency is measured by its absolute value at that frequency. In MIMO systems, the signal is a vector and its size at a particular frequency is measured by its Euclidian norm at that frequency. When it comes to a system represented by its transfer matrix $\boldsymbol{G}(s)$, its size is judged by the singular values associated with the system. Thus $\boldsymbol{G}(j\omega)$ is considered 'small' if its largest singular value $\overline{\sigma}[\boldsymbol{G}(j\omega)]$ is small. It will be considered 'large' if its smallest singular value $\underline{\sigma}[\boldsymbol{G}(j\omega)]$ is large. For a detailed discussion on singular values refer to Appendix C.

Earlier we showed that if $\boldsymbol{G} \in \mathcal{L}_\infty$ then $\boldsymbol{G}\mathcal{L}_2 \subset \mathcal{L}_2$. The converse of this statement that if an $\mathcal{L}_2$ input yields an $\mathcal{L}_2$ output, then the transfer matrix $\boldsymbol{G} \in \mathcal{L}_\infty$ is also true. The proof is omitted.

$\mathcal{H}_\infty$ space of transfer matrices

Consider a linear time invariant system with transfer matrix $\boldsymbol{G}$. Let $z = \boldsymbol{G}w$ where both z and w belong to $\mathcal{H}_2$. Then $\boldsymbol{G}$ defines a stable system. Let us now evaluate the $\mathcal{H}_2$ norm of the signal $\boldsymbol{G}w$. We have from (3.6)

$$\begin{aligned}
||\boldsymbol{G}w||_2^2 &= \sup_{\alpha>0} \frac{1}{2\pi} \int_{-\infty}^{\infty} ||\boldsymbol{G}(\alpha + j\omega)w(\alpha + j\omega)||^2 \, d\omega \\
&\leq \sup_{\alpha>0} \frac{1}{2\pi} \int_{-\infty}^{\infty} \overline{\sigma}[\boldsymbol{G}(\alpha + j\omega)]^2 ||w(\alpha + j\omega)||^2 \, d\omega \\
&\leq \sup_{\alpha>0} \sup_{\omega} \overline{\sigma}[\boldsymbol{G}(\alpha + j\omega)]^2 \sup_{\alpha>0} \frac{1}{2\pi} \int_{-\infty}^{\infty} ||w(\alpha + j\omega)||^2 \, d\omega \\
&= \sup_{\alpha>0} \sup_{\omega} \overline{\sigma}[\boldsymbol{G}(\alpha + j\omega)]^2 ||w||_2^2
\end{aligned}$$

The output $\boldsymbol{G}w$ will be finite only if the supremum term in the above inequality is finite for $w \in \mathcal{H}_2$. Of course $\boldsymbol{G}$ should be analytic in the open right half plane to guarantee that z belongs to $\mathcal{H}_2$. We therefore define the $\mathcal{H}_\infty$ space of matrix transfer functions as follows

$$\mathcal{H}_\infty = \{\boldsymbol{G} : \boldsymbol{G} \text{ is analytic in } \operatorname{Re} s > 0 \text{ and } ||\boldsymbol{G}||_\infty < \infty\} \tag{3.19}$$

where

$$||\boldsymbol{G}||_\infty = \sup_{\alpha>0} \left\{ \sup_{\omega} \bar{\sigma} \left[\boldsymbol{G}(\alpha + \jmath\omega)\right] \right\} \tag{3.20}$$

Clearly, a system that has its transfer matrix in $\mathcal{H}_\infty$ is a stable system and conversely if a system is stable it must belong to $\mathcal{H}_\infty$.

Now define a boundary function $\boldsymbol{G}_b(\jmath\omega) = \lim_{\alpha \to 0_+} \boldsymbol{G}(\alpha + \jmath\omega)$. If $\boldsymbol{G} \in \mathcal{H}_\infty$ then $\boldsymbol{G}_b(\jmath\omega)$ exists for almost all ω. Evaluation of $||\boldsymbol{G}||_\infty$ is simplified by evaluating

$$||\boldsymbol{G}||_\infty = \sup_{\omega} \bar{\sigma}[\boldsymbol{G}_b(\jmath\omega)]$$

Since $\boldsymbol{G}_b(\jmath\omega)$ is equal to $\boldsymbol{G}(\jmath\omega)$ almost everywhere on the imaginary axis, we may replace $\boldsymbol{G}_b(\jmath\omega)$ by $\boldsymbol{G}(\jmath\omega)$ and obtain

$$||\boldsymbol{G}||_\infty = \sup_{\omega} \bar{\sigma}[\boldsymbol{G}(\jmath\omega)] \tag{3.21}$$

Transfer matrices belonging to $\mathcal{H}_\infty$ space differ from transfer matrices belonging to $\mathcal{L}_\infty$ space in that the former are not only bounded on the imaginary axis (as required in the case of L_∞ spaces) but are also analytic and bounded in the open right half plane. In actual fact, $\mathcal{H}_\infty$ space is a closed sub space of $\mathcal{L}_\infty$ space.

$\mathcal{RL}_\infty$ and $\mathcal{RH}_\infty$ spaces

The following spaces are frequently referred to in our discussions.

1. $\mathcal{RL}_\infty$ space – space of matrices in $\mathbb{R}(s)^{p\times q}$ which have no poles on the imaginary axis (including the point at ∞)
2. $\mathcal{RH}_\infty$ space – Sub space of $\mathcal{RL}_\infty$ matrices which have no poles in $\operatorname{Re} s > 0$
3. $\mathcal{RH}_\infty^-$ space – sub space of $\mathcal{RL}_\infty$ matrices which have no poles in $\operatorname{Re} s < 0$.

All the above spaces are normed spaces but they are not complete with respect to the metric induced by their respective norms. Hence they are not Banach spaces.

The prefix $\mathcal{R}$ attached to $\mathcal{L}_\infty$ and $\mathcal{H}_\infty$ refers to the fact that we are dealing with real, rational matrices.

3.3.3 Induced Norms

Depending upon the norms employed for measuring the size of the input and output signals, the amplification changes and this leads to the definition of different types of induced norms.

Before going into details we shall discuss some preliminaries.

Consider a signal

$$u = \begin{bmatrix} u_1(t) \\ u_2(t) \\ \vdots \\ u_m(t) \end{bmatrix} \in \mathbb{R}^m$$

Its *Auto correlation function* is defined as

$$R_{uu}(\tau) = \lim_{T\to\infty} \frac{1}{2T} \int_{-T}^{T} u(t+\tau)u'(t)\,\mathrm{d}t \tag{3.22}$$

where we assume that the indicated limit in (3.22) exists. It can be shown that

$$R_{uu}(\tau) = R_{uu}(-\tau) \geq 0$$

The Fourier transform of $R_{uu}(\tau)$ (if it exists) is denoted by $S_{uu}(\jmath\omega)$. It is called the *spectral density of u.*

$$S_{uu}(\jmath\omega) = \int_{-\infty}^{\infty} R_{uu}(\tau)e^{-\jmath\omega\tau}\,\mathrm{d}\tau \tag{3.23}$$

$R_{uu}(\tau)$ could be obtained from $S_{uu}(\jmath\omega)$ by taking its inverse Fourier transform. Thus

$$R_{uu}(\tau) = \frac{1}{2\pi} \int_{-\infty}^{\infty} S_{uu}(\jmath\omega)e^{\jmath\omega\tau}\,\mathrm{d}\omega \tag{3.24}$$

Spectral density matrices are hermitian ($S_{uu}(\jmath\omega) = S_{uu}^*(\jmath\omega)$) and positive semi definite ($S_{uu}(\jmath\omega) \geq 0$)

Recall from Section 3.2.4 that the semi-norm associated with a bounded power signal is

$$||u||_{\mathcal{P}} = \left\{ \lim_{T\to\infty} \frac{1}{2T} \int_{-T}^{T} ||u(t)||^2 \mathrm{d}t \right\}^{\frac{1}{2}}$$

It therefore follows from (3.22) that

$$||u||_{\mathcal{P}} = [\text{trace } R_{uu}(0)]^{\frac{1}{2}} \tag{3.25}$$

from (3.24) we have for $\tau = 0$

$$R_{uu}(0) = \frac{1}{2\pi} \cdot \int_{-\infty}^{\infty} S_{uu}(\jmath\omega)\, d\omega$$

Hence the power norm of a bounded power signal can also be computed from its spectral power density function as follows

$$||u||_{\mathcal{P}} = \left[\frac{1}{2\pi} \int_{-\infty}^{\infty} \text{trace } [S_{uu}(\jmath\omega)]\, d\omega\right]^{\frac{1}{2}} \tag{3.26}$$

It can be deduced from the definitions of power norm and infinity norm that

$$||u||_{\mathcal{P}} \leq ||u||_{\infty}$$

Let $g(t)$ be the inverse Laplace transform of $\boldsymbol{G}(s)$ and let u and z be the input and output signals respectively.

The following standard properties (given without proof) are useful while considering the propagation of random signals through systems. (star multiplication in the formula indicates convolution)

$$R_{zu}(\tau) = g(\tau) * R_{uu}(\tau) \tag{3.27}$$

$$R_{zz}(\tau) = g(\tau) * R_{uu}(\tau) * g'(-\tau) \tag{3.28}$$

$$S_{zu}(\jmath\omega) = \boldsymbol{G}(\jmath\omega) S_{uu}(\jmath\omega) \tag{3.29}$$

$$S_{zz}(\jmath\omega) = \boldsymbol{G}(\jmath\omega) S_{uu}(\jmath\omega) \boldsymbol{G}^*(\jmath\omega) \tag{3.30}$$

2-norm of a system

With the above preliminaries out of the way we define the 2-norm of a system as the expected Root Mean Square (RMS) value of the output, when the input to the system is the realization of a unit variance white noise process.

If the input is a unit invariance white noise process then $S_{uu}(\jmath\omega) = I$ and from (3.30)

$$S_{zz}(\jmath\omega) = \boldsymbol{G}(\jmath\omega)\boldsymbol{G}^*(\jmath\omega)$$

and

$$\mathcal{E}\left\{\lim_{T\to\infty} \frac{1}{2T} \int_{-T}^{T} z'(t)z(t)\, dt\right\} = \frac{1}{2\pi} \int_{-\infty}^{\infty} \text{trace } [S_{zz}(\jmath\omega)\, d\omega$$

where $\mathcal{E}$ is the ‘expectation operator’.

Substituting for $S_{zz}(j\omega)$ in the above equation and applying the definition of 2-norm we obtain

$$||G||_2 = \left[\frac{1}{2\pi}\int_{-\infty}^{\infty} \text{trace}\,[G(j\omega)G^*(j\omega)]\,\mathrm{d}\omega\right]^{\frac{1}{2}} \tag{3.31}$$

The 2-norm of a matrix transfer function can also be characterized in terms of input-output experiments. Let e_i be the ith standard basis vector. Apply the impulse input $\delta(t)e_i$ and denote the output by $z_i(t)$. Then $z_i(t) \in \mathcal{L}_2[0, \infty)$ (note that the 2-norm diverges if G is not strictly proper) and we have

$$||G||_2^2 = \sum_i ||z_i||_2^2 \tag{3.32}$$

where $||z_i||_2^2$ is defined as in (3.3)

Induced norms

Let S_1 and S_2 be two normed spaces and let G be a mapping $G : S_1 \to S_2$

The induced norm of G is defined as

$$||G|| = \sup_{w \neq 0} \frac{||Gw||_{S_1}}{||w||_{S_2}} \tag{3.33}$$

It can be verified that the induced norm satisfies all properties associated with norms. Further it also possesses the sub-multiplicative property already explained earlier.

Let $z = Gw$ where $w \in \mathcal{L}_2$. According to (3.15)

$$||Gw||_2 \leq ||G||_\infty ||w||_2$$

Hence

$$\frac{||Gw||_2}{||w||_2} \leq ||G||_\infty$$

It can be shown that $||G||_\infty$ is not merely an upper bound but represents the least upper bound. Hence

$$||G|| = \sup_{w \neq 0} \frac{||Gw||_2}{||w||_2} = ||G||_\infty$$

Hence if we use 2-norms for the input and output measurement, then the supremum of the quotient is given by $||G||_\infty$. Further from (3.17) we know that

$$||G||_\infty = \sup_\omega \overline{\sigma}[G(j\omega)]$$

Note that the 2-norm of G viz. $||G||_2$ is not an induced norm. We have to make a distinction between the norm of a matrix transfer function induced by the 2-norm and the 2-norm of the transfer function. These two are entirely different quantities. Other possible norms could be assigned for inputs and outputs to obtain various induced norms.

For example we may consider the case where both input and output are bounded power signals. Representing by $\mathcal{P}$ the class of such signals we have $G : \mathcal{P} \to \mathcal{P}$.

From (3.26) and (3.30) it follows that

$$
\begin{aligned}
||z||_{\mathcal{P}}^2 &= \frac{1}{2\pi} \int_{-\infty}^{\infty} \text{trace } [S_{zz}(j\omega)] \, d\omega \\
&= \frac{1}{2\pi} \int_{-\infty}^{\infty} \text{trace } [G(j\omega) S_{ww}(j\omega) G^*(j\omega)] \, d\omega
\end{aligned}
$$

Noting that trace $(AB') =$ trace $(B'A)$ for all A, B matrices of the same dimension, we have

$$
\begin{aligned}
||z||_{\mathcal{P}}^2 &= \frac{1}{2\pi} \int_{-\infty}^{\infty} \text{trace } [G^*(j\omega) G(j\omega) S_{ww}(j\omega)] \, d\omega \\
&\leq \frac{1}{2\pi} \int_{-\infty}^{\infty} \text{trace } \{\overline{\sigma}[G(j\omega)]^2 S_{ww}(j\omega)\} \, d\omega \\
&\leq ||G||_{\infty}^2 \frac{1}{2\pi} \int_{-\infty}^{\infty} \text{trace } [S_{ww}(j\omega)] \, d\omega \\
&= ||G||_{\infty}^2 ||w||_{\mathcal{P}}^2
\end{aligned}
$$

It can be shown that $||G||_{\infty}$ represents the least upper bound. Hence the induced norm when the input and output signals are measured using the power norm is equal to $||G||_{\infty}$

3.4 Computation of $\mathcal{H}_2$ Norm

In this section we describe a method for computing the $\mathcal{H}_2$ norm of a stable, strictly proper rational transfer function $G(s)$.

Let (A, B, C) be the realization of transfer matrix $\boldsymbol{G}(s)$ with A assumed to be stable. For zero initial conditions we have

$$x(t) = \int_0^t e^{A(t-\tau)} Bu(\tau)\,d\tau$$

$$y(t) = \int_0^t Ce^{A(t-\tau)} Bu(\tau)\,d\tau$$

where x is the state, u the input and y the output associated with the system. We have

$$\boldsymbol{G}(s) = C(sI - A)^{-1}B \quad \text{and}$$

$$\boldsymbol{G}(t) = \begin{cases} Ce^{At}B & \text{for } t \geq 0 \\ 0 & \text{for } t < 0 \end{cases}$$

where $\boldsymbol{G}(t)$ is the inverse Laplace transform of $\boldsymbol{G}(s)$

By definition

$$\|\boldsymbol{G}\|_2^2 = \frac{1}{2\pi}\int_{-\infty}^{\infty} \text{trace}\,[\boldsymbol{G}(j\omega)\boldsymbol{G}^*(j\omega)]\,d\omega$$

$$= \int_{-\infty}^{\infty} \text{trace}\,[\boldsymbol{G}(t)\boldsymbol{G}'(t)]\,dt$$

The last step follows from Parseval's theorem.

Substituting for $\boldsymbol{G}(t)$ we have

$$\|\boldsymbol{G}\|_2^2 = \int_{-\infty}^{\infty} \text{trace}\,[Ce^{At}BB'e^{A't}C']\,dt$$

$$= \text{trace}\,[CPC']$$

where $P = \int_{-\infty}^{\infty} e^{At}BB'e^{A't}dt$, is the controllability gramian which is finite because $e^{At} \to 0$ as $t \to \infty$ on account of A being stable. We have thus proved that.

$$\|\boldsymbol{G}\|_2 = \sqrt{\text{trace}\,(CPC')} \tag{3.34}$$

Incidentally, we also demonstrate that P satisfies the Lyapunov equation

$$AP + PA' + BB' = 0$$

To show this, we note that

$$\frac{\mathrm{d}}{\mathrm{d}t}[e^{At}BB'e^{A't}] = Ae^{At}BB'e^{A't} + e^{At}BB'e^{A't}A'$$

Integrating both sides of the equation between the indicated limits, we have

$$e^{At}B \cdot B'e^{A't}\Big|_0^{\infty} = A\int_0^{\infty} e^{At}BB'e^{A't}\mathrm{d}t + \int_0^{\infty} e^{At}BB'e^{A't}\mathrm{d}tA'$$

Since e^{At} vanishes at $t = \infty$ (because A is stable) we have on simplification of the above equation

$$-BB' = AP + PA'.$$

and the result follows.

To summarise, $||\boldsymbol{G}||_2$ may be evaluated in two steps.
Step 1: Solve the matrix Lyapunov equation

$$AP + PA' + BB' = 0$$

for the symmetric matrix P.
Step 2: Evaluate $[\text{trace } (CPC')]^{\frac{1}{2}}$ which yields $||\boldsymbol{G}||_2$.
In an identical manner it can be shown that

$$||\boldsymbol{G}||_2 = \sqrt{\text{trace } B'QB} \tag{3.35}$$

where Q is the observability gramian satisfying the Lyapunov matrix equation

$$A'Q + QA + C'C = 0$$

Note that the 2-norm of a transfer matrix may be obtained in a finite number of steps

3.5 Computation of $\mathcal{H}_\infty$ Norm

In this section we describe a method for computing the $\mathcal{H}_\infty$ norm of a stable proper rational transfer matrix.

Recall that by definition, the $\mathcal{H}_\infty$ norm of a transfer matrix $\boldsymbol{G}(s)$ is given by

$$||\boldsymbol{G}||_\infty = \sup_{\omega \in (-\infty,\infty)} \bar{\sigma}[\boldsymbol{G}(j\omega)]$$

Calculating the $\mathcal{H}_\infty$ norm making use of the above definition is impractical as it involves a search over an infinite range of frequencies. In the case of

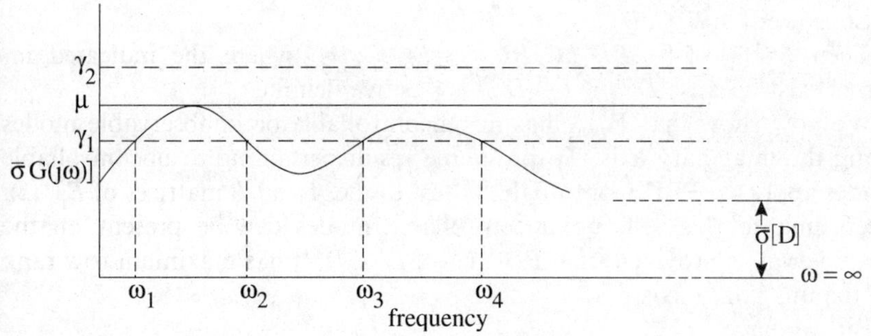

Figure 3.1. Variation of $\bar{\sigma}[\boldsymbol{G}(j\omega)]$with frequency.

SISO systems, the $\mathcal{H}_\infty$ norms can be directly read from the Nyquist plot, it being the radius of the smallest circle that circumscribes the plot. Generalizing this notion to MIMO systems, in Figure 3.1 we give a hypothetical plot of the maximum singular value of $\boldsymbol{G}(j\omega)$ versus ω as ω varies from zero to infinity.

Let (A, B, C, D) be a realization of $\boldsymbol{G}(s)$ with A assumed stable.

From the figure, $||\boldsymbol{G}||_\infty$ may be directly read. For the given plot its magnitude is μ. Let γ_1 and γ_2 be two real positive numbers such that $\gamma_1 < \mu < \gamma_2$

A line drawn parallel to the horizontal axis at γ_1 intersects the plot at a few points. A similar line drawn at γ_2 parallel to the horizontal axis does not intersect the plot at all. Since $\lim_{\omega \to \infty} \boldsymbol{G}(s) = D$, the singular value at $\omega = \infty$ is given by $\bar{\sigma}(D)$. Clearly, $||\boldsymbol{G}||_\infty \geq \bar{\sigma}(D)$. Assume for the time being that the chosen γ is greater than μ. We then have

$$\gamma^2 \boldsymbol{I} - \boldsymbol{G}^*(j\omega)\boldsymbol{G}(j\omega) > 0 \quad \forall \omega \in (-\infty, \infty)$$

with the above choice of γ we also have

$$\gamma^2 I - D'D > 0.$$

Let $\boldsymbol{F}(s) = \gamma^2 \boldsymbol{I} - \boldsymbol{G}'(-s)\boldsymbol{G}(s)$.
The state space realization if $\boldsymbol{F}(s)$can be shown to be

$$\boldsymbol{F}(s) = \left[\begin{array}{cc|c} A & 0 & -B \\ -C'C & -A' & C'D \\ \hline D'C & B' & \gamma^2 I - D'D \end{array}\right] \tag{3.36}$$

The next step is to obtain the state space realization of $\boldsymbol{F}^{-1}(s)$.

Let $\boldsymbol{F}(s) : (\tilde{A}, \tilde{B}, \tilde{C}, \tilde{D})$.

Then $\boldsymbol{F}^{-1}(s) : (\tilde{A} - \tilde{B}\tilde{D}^{-1}\tilde{C}, \tilde{B}\tilde{D}^{-1}, -\tilde{D}^{-1}\tilde{C}, \tilde{D}^{-1})$where the indicated inverses exist because $\tilde{D} = \gamma^2 I - D'D$ is positive definite.

We now show that $\boldsymbol{F}^{-1}(s)$has no uncontrollable or unobservable modes along the imaginary axis. To prove the result pertaining to uncontrollable modes apply the PBH Controllability test to the A and B matrices of $\boldsymbol{F}^{-1}(s)$. According to this test, no uncontrollable modes can be present on the imaginary axis provided $[\tilde{A} - \tilde{B}\tilde{D}^{-1}\tilde{C} - \lambda I \quad \tilde{B}\tilde{D}^{-1}]$ has maximum row rank on the imaginary axis.

But we have

$$\text{Rank } [\tilde{A} - \tilde{B}\tilde{D}^{-1}\tilde{C} - \lambda I \qquad \tilde{B}\tilde{D}^{-1}] = [\tilde{A} - \lambda I \qquad \tilde{B}\tilde{D}^{-1}]$$

This can be seen by post multiplying the second block column of left hand side matrix by $\tilde{C}$ and adding the same to its first block column. This being an elementary column operation, rank of the matrix remains unchanged. We have

$$[\tilde{A} - \gamma I] = \begin{bmatrix} A - \gamma I & 0 \\ -C'C & -A' - \gamma I \end{bmatrix}$$

But det $(\tilde{A} - \lambda I)$ which is equal to $-$det $(A - \lambda I)$ det $(A' + \lambda I)$does not vanish on the imaginary axis because A is stable.

Hence there are no uncontrollable modes for the inverse system on the imaginary axis. The results pertaining to unobservable modes could be similarly proved.

Hence the inverse system has no uncontrollable and unobservable modes on the imaginary axis. But it could still have controllable and observable modes located on the imaginary axis. But these modes are precisely the poles of $\boldsymbol{F}^{-1}(s)$. However, the poles of $\boldsymbol{F}^{-1}(s)$ are the zeros of $\boldsymbol{F}(s)$. If $\gamma^2 I - \boldsymbol{G}^*(\jmath\omega)\boldsymbol{G}(\jmath\omega) > 0$ (which is implied by the statement $||\boldsymbol{G}||_\infty < \gamma$), det $\boldsymbol{F}(s)$does not vanish on the imaginary axis and consequently $\boldsymbol{F}(s)$has no zeros there. Let the realization of $\boldsymbol{F}^{-1}(s)$ be denoted by $(\hat{A}, \hat{B}, \hat{C}, \hat{D})$. From the above reasoning we may conclude that provided $||\boldsymbol{G}||_\infty < \gamma$, none of the eigen values of $\hat{A}$ are located on the imaginary axis. The converse is easily seen to be true i.e. if none of the eigen values of $\hat{A}$ are located on the imaginary axis, then $||\boldsymbol{G}||_\infty < \gamma$. Spelling out the $\hat{A}$ matrix in fuller detail we have

$$\hat{A} = \begin{bmatrix} A & 0 \\ -C'C & -A' \end{bmatrix} - \begin{bmatrix} -B \\ C'D \end{bmatrix} \tilde{D}^{-1} [D'C \qquad B'] \tag{3.37}$$

which on simplification yields

$$\hat{A} = \begin{bmatrix} A + B\tilde{D}^{-1}D'C & B\tilde{D}^{-1}B' \\ -C'(I + D\tilde{D}^{-1}D')C & -(A + B\tilde{D}^{-1}D'C)' \end{bmatrix} \tag{3.38}$$

Note that the diagonal block matrices of $\hat{A}$ are transposes of each other and the off diagonal block matrices are symmetric. Such a matrix is known as the Hamiltonian matrix and has the generic form

$$H = \begin{bmatrix} A & R \\ -Q & -A' \end{bmatrix} \tag{3.39}$$

where R and Q are symmetric matrices. The role of H in the solution of algebraic Riccati equation and its properties are discussed in Chapter 8. Based on the theory developed above, the H_∞ norm of a transfer matrix may be obtained as follows.

1. Let $||\boldsymbol{G}||_\infty$ be denoted by μ. Obtain an upper bound γ_{ub} and a lower bound γ_{lb} for μ. A method of arriving at these bounds will be presented later.
2. Choose a value of $\gamma = \dfrac{\gamma_{ub} + \gamma_{lb}}{2}$
3. Compute the eigen values of $\hat{A}$. Note that they are symmetric about the imaginary axis because $\hat{A}$ is a Hamiltonian matrix.
4. If there are no eigen values on the imaginary axis, decrease γ. If the opposite is true then increase γ.
5. Based on a bisection search procedure estimate μ to any desired value of accuracy.

3.5.1 Upper and Lower Bounds of $||\boldsymbol{G}||_\infty$

From Figure 3.1 it is clear that

$$||\boldsymbol{G}||_\infty \geq \overline{\sigma}(\boldsymbol{D})$$

It can be shown that

$$||\boldsymbol{G}||_\infty \geq ||\boldsymbol{G}||_H$$

where $||\boldsymbol{G}||_H$ refers to the maximum Hankel singular value of $\boldsymbol{G}$ (Hankel singular values and their properties are discussed in Chapter 7).

We therefore have

$$\gamma_{lb} = \max[||\boldsymbol{G}||_H, \overline{\sigma}(\boldsymbol{D})] \tag{3.40}$$

In order to calculate the upper bound, we note that

$$\boldsymbol{G}(s) = \boldsymbol{G}_1(s) + \boldsymbol{D}$$

and

$$||\boldsymbol{G}||_\infty \leq ||\boldsymbol{G}_1||_\infty + ||\boldsymbol{D}||_\infty \quad \text{(triangle inequality)}$$

where G_1 is the strictly proper part of G. We also have

$$||G_1||_\infty \leq 2(\sigma_1 + \sigma_2 + \cdots + \sigma_r)$$

where $\sigma_1, \sigma_2, \ldots \sigma_r$ are the Hankel singular values of G_1. (See Glover 1984 for proof) It immediately follows that

$$\gamma_{ub} = 2(\sigma_1 + \sigma_2 + \cdots + \sigma_r) + \overline{\sigma}(D) \tag{3.41}$$

The choice of γ can be narrowed down to an even closer interval by nothing that if $\gamma < \mu$, the imaginary axis eigen values of $\hat{A}$ occur at frequencies where $\overline{\sigma}(G) = \gamma$. In a typical case shown in Figure 3.1, this occurs at four frequencies $\omega_1, \omega_2, \omega_3$ and ω_4. Let the mid frequencies of the corresponding frequency intervals be ω_{12}, ω_{23} and ω_{34}. The values of γ may then be chosen from max $\{\overline{\sigma}(G(j\omega_{12})), \overline{\sigma}(G(j\omega_{23})), \overline{\sigma}(G(j\omega_{34}))\}$

3.6 $\mathcal{H}_\infty$ Norm and Associated Algebraic Relationship

The central issue in this book is the minimizations of the $\mathcal{H}_\infty$ norm of a transfer matrix using appropriate stabilizing controllers. In our quest for arriving at a solution to this problem, we invariably, come across questions such as existence of solutions to the Riccati equation, its uniqueness and its ability to stabilize the system. We now state and prove a theorem which neatly provides answers to most of these questions raised. Explanations for various technical terms used such as Ric operator, domain of Ric operator (written as *dom* (Ric)) and the solution of the Riccati equation (which is a function of the Hamiltonian matrix H and written as *Ric* (H)) may be found in Chapter 8.

THEOREM 3.1
Suppose $G = C(sI - A)^{-1}B + D$ with A asymptotically stable. Let γ be a real number such that $S = \gamma^2 I - D'D > 0$. Then the following conditions are equivalent

(a) $||G||_\infty < \gamma$

(b) $H = \begin{bmatrix} A & 0 \\ -C'C & -A' \end{bmatrix} - \begin{bmatrix} -B \\ C'D \end{bmatrix} S^{-1} [D'C \quad B']$

has no eigen values on the imaginary axis

(c) $H \in dom$ (Ric)

(d) $H \in dom$ (Ric) and $Ric\ (H) \geq 0$

Proof: The equivalence of (*a*) and (*b*) has already been proved in Section 3.5.
The equivalence of (*b*) and (*c*) is established as follows.
To show that (b) ⇒ (c)
Let

$$H = \begin{bmatrix} \hat{A} & R \\ -Q & -\hat{A} \end{bmatrix}$$

where

$$\hat{A} = A + BS^{-1}D'C; R = BS^{-1}B'$$

$$Q = C'(I + DS^{-1}D')C$$

To show that $H \in dom$ (Ric), three conditions have to be satisfied (see lemma 8.2 in Chapter 8). These are: (i) H has no eigen values on the imaginary axis; (ii) R is positive or negative semi-definite; (iii) $(\hat{A}, R)$ is stabilizable. If we assume (b), then H has no eigen values on the imaginary axis. Because of the non-singularity of S, the matrix R is sign-definite. It remains to be shown that $(\hat{A}, R)$ is stabilizable (Note that A is stable by assumption). This can be shown if there exists a matrix F such that
$A + BS^{-1}D'C + BS^{-1}B'F$ is stable. The proof is omitted for the sake of brevity. With the three conditions cited above satisfied $H \in dom$ (Ric). Hence the following Riccati equation is satisfied by X (denoted as Ric (H)). We have

$$X\hat{A} + \hat{A}'X + XRX + Q = 0$$

on substituting for the matrix $\hat{A}, R$ and Q we have

$$X(A + BS^{-1}D'C) + (A + BS^{-1}D'C)'X$$

$$+XBS^{-1}B'X + C'(I + DS^{-1}D')C = 0 \tag{3.42}$$

Further $(\hat{A} + RX)$ is stable.

We show that (c) ⇒ (d) as follows. We start with the hypothesis that $H \in dom$ (Ric) and $X = Ric(H)$. Hence X satisfies (3.42). This equation may be re-written as

$$XA + A'X + (D'C + B'X)'S^{-1}(D'C + B'X) + C'C = 0 \tag{3.43}$$

Noting that A is stable and that $S^{-1} > 0$, it follows that (3.43) is the Lyapunov equation associated with the observability gramian of

$$\left(\begin{bmatrix} C \\ S^{-\frac{1}{2}}(D'C + B'X) \end{bmatrix}, A \right)$$

Invoking the well known property of Lyapunov equation (see Theorem 3.2), a stable A-matrix gives rise to a solution X for 3.43 which is positive semi-definite. The proof is now complete.

The importance of Theorem 3.1 rests on the fact that it lays down an equivalent algebraic condition in state space for the statement $||G||_\infty < \gamma$ in the frequency domain.

3.7 All Pass Systems

By an *all pass* system we mean a system which permits signals of all frequencies to pass through without their undergoing any change in magnitude. In contrast to this, we have a *band pass* system which allows only signals in a particular frequency band to pass through, cutting off all other signals. Similarly *low pass* and *high pass* systems do the same thing to signals in the low frequency and high frequency ranges. Let G be a linear mapping from a signal space S_1 to a signal space S_2 (both are normed spaces). G is said to be all pass if

$$||Gx||_{S_2} = ||x||_{S_1} \quad \forall x \in S_1$$

From a geometric viewpoint, the signal gets merely rotated, with its magnitude remaining unchanged.

If the mapping is from one Hilbert space to another then it can be shown that for an all pass system, the inner products in the two spaces are preserved. Thus

$$< Gx, Gy >_{S_2} = < x, y >_{S_1} \quad \forall x, y \in S_1$$

But $\langle Gx, Gy \rangle_{S_2} = \langle x, G^{\sim} Gy \rangle_{S_1}$, where $G^{\sim}$ is a unique operator known as the adjoint operator.
Hence

$$\langle x, G^{\sim} Gy \rangle_{S_1} = \langle x, y \rangle_{S_1}$$

It therefore follows that for an all pass system

$$G^{\sim} G = I$$

If the spaces S_1 and S_2 are of the same dimension then $G^{\sim} = G^{-1}$.

All pass systems in state-space can be characterized with the help of the following theorem.

THEOREM 3.2

Let $G = \left[\begin{array}{c|c} A & B \\ \hline C & D \end{array}\right]$ with (C, A) detectable and let $Q = Q'$ satisfy the Lyapunov equation

$$A'Q + QA + C'C = 0$$

Then the following hold.

a) $Q \geq 0$ if and only if A is stable
b) $D'C + B'Q = 0$ implies $\mathbf{G}^{\sim}\mathbf{G} = D'D$
c) $Q \geq 0$, (A, B) controllable and $\mathbf{G}^{\sim}\mathbf{G} = D'D$ implies $D'C + B'Q = 0$.

Proof: Detailed proof is omitted. Instead the following heuristic explanation is offered.

Explanation: We note that the observability gramian Q is defined as

$$Q = \int_0^\infty e^{A't} C' C e^{At} \, dt$$

Further, the output y (with initial condition x_0), corresponding to zero input response is given by

$$y = Ce^{At} x_0$$

Using the spectral expansion of e^{At} , one can easily see that the eigen vectors corresponding to the unobservable modes are anihilated by C. Thus only the observable modes are present in y. Further we have

$$\int_0^\infty y'y \, dt = x_0' \int_0^\infty e^{A't} C' C e^{At} dt \, x_0 = x_0' Q x_0$$

If A is stable then the above integral converges and Q is finite. If (C, A) is detectable, then all the unobservable modes are located in the open left-half plane. The observable modes (all of which are present in y) influence the observability grammian Q. Since Q is obtained as an integral of $y'y$, it is indeed positive semi-definite and is finite only if there are no unstable observable modes. Hence $Q \geq 0$ implies that A is stable provided there are no unobservable unstable modes i.e. (C, A) is detectable. The converse is straightforward, because if A is stable then the integral will converge and $Q \geq 0$. In fact we may even tighten the hypothesis further and stipulate that (C, A) be observable instead of merely being detectable. In that case y contains every mode of e^{At} and $\int_0^\infty y'y \, dt$ can be zero if and only if $x_0 = 0$. This implies Q is positive definite provided A is stable. The converse is also true. If in the above theorem, detectability of (C, A) assumption is omitted, then A being stable implies $Q \geq 0$ but $Q \geq 0$ does not imply that A is stable.

(b) is proved as follows:
Noting that $G^{\sim}(s) = G'(-s)$ we have

$$\begin{aligned} G^{\sim}G &= \{D' + B'(-sI - A')^{-1}C'\}\{D + C(sI - A)^{-1}B\} \\ &= D'D + B'(-sI - A')^{-1}C'D + D'C(sI - A)^{-1}B \\ &\quad + B'(-sI - A')^{-1}C'C(sI - A)^{-1}B \end{aligned}$$

Now substitute $-(A'Q + QA)$ for $C'C$ in the above equation. We then have

$$\begin{aligned} G^{\sim}G &= D'D + B'(-sI - A')^{-1}C'D + D'C(sI - A)^{-1}B \\ &\quad + B'(sI - A')^{-1}(-A'Q - QA)(sI - A)^{-1}B \\ &= D'D + B'(-sI - A')^{-1}C'D + D'C(sI - A)^{-1}B \\ &\quad + B'(-sI - A')^{-1}[Q(sI - A) + (-sI - A')Q](sI - A)^{-1}B \\ &= D'D + B'(-sI - A')^{-1}C'D + D'C(sI - A)^{-1}B \\ &\quad + B'(-sI - A')^{-1}QB + B'Q(sI - A)^{-1}B \\ &= D'D + B'(sI - A)^{-1}(C'D + QB) + (D'C + B'Q)(sI - A)^{-1}B \end{aligned}$$

But $D'C + B'Q = 0$ by hypothesis.
Hence $G^{\sim}G = D'D$

3.8 The Adjoint Operator

Consider two Hilbert spaces S_1 and S_2 and let G be a linear operator mapping S_1 and S_2. Then it can be shown that there exists a unique linear operator $G^{\sim}$ mapping S_2 to S_1 such that.

$$\langle Gw, y\rangle_{S_2} = \langle w, G^{\sim}y\rangle_{S_1}$$

where $w \in S_1$ and $y \in S_2$

$G^{\sim}$ is known as the adjoint of the operator G. Let the state space realization of the system be

$$G : (A, B, C, D)$$

Then we have

$$G(s) = C(sI - A)^{-1}B + D$$

and

$$G(t) = Ce^{At}B + D\delta(t)$$

where $G(t)$ is the inverse Laplace transform of $G(s)$ and $\delta(t)$ is the unit impulse function. Let $w(t) \in \mathcal{L}_2[0, \infty)$ be the system input.

Then the system output via the convolution integral is given by

$$(\boldsymbol{G}w)(t) = \int_0^{\infty} [Ce^{A(t-\tau)}B + D\delta(t-\tau)]w(\tau)\,\mathrm{d}\tau \quad \text{for } t \geq \tau$$

For any $y \in \mathcal{L}_2[0, \infty)$ we have

$$\langle \boldsymbol{G}w, y\rangle = \int_0^{\infty} y'(t)\,\mathrm{d}t \int_0^{\infty} [Ce^{A(t-\tau)}B + D\delta(t-\tau)]w(\tau)\,\mathrm{d}\tau$$
$$= \int_0^{\infty} \mathrm{d}t \int_0^{\infty} [\{B'e^{A'(t-\tau)}C' + D'\delta(t-\tau)\}y(t)]'w(\tau)\,\mathrm{d}\tau$$

Interchanging the order of integration, we have

$$\langle \boldsymbol{G}w, y\rangle = \int_0^{\infty} \mathrm{d}\tau \int_0^{\infty} [\{B'e^{A'(t-\tau)}C' + D'\delta(t-\tau)\}y(t)]'\,\mathrm{d}t w(\tau)$$

Let $\eta(\tau) = \int_0^{\infty} \{B'e^{A'(t-\tau)}C' + D'\delta(t-\tau)\}y(t)\,\mathrm{d}t$

Then

$$< \boldsymbol{G}w, y > = < w, \eta >$$

Hence

$$(\boldsymbol{G}^{\sim}y)(\tau) = \int_0^{\infty} [B'e^{A'(t-\tau)}C' + D'\delta(t-\tau)]y(t)\mathrm{d}t$$

and

$$(\boldsymbol{G}^{\sim}y)(t) = -\int_{\infty}^{0} [B'e^{-A'(t-\tau)}C' + D'\delta(t-\tau)]y(\tau)d\tau \qquad \text{for } \tau \geq t$$

The adjoint operator may be associated with a system with input $y(t)$ and output $\eta(t)$. The system realization follows from the above equation. We have

$$\boldsymbol{G}^{\sim} : (-A', -C', B', D')$$

whose state space representation is

$$\dot{p}(t) = -A'p(t) - C'y(t) \quad p(t)|_{t=\infty} = 0$$
$$\eta(t) = B'p(t) + D'y(t)$$

The transfer matrix of the adjoint system is

$$G^{\sim}(s) = -B'(sI + A')C' + D'$$

which is nothing but $G'(-s)$. Hence

$$G^{\sim}(s) = G'(-s)$$

We now show that the linear operator G and its adjoint $G^{\sim}$ have identical norms. Recall from (3.31) that the 2-norm of the transfer matrix G is

$$\|G\|_2^2 = \frac{1}{2\pi}\int_{-\infty}^{\infty} \text{trace } [G(j\omega)G^*(j\omega)]\, d\omega$$

Applying this formula to obtain $\|G^{\sim}\|_2^2$ we have

$$\|G^{\sim}\|_2^2 = \frac{1}{2\pi}\int_{-\infty}^{\infty} \text{trace } [G^*(j\omega)G(j\omega)]\, d\omega$$

(Note that along the imaginary axis $G^{\sim}(j\omega) = G^*(j\omega)$)

Since trace $[G^*(j\omega)G(j\omega)] =$ trace $[G(j\omega)G^*(j\omega)]$.

We have $\|G\|_2 = \|G^{\sim}\|_2$.

The ∞ -norm of transfer matrix G is given by (see (3.17))

$$\|G\|_\infty = \sup_\omega \bar{\sigma}[G(j\omega)]$$

But $\bar{\sigma}[G(j\omega)] = \lambda_{\max}^{\frac{1}{2}}[G^*(j\omega)G(j\omega)] = \lambda_{\max}^{\frac{1}{2}}[G(j\omega)G^*(j\omega)]$ where $\lambda_{\max}^{\frac{1}{2}}\,[\cdot]$ denotes the positive square root of the maximum eigen value of the argument.

Hence $\|G\|_\infty = \|G^{\sim}\|_\infty$

3.9 Inverse Systems

Let system $G : (A, B, C, D)$ be characterized by the state equations

$$\dot{x} = Ax + Bu$$

$$y = Cx + Du$$

where x is the state, u is the input and y is the output. The required inverse system (if one exists) has y as its input and u as its output. This system operating in tandem with the original system yields a composite system with transfer matrix unity.

To obtain the inverse system we proceed as follows. From the output equation of the original system we obtain, provided D is non-singular,

$$u = -D^{-1}Cx + D^{-1}y$$

Substituting for u in the state equation of the original system we have

$$\dot{x} = (A - BD^{-1}C)x + BD^{-1}y$$

Hence the inverse system G^{-1} has the state space realization

$$G^{-1}(A - BD^{-1}C, BD^{-1}, -D^{-1}C, D^{-1}) \tag{3.44}$$

So long as $G(s)$ is square and $G^{-1}(s)$ exists, it is immaterial if the inverse system is placed before or after the original system because $GG^{-1} = G^{-1}G = I$. However for non square systems with m inputs and p outputs, one has to distinguish between left and right inverses. If $p \leq m$, then the right inverse $G^{\dagger}$ satisfies the relation $GG^{\dagger} = I_p$. The right inverse exists only if rank of $D = p$. If $p \geq m$ and the rank of $D = m$ then we have the left inverse $G^{\dagger}G = I_m$

When $p = m$ then we have the case of the square matrix discussed earlier. Whether we are considering the right or the left inverse, the realization of the inverse is always given by (3.44). However for non-square transfer matrices D^{-1} may be replaced by $D^{\dagger}$ where $D^{\dagger}$ is the pseudo-inverse of D. We have $DD^{\dagger} = I_p$ in the case of the right inverse and $D^{\dagger}D = I_m$ in the case of the left inverse. That the inverse system thus obtained leads to unit transfer matrix when cascaded with the original system can be verified. For proving this we recall the following standard result.

If $G_1 : (A_1, B_1, C_1, D_1)$ and $G_2 : (A_2, B_2, C_2, D_2)$ are cascaded in such a way that the composite transfer function is G_1G_2, then its state space realization is

$$G_1G_2 : \left\{ \begin{bmatrix} A_1 & B_1C_2 \\ 0 & A_2 \end{bmatrix}, \begin{bmatrix} B_1D_2 \\ B_2 \end{bmatrix}, [C_1 \quad D_1C_2], [D_1D_2] \right\}$$

If for example the system possesses a right inverse, applying the above formula, we get

$$GG^{\dagger} = \begin{bmatrix} A & B \\ C & D \end{bmatrix} * \begin{bmatrix} A - BD^{\dagger}C & BD^{\dagger} \\ -D^{\dagger}C & D^{\dagger} \end{bmatrix}$$

$$= \left[\begin{array}{cc|c} A & -BD^{\dagger}C & BD^{\dagger} \\ 0 & A - BD^{\dagger}C & BD^{\dagger} \\ \hline C & -C & I_p \end{array}\right]$$

Let $GG^{\dagger} : (\hat{A}, \hat{B}, \hat{C}, \hat{D})$.

Now apply a similarity transformation

$$T = \begin{bmatrix} I & I \\ 0 & I \end{bmatrix}$$

The transfer matrix does not change under such a transformation. Hence

$$\boldsymbol{G}\boldsymbol{G}^{\dagger} = (T^{-1}\hat{A}T, T^{-1}\hat{B}, \hat{C}T, \hat{D})$$

We have

$$T^{-1}\hat{A}T = \begin{bmatrix} A & 0 \\ 0 & A - BD^{\dagger}C \end{bmatrix}; \quad T^{-1}\hat{B} = \begin{bmatrix} 0 \\ BD^{\dagger} \end{bmatrix}$$
$$\hat{C}T = [C \quad 0]; \quad \hat{D} = I_p$$

with the above coordinate transformation, $\boldsymbol{G}\boldsymbol{G}^{\dagger}$ can be easily calculated and we have $\boldsymbol{G}\boldsymbol{G}^{\dagger} = I_p$ which proves that $\boldsymbol{G}^{\dagger}$ indeed represents the right inverse of $\boldsymbol{G}$. In the above formula $D^{\dagger}$ can be obtained from the singular value decomposition of D as explained in Appendix C.

3.10 Summary

In science and engineering we deal with a wide variety of signals. The signals of primary interest are the $\mathcal{L}_2(-\infty, \infty), \mathcal{L}_2[0, \infty)$ and $\mathcal{L}_2(-\infty, 0]$ signals. They are important because all of them possess finite energy and are close to the signals we come across in nature. These signal spaces are Hilbert spaces where the metric is induced by the respective inner products. $\mathcal{L}_2(-\infty, 0]$ and $\mathcal{L}_2[0, \infty)$ are closed sub spaces of $\mathcal{L}_2(-\infty, \infty)$. Further they are orthogonal sub spaces. We have

$$\mathcal{L}_2(-\infty, \infty) = \mathcal{L}_2(-\infty, 0] \overset{\perp}{\oplus} \mathcal{L}_2[0, \infty)$$

The familiar Fourier transform is a Hilbert-space isomorphism from the time domain 2-space $\mathcal{L}_2(-\infty, \infty)$ to the frequency domain 2-space $\mathcal{L}_2$. This leads us logically to the so called Hardy spaces $\mathcal{H}_2$ and $\mathcal{H}_\infty$.

$\mathcal{H}_2$ space consists of functions which are analytic and bounded in the open right, half plane and bounded on the imaginary axis. The norm employed for testing boundedness is the 2-norm. $\mathcal{H}_2$ space is a closed sub space of $\mathcal{L}_2$ space. We may in an identical fashion define $\mathcal{H}_2^{\perp}$ space, as a space analytic and bounded in the open left half plane and bounded on the imaginary axis. $\mathcal{H}_2^{\perp}$ is an orthogonal complement of $\mathcal{H}_2$ in $\mathcal{L}_2$ space.

$\mathcal{H}_\infty$ space consists of functions which are analytic and bounded (in the ∞ - norm sense) in the open right half plane and bounded on the imaginary axis.

Similarly $\mathcal{H}_\infty^-$ space consists of functions which are analytic and bounded in the open left half plane and bounded on the imaginary axis. They are sub spaces of $\mathcal{L}_\infty$ space. The Laplace transform is a Hilbert space isomorphism from the time domain 2-space $\mathcal{L}_2[0,\infty)$ to the frequency domain 2-space $\mathcal{H}_2$.

The concept of norms applied to signals can be extended to cover systems represented by matrix valued transfer functions. The norms often used in control theory are the $\mathcal{H}_2$ and $\mathcal{H}_\infty$ norms. Both of them are defined for stable transfer matrices. They are respectively,

$$||\boldsymbol{G}||_2^2 = \frac{1}{2\pi}\int_{-\infty}^{\infty} \text{trace } [\boldsymbol{G}^*(\jmath\omega)\cdot\boldsymbol{G}(\jmath\omega)]\, d\omega$$

and

$$||\boldsymbol{G}||_\infty = \sup_\omega \overline{\sigma}[\boldsymbol{G}(\jmath\omega)]$$

In simple words, $\mathcal{H}_2$ norm is the square root of the integral of the squared magnitude of the transfer function on the imaginary axis. It measures the output power, assuming that the input power is a white Gaussian stochastic process of unit intensity. The $\mathcal{H}_\infty$ norm is the supremum of the magnitude of the transfer function evaluated on the imaginary axis. It measures maximal energy gain.

Depending upon the norms employed for measuring input and output signals, different induced system gains can be obtained. Thus the induced norms for energy to energy and power to power are both ∞-norms. The 2-norm can be calculated from the following formula

$$||G||_2 = \sqrt{\text{trace } (CPC')} = \sqrt{\text{trace } (B'QB)}$$

where P and Q are controllability and observability gramians.

The ∞-norm of a transfer matrix can be evaluated to any desired degree of accuracy by iterative methods. The algorithm proposed is based on the result that $||\boldsymbol{G}||_\infty < \gamma$ is equivalent to the associated Hamiltonian matrix having no eigen values on the imaginary axis. It is possible to fix upper and lower bounds for $||\boldsymbol{G}||_\infty$ and this helps in the initial choice of γ. An important result which throws much light on the solution of the Riccati equation associated with the problem $||\boldsymbol{G}||_\infty < \gamma$ is contained in Theorem 3.1. This theorem is similar to the so called bounded real lemma referred to in literature (see Green and Limebeer 1995). The closing part of this chapter deals with three important system categories – the all pass system, the adjoint system and the inverse system. The all pass system allows signals of all frequencies to pass through without any change in magnitude.

Given a system $\boldsymbol{G} : (A, B, C, D)$ its adjoint $\boldsymbol{G}^\sim$ has a realization $\boldsymbol{G}^\sim : (-A', -C', B', D')$. This property is frequently invoked in our discussions in subsequent chapters. The last topic covered, namely inverse systems has no practical use in control theory. This is mainly because their introduction may produce unstable pole-zero cancellation. Their main usefullness is in theoretical studies as demonstrated in this chapter, were we derived the Hamiltonian matrix as the A-matrix of the inverse of the system $\gamma^2 I - G^{\sim}(s)G(s)$.

Notes and Additional References

For a detailed coverage of Hardy spaces the reader may refer to two books – one by Koosis (1980) and the other by Duren (1970). A very readable account of Hilbert spaces and their properties is given in the book by Young (1988).

Exercises

1. Let u be a bounded power signal and let $||u||_{\mathcal{P}} < \infty$. Show that $||u||_{\mathcal{P}} < ||u||_{\infty}$.
2. The 1-norm of a signal $g(t)$ is given by $||g||_1 = \int_{-\infty}^{\infty} |g(t)| \mathrm{d}t$

   If $g(s)$ is stable and strictly proper, show that $||g||_1 < \infty$ and obtain an inequality connecting $||g||_{\infty}$ and $||g||_1$.
3. Let $g(s)$ be a scalar transfer function which is stable and proper. Obtain the induced system gain when the input u and the output y are measured by their respective power norms $||u||_{\mathcal{P}}$ and $||y||_{\mathcal{P}}$.
4. Let $g(s)$ be strictly proper with no poles on the imaginary axis. Then we have

$$||g||_2^2 = \frac{1}{2\pi} \int_{-\infty}^{\infty} g^*(j\omega) g(j\omega) \, \mathrm{d}\omega$$

$$= \frac{1}{2\pi j} \int_{-j\infty}^{j\infty} g(-s) g(s) \, \mathrm{d}s$$

$$= \frac{1}{2\pi j} \oint g(-s) g(s) \, \mathrm{d}s$$

where the contour integration is taken along the imaginary axis and around a semi-circle of infinite radius in the left half plane. The contribution to the integral from the semi-circle is zero as $g(s)$ is strictly proper. By residue theorem $||g||_2^2$ is equal to the sum of the residues of $g(-s)g(s)$ at its poles in the left half plane. Using the above principle calculate the $||g||_2$ norm of the transfer function $g(s) = \dfrac{s+1}{(s+2)(s-3)}$.

5. Let A and B be transfer matrices of compatible order. Prove that $||AB||_\infty \leq ||A||_\infty ||B||_\infty$. This is known as the sub-multiplicative property. Show that the 2-norm of transfer matrices does not satisfy this property.
6. Show that a linear system and its adjoint system have the same induced norm.
7. The transfer function of a system is given by

$$g(s) = \frac{6}{(s+1)(s+2)(s+3)}$$

obtain a state-space realization of the transfer function and hence calculate $||g||_2$ by using the formula $||g||_2 = \sqrt{\text{trace } CPC'}$ where P is the controllability gramian.

8. Consider the transfer function $g(s) = \dfrac{1}{s+4}$. Check whether $||g||_\infty < 5$ by first obtaining the Hamiltonian matrix and verifying whether any of its eigen values are located on the imaginary axis. Repeat the same procedure for $||g||_\infty < 0.4$.
9. Given $g(s) = \dfrac{1}{s+a}$ where a is a positive constant. What is the lower bound of γ in the inequality $||g||_\infty < \gamma$ so that the Riccati equation (3.42) yields a stabilizing solution i.e. a solution X which makes $\hat{A} + RX$ stable. $\hat{A}$ and R are as defined in Theorem 3.1.
10. If $[G_1 \quad G_2] = \left[\begin{array}{c|cc} A & B_1 & B_2 \\ \hline C & D_1 & D_2 \end{array}\right]$ obtain a state space realization of $G_1 G_2^{-1}$. Assume D_1 and D_2 to be nonsingular.
11. Let Q satisfy the Lyapunov matrix equation

$$A'Q + QA + C'C = 0$$

Prove the following:

i) A being asymptotically stable and (C, A) observable implies $Q > 0$.

ii) $Q \geq 0$ implies that unstable modes of A (if any) are unobservable.

4

Matrix Fraction Description

4.1 Introduction

The state space and transfer function description of linear time invariant-systems are quite familiar to control engineers. There is yet another type of description known as the Matrix Fraction Description (MFD) which is now being increasingly used in control theory literature. This chapter deals with the basics of MFD of systems. Section 4.2 introduces the notion of equivalence of polynomial matrices. Using the equivalence concept it is shown how Hermite and Smith forms could be generated via unimodular transformations. In Section 4.3, the coprimeness property is explained with reference to polynomial matrices and tests for coprimeness are established. In Section 4.4 the subject of study pertains to matrix valued rational functions. Factorization of these matrices in terms of polynomial matrices is investigated. The well known Smith–McMillan form is introduced in Section 4.5. In Section 4.6, results obtained in the earlier sections are extended to cover factorization of a transfer matrix over a ring of stable matrices. This section further discusses the fractional representation approach from a more general viewpoint. In Section 4.7 it is shown how, given the state space realization of a transfer matrix, it is possible to obtain the state space description of right and left-coprime factors associated with the transfer matrix.

4.2 Factorization of Polynomial Matrices

A study of polynomial matrix factorization is important because it is a basic building block in the analysis and synthesis of systems. We begin with some definitions.

DEFINITION 4.1 (Right divisor, left multiple)
Let $\boldsymbol{A}(s) = \boldsymbol{B}(s)\boldsymbol{C}(s)$ where $\boldsymbol{A}(s), \boldsymbol{B}(s)$ and $\boldsymbol{C}(s)$ are polynomial matrices. Then $\boldsymbol{C}(s)$ is called the *right divisor* of $\boldsymbol{A}(s)$ and $\boldsymbol{A}(s)$ the *left multiple* of $\boldsymbol{C}(s)$.

Analogously we may define $\boldsymbol{B}(s)$ as the *left divisor* of $\boldsymbol{A}(s)$ and $\boldsymbol{A}(s)$ as the *right multiple* of $\boldsymbol{B}(s)$

DEFINITION 4.2 (Common right divisor)
Let $\boldsymbol{A}(s) = \boldsymbol{A}_1(s)\boldsymbol{R}(s)$ and $\boldsymbol{B}(s) = \boldsymbol{B}_1(s)\boldsymbol{R}(s)$ Then $\boldsymbol{R}(s)$ is called the *Common right divisor* of $\boldsymbol{A}(s)$ and $\boldsymbol{B}(s)$

Analogously if $\boldsymbol{A}(s) = \boldsymbol{L}(s)\boldsymbol{A}_2(s)$ and $\boldsymbol{B}(s) = \boldsymbol{L}(s)\boldsymbol{B}_2(s)$, than $\boldsymbol{L}(s)$ is called the *Common left divisor* of $\boldsymbol{A}(s)$ and $\boldsymbol{B}(s)$

DEFINITION 4.3 (Greatest common right divisor)
Let $\boldsymbol{R}(s)$ in Definition 4.2 be the left-multiple of every common right divisor of $\boldsymbol{A}(s)$ and $\boldsymbol{B}(s)$. Then $\boldsymbol{R}(s)$ is called the *greatest-common right-divisor (g.c.r.d) of $\boldsymbol{A}(s)$ and $\boldsymbol{B}(s)$.*

Similarly we may define the *greatest-common left-divisor* (g.c.l.d) of $\boldsymbol{A}(s)$ and $\boldsymbol{B}(s)$

Let $\boldsymbol{R}(s)$ be the g.c.r.d. of $\boldsymbol{A}(s)$ and $\boldsymbol{B}(s)$. Then it can be shown that $\boldsymbol{U}(s)\boldsymbol{R}(s)$ is also a g.c.r.d of $\boldsymbol{A}(s)$ and $\boldsymbol{B}(s)$ where $\boldsymbol{U}(s)$ is a unimodular matrix (Square polynomial matrices whose determinant is a non-zero constant) The g.c.r.d thus obtained is unique up to a left unimodular factor. Similarly g.c.l.d is unique up to a right unimodular factor

4.2.1 Equivalent Matrices and Standard Forms

Postmultiplying and premultiplying polynomial matrices by appropriate unimodular matrices it is possible to reduce them to the so called standard forms. These unimodular matrices are obtained as a product of a succession of elementary operations

DEFINITION 4.4 (Unimodular matrix)
A polynomial matrix $\boldsymbol{M} \in \mathbb{R}\,[s]^{n\times n}$ is *unimodular* or invertable if $\det \boldsymbol{M} = k$, a non-zero constant. If $\boldsymbol{M}$ is unimodular $\boldsymbol{M}^{-1}$ exists and belongs to $\mathbb{R}\,[s]^{n\times n}$

Elementary row and column operations: Elementary row operations on a polynomial matrix comprise:

1. Interchange of ith row (ρ_i) with j^{th} row (ρ_j)
2. Multiplication of a row by a non-zero constant k. Thus ρ_i is replaced by $k\rho_i$

3. For $j \neq i$, add r times the jth row (where $r \in \mathbb{R}[s]$) to the ith row i.e. ρ_i replaced by $\rho_i + r\rho_j$

Elementary row operations on $\boldsymbol{M} \in \mathbb{R}[s]^{m\times n}$ are performed by first carrying out the relevant row operations on the unit matrix of appropriate order and them premultiplying $\boldsymbol{M}$ by this matrix. It is easily verified that elementary row operations on a polynomial matrix leave its normal rank unchanged.

Elementary column operations may be similarly defined. The corresponding operations are:

1. Interchange of column γ_i with column γ_j
2. Multiplication of column γ_i by a non-zero constant k
3. Adding to column γ_i, r times column γ_j where $r \in \mathbb{R}[s]$

Elementary column operations on $\boldsymbol{M}$ are equivalent to postmultiplying $\boldsymbol{M}$ by an elementary matrix obtained from the unit matrix by performing the column operations on it.

Definition 4.5 (Equivalent matrices)
$\boldsymbol{A}(s)$ and $\boldsymbol{B}(s) \in \mathbb{R}[s]^{m\times n}$ are said to be equivalent if there exists unimodular matrices $\boldsymbol{L} \in \mathbb{R}[s]^{m\times m}$ and $\boldsymbol{R} \in \mathbb{R}[s]^{n\times m}$ such that

$$\boldsymbol{A} = \boldsymbol{LBR}$$

$\boldsymbol{A}$ equivalent to $\boldsymbol{B}$ is represented as $\boldsymbol{A} \sim \boldsymbol{B}$.
$\boldsymbol{A}$ right equivalent to $\boldsymbol{B}$ is represented as $\boldsymbol{A} \underset{r}{\sim} \boldsymbol{B}$.
$\boldsymbol{A}$ left equivalent to $\boldsymbol{B}$ is represented as $\boldsymbol{A} \underset{l}{\sim} \boldsymbol{B}$.
Note that $\boldsymbol{A} \sim \boldsymbol{A}$; $\boldsymbol{A} \sim \boldsymbol{B} \Rightarrow \boldsymbol{B} \sim \boldsymbol{A}$ and $\boldsymbol{A} \sim \boldsymbol{B}$ and $\boldsymbol{B} \sim \boldsymbol{C} \Rightarrow \boldsymbol{A} \sim \boldsymbol{B} \sim \boldsymbol{C}$.

Standard Forms
Through the application by elementary operations polynomial matrices may be reduced to standard forms. The advantage of standard form is that it reveals the basic structure of the matrix and serves as a typical representation of a whole class of equivalent matrices. Two commonly used standard forms are 1. Hermite form; 2. Smith form

Theorem 4.1 (Hermite form)
Let $\boldsymbol{M} \in \mathbb{R}[s]^{m\times n}$. Then there exists a unimodular matrix $\boldsymbol{L} \in \mathbb{R}[s]^{m\times m}$ (obtained from elementary row operations) such that

$$\boldsymbol{LM} = \boldsymbol{H}$$

where H is the Hermite form. This form is upper quasi-triangular and has the following properties:

1. Rank H = Rank $M = r \leq (\min(m, n))$.
2. If $m > r$, then the $m - r$ rows are identically zero.
3. In column $j, 1 \leq j \leq r$, the diagonal element is monic and is of higher degree than any other non-zero element about it.
4. In column $j, 1 \leq j \leq r$, if the diagonal element is unity, all elements directly above it are zero.
5. If $n > r$, no special properties can be attributed to elements in the last $n - r$ columns and the first r rows.

■

By postmultiplying M by $R \in \mathbb{R}[s]^{n \times n}$ where R is a unimodular matrix, one may obtain

$$MR = H$$

in Hermite column form. It is lower quasi-triangular and its structure could be worked out as in the case of Hermite row form.

By performing both the row and column operations on M it is possible to obtain a matrix S whose only non zero entries are diagonal. This is the *Smith form* whose structure is described by the following theorem.

THEOREM 4.2 (Smith form)
Let $M \in \mathbb{R}[s]^{m \times n}$. Then there exist unimodular matrices $L \in \mathbb{R}[s]^{m \times m}$ and $R \in \mathbb{R}[s]^{n \times n}$ such that

$$L\,M\,R = S = \left[\begin{array}{cccc|c} \lambda_1(s) & & & & \\ & \lambda_2(s) & & & 0 \\ & & \ddots & & \\ & & & \lambda_r(s) & \\ \hline & & 0 & & 0 \end{array}\right] \tag{4.1}$$

S is called the Smith form.

The polynomials $\lambda_i(s)(i = 1, 2, \ldots r)$ are called the *invariant polynomials* of M

They are monic, uniquely defined and exhibit the division property

$$\lambda_i | \lambda_{i+1} \qquad i = 1, 2, \ldots, r - 1$$

i.e. λ_i divides λ_{i+1} without remainder

■

Let Δ_i be the monic greatest-common divisor of $i \times i$ minors of $\boldsymbol{M}$ and let $\Delta_0 = 1$ These polynomials are called the determinatal *divisors* of $\boldsymbol{M}$. They are related to the invariant polynomials of $\boldsymbol{M}$ by

$$\lambda_i = \Delta_i / \Delta_{i-1} \; i = 1, 2, \ldots, r$$

The product polynomial

$$\Lambda(s) = \lambda_1(s)\lambda_2(s)\ldots\lambda_r(s)$$

is called the zero polynomial of $\boldsymbol{S}$. The roots of $\Lambda(s)$ are called the zeros of $\boldsymbol{S}$. The integer r is called the normal rank of S. It is easy to see that the normal rank of $\boldsymbol{S}$ is equal to rank of $\boldsymbol{S}(s)$ for all $s \in \mathbb{C}$ if and only if S is unimodularly equivalent to $\begin{bmatrix} I_r & 0 \\ 0 & 0 \end{bmatrix}$.

Smith form is a unique standard form under equivalence. Hence polynomial matrices are equivalent if and only if they have the same Smith form

4.3 Testing For Coprimeness

DEFINITION 4.6 (Coprimeness)

Let $\boldsymbol{A} \in \mathbb{R}[s]^{p\times m}$ and $\boldsymbol{B} \in \mathbb{R}[s]^{q\times m}$ be two polynomial matrices having the same number of columns. The two matrices are said to be *right coprime* if the g.c.r.d. of $\boldsymbol{A}$ and $\boldsymbol{B}$ is unimodular.

Two polynomial matrices having the same number of rows are said to be *left coprime* if the g.c.l.d. of A and B is unimodular.

THEOREM 4.3

Let $\boldsymbol{A} \in \mathbb{R}[s]^{p\times m}$ and $\boldsymbol{B} \in \mathbb{R}[s]^{q\times m}$ have the same number of columns. Then $\boldsymbol{A}$ and $\boldsymbol{B}$ will always have a g.c.r.d. If $\boldsymbol{D} \in \mathbb{R}[s]^{m\times m}$ is the g.c.r.d. of $\boldsymbol{A}$ and $\boldsymbol{B}$, then there exists an $\boldsymbol{X} \in \mathbb{R}[s]^{m\times p}$ and $\boldsymbol{Y} \in \mathbb{R}[s]^{m\times q}$ such that

$$\boldsymbol{XA} + \boldsymbol{YB} = \boldsymbol{D}$$

Proof: Let $\boldsymbol{F} = \begin{bmatrix} \boldsymbol{A} \\ \boldsymbol{B} \end{bmatrix}$

Consider this case when $\boldsymbol{F}$ has at least as many rows as columns. The rank of $\boldsymbol{F}$ therefore cannot exceed m. Hence there exists a unimodular matrix $\boldsymbol{T}$ of appropriate size such that

$$\begin{bmatrix} \boldsymbol{T}_{11} & \boldsymbol{T}_{12} \\ \boldsymbol{T}_{21} & \boldsymbol{T}_{22} \end{bmatrix} \begin{bmatrix} \boldsymbol{A} \\ \boldsymbol{B} \end{bmatrix} = \begin{bmatrix} \boldsymbol{D} \\ \boldsymbol{0} \end{bmatrix}$$

where $\boldsymbol{D}$ is a $m \times m$ polynomial matrix which could be in Hermite form though this is unnecessary for our purpose. From the above relationship we may write

$$\boldsymbol{T}_{11}\boldsymbol{A} + \boldsymbol{T}_{12}\boldsymbol{B} = \boldsymbol{D} \tag{4.2}$$

Let

$$\begin{bmatrix} \boldsymbol{T}_{11} & \boldsymbol{T}_{12} \\ \boldsymbol{T}_{21} & \boldsymbol{T}_{22} \end{bmatrix}^{-1} = \begin{bmatrix} \boldsymbol{S}_{11} & \boldsymbol{S}_{12} \\ \boldsymbol{S}_{22} & \boldsymbol{S}_{22} \end{bmatrix}$$

Hence

$$\begin{bmatrix} \boldsymbol{A} \\ \boldsymbol{B} \end{bmatrix} = \begin{bmatrix} \boldsymbol{S}_{11} & \boldsymbol{S}_{12} \\ \boldsymbol{S}_{21} & \boldsymbol{S}_{22} \end{bmatrix} \begin{bmatrix} \boldsymbol{D} \\ \boldsymbol{O} \end{bmatrix}$$

$$= \begin{bmatrix} \boldsymbol{S}_{11}\boldsymbol{D} \\ \boldsymbol{S}_{21}\boldsymbol{D} \end{bmatrix} \tag{4.3}$$

from (4.3) we note that $\boldsymbol{D}$ is the common right divisor of $\boldsymbol{A}$ and $\boldsymbol{B}$ i.e., $\boldsymbol{A} = \boldsymbol{S}_{11}\boldsymbol{D}$ and $\boldsymbol{B} = \boldsymbol{S}_{21}\boldsymbol{D}$

Let $\boldsymbol{C}$ be another common right divisor of $\boldsymbol{A}$ and $\boldsymbol{B}$ i.e. $\boldsymbol{A} = \boldsymbol{A}_1\boldsymbol{C}$; $\quad \boldsymbol{B} = \boldsymbol{B}_1\boldsymbol{C}$.

Substitute in (4.2), for $\boldsymbol{A}$ and $\boldsymbol{B}$ as given above. We have

$$(\boldsymbol{T}_{11}\boldsymbol{A}_1 + \boldsymbol{T}_{12}\boldsymbol{B}_1)\boldsymbol{C} = \boldsymbol{D}$$

Hence $\boldsymbol{D}$ is the left multiple of every common right divisor of $\boldsymbol{A}$ and $\boldsymbol{B}$ which proves that $\boldsymbol{D}$ is the g.c.r.d. of $\boldsymbol{A}$ and $\boldsymbol{B}$.

Consider now the case when $p + q < m$. We assert that the g.c.r.d. of $\boldsymbol{A}$ and $\boldsymbol{B}$ in this case is

$$\boldsymbol{D} = \begin{bmatrix} \boldsymbol{A} \\ \boldsymbol{B} \\ 0_{k \times m} \end{bmatrix} \quad \text{where } k = m - (p + q)$$

To show that this statement is true, we first observe that $\boldsymbol{D}$ is the common right divisor of $\boldsymbol{A}$ and $\boldsymbol{B}$ because

$$\boldsymbol{A} = [I_p \quad 0 \quad 0]\boldsymbol{D} \text{ and } \boldsymbol{B} = [0 \quad I_q \quad 0]\boldsymbol{D}$$

Secondly

$$\begin{bmatrix} I_p \\ 0 \\ 0 \end{bmatrix} \boldsymbol{A} + \begin{bmatrix} 0 \\ I_q \\ 0 \end{bmatrix} \boldsymbol{B} = \boldsymbol{D}$$

From the above statements it can be easily shown that D is the left multiple of every common right divisor of A and B. Hence D is the g.c.r.d. of A at B.

■

If D is unimodular, then D^{-1} is a polynomial matrix and (4.2) implies that

$$D^{-1}T_{11}A + D^{-1}T_{12}B = I$$
$$XA + YB = I \tag{4.4}$$

where X and Y are polynomial matrices.

If g.c.r.d. of A and B is unimodular, then (4.4) is satisfied. This could be used as a test to establish coprimeness as stated by the following theorem.

THEOREM 4.4 (Bezout identity)
The matrices $A \in \mathbb{R}[s]^{p\times m}$ at $B \in \mathbb{R}[s]^{q\times m}$ are right coprime if and only if there exists an $X \in \mathbb{R}[s]^{m\times p}$ and a $Y \in \mathbb{R}[s]^{m\times q}$ such that

$$XA + YB = I$$

The above equation is known as the *right Bezout identity*

If A and B have the same number of rows, then using identical arguments one can establish the *left Bezout identity* according to which

$$A\tilde{X} + B\tilde{Y} = I$$

The next theorem states that g.c.r.d.s are unique up to a unimodular matrix

THEOREM 4.5 (Non uniqueness of g.c.r.d)
Let $A \in \mathbb{R}[s]^{p\times m}$ and $B \in \mathbb{R}[s]^{q\times m}$

Let D be the g.c.r.d of A and B. Then UD is also a g.c.r.d. for any unimodular polynomial matrix U of appropriate dimension. Further if D_1 is another g.c.r.d. of A and B, $D_1 = UD$

Proof: Let $A = A_1D$ and $B = B_1D$

Hence $A = (A_1U^{-1})(UD)$ and $B = (B_1U^{-1})(UD)$ for unimodular U

Therefore UD is the c.r.d. of A and B. Since UD is the left multiple of D, which by virtue of it being a g.c.r.d. is a left multiple of every c.r.d. of A and B, it follows that UD is the left multiple of every c.r.d. of A and B and the result follows.

We now prove the second part of the theorem for the case when A is square and non singular. we have

$$T\begin{bmatrix} A \\ B \end{bmatrix} = \begin{bmatrix} D \\ 0 \end{bmatrix}$$

where $\boldsymbol{T}$ is a unimodular matrix. Since $\boldsymbol{A}$ is non singular it follows that $\boldsymbol{D}$ is also non-singular. Further, it is the g.c.r.d. of $\boldsymbol{A}$ and $\boldsymbol{B}$. Let $\boldsymbol{D}_1$ be another g.c.r.d. of $\boldsymbol{A}$ and $\boldsymbol{B}$. Then by definition of g.c.r.d.

$$\boldsymbol{D}_1 = \boldsymbol{L}_1\boldsymbol{D}$$

similar argument leads to the equation

$$\boldsymbol{D} = \boldsymbol{L}\boldsymbol{D}_1$$

combining the two we obtain

$$\boldsymbol{D} = \boldsymbol{L}\boldsymbol{L}_1\boldsymbol{D}.$$

Since $\boldsymbol{D}$ is non singular $\det \boldsymbol{L} \det \boldsymbol{L}_1 = 1$ Hence both $\boldsymbol{L}$ and $\boldsymbol{L}_1$ are unimodular. $\boldsymbol{D}$ is therefore related to $\boldsymbol{D}_1$ by a unimodular transformation The above result is true even if $\boldsymbol{D}$ is singular. For proof refer to Vidyasagar (1985)

4.4 Rational Functions – Matrix Fraction Description

In the case of transfer matrices obtained from state-space description, the elements of the matrix are not polynomials but rational functions of s. This provides the motivation for studying MFD of transfer matrices $\boldsymbol{G}(s) \in \mathbb{R}\ (s)^{p\times q}$

DEFINITION 4.7 (Coprime fraction)
Let $\boldsymbol{G}(s) \in \mathbb{R}\ (s)^{q\times q}$. We say that the matrix pair $(\boldsymbol{N}_r, \boldsymbol{D}_r)$ with $\boldsymbol{N}_r \in \mathbb{R}\ [s]^{p\times q}$ and $\boldsymbol{D}_r \in \mathbb{R}\ [s]^{q\times q}$ is a *right coprime fraction* of $\boldsymbol{G}$ if and only if

1. Det $\boldsymbol{D}_r \neq 0$
2. $\boldsymbol{G} = \boldsymbol{N}_r\boldsymbol{D}_r^{-1}$
3. $(\boldsymbol{N}_r, \boldsymbol{D}_r)$ is right coprime

If condition 3 is not satisfied then $(\boldsymbol{N}_r, \boldsymbol{D}_r)$ is merely known as a *right fraction.*

Similarly the matrix pair $(\boldsymbol{D}_l, \boldsymbol{N}_l)$ with $\boldsymbol{N}_l \in \mathbb{R}\ [s]^{q\times q}$ and $\boldsymbol{D}_l \in \mathbb{R}\ [s]^{p\times p}$ is a *left coprime fraction* if, and only if,

1. Det $D_l \neq 0$
2. $\boldsymbol{G} = \boldsymbol{D}_l^{-1}\boldsymbol{N}_l$
3. $(\boldsymbol{D}_l, \boldsymbol{N}_l)$ is left coprime

If condition 3 is not satisfied then $(\boldsymbol{D}_l, \boldsymbol{N}_l)$ is merely known as a *left fraction*

■

We now discuss a procedure for obtaining the relevant factorization. Given $\boldsymbol{G}(s) \in \mathbb{R}\ (s)^{p \times q}$ compute the least common denominator (l.c.d) of elements column wise. If any column has all its elements zero, define l.c.d. pertaining to that column as unity.

We write

$$\begin{aligned}\boldsymbol{G}(s) &= \boldsymbol{G}(s)\ \text{diag}\ (d_1 \ldots d_q)\ \text{diag}\ (1/d_1 \ldots 1/d_q) \\ &= \overline{\boldsymbol{N}}_r(s)\overline{\boldsymbol{D}}_r(s)^{-1}\end{aligned}$$

Note that $\overline{\boldsymbol{N}}_r$ and $\overline{\boldsymbol{D}}_r$ are polynomial matrices with $\overline{\boldsymbol{D}}_r$ non-singular.

Now obtain the g.c.r.d. of $(\overline{D}_r, \overline{N}_r)$ via a unimodular transformation $\boldsymbol{T}$ as follows

$$T\begin{bmatrix}\overline{\boldsymbol{D}}_r \\ \overline{\boldsymbol{N}}_r\end{bmatrix} = \begin{bmatrix}\boldsymbol{R} \\ O\end{bmatrix}$$

From (4.3) it is seen that

$$\overline{\boldsymbol{D}}_r = \boldsymbol{S}_{11}\boldsymbol{R} \text{ and } \overline{\boldsymbol{N}}_r = \boldsymbol{S}_{21}\boldsymbol{R}$$

where $\boldsymbol{S}_{11}$ and $\boldsymbol{S}_{21}$ are the (1, 1) and (2, 1) block matrices of the inverse of $\boldsymbol{T}$. Note that $\boldsymbol{R}$ is non-singular. We have

$$\overline{\boldsymbol{N}}_r\overline{\boldsymbol{D}}_r^{-1} = \boldsymbol{S}_{21}\boldsymbol{R}\boldsymbol{R}^{-1}\boldsymbol{S}_{11}^{-1} = \boldsymbol{S}_{21}\boldsymbol{S}_{11}^{-1}$$

The coprime factors $(\boldsymbol{N}_r, \boldsymbol{D}_r)$ are thus identified as $(\boldsymbol{S}_{21}, \boldsymbol{S}_{11})$.

Let $\boldsymbol{T}$ and $\boldsymbol{T}^{-1}$ be partitioned as block matrices as indicated below

$$\boldsymbol{T} = \begin{bmatrix}\boldsymbol{Y}_r & \boldsymbol{X}_r \\ -\boldsymbol{N}_l & \boldsymbol{D}_l\end{bmatrix} \qquad \boldsymbol{T}^{-1} = \begin{bmatrix}\boldsymbol{D}_r & -\boldsymbol{X}_l \\ \boldsymbol{N}_r & \boldsymbol{Y}_l\end{bmatrix}$$

The reasons for the nomenclature adopted for the partitioned matrices will be clear from later discussions. Since $\boldsymbol{T}\boldsymbol{T}^{-1} = I$, carrying out the indicated matrix multiplication we have

$$\boldsymbol{Y}_r\boldsymbol{D}_r + \boldsymbol{X}_r\boldsymbol{N}_r = I$$

which is the right Bezout identity. Further,

$$-\boldsymbol{N}_l\boldsymbol{D}_r + \boldsymbol{D}_l\boldsymbol{N}_r = 0$$

Since $\overline{D}_r$ is non-singular (by hypothesis) and $\overline{\boldsymbol{D}}_r = \boldsymbol{D}_r\boldsymbol{R}$ where $\boldsymbol{R}$ is non-singular, it follows that $\boldsymbol{D}_r$ is non-singular. Similarly it can be shown that $\boldsymbol{D}_l$ is also non-singular and it follows from the above equation that

$$\boldsymbol{G} = \boldsymbol{N}_r\boldsymbol{D}_r^{-1} = \boldsymbol{D}_l^{-1}\boldsymbol{N}_l$$

Further equating the (2, 2) term of $TT^{-1} = I$ on either side yields

$$N_l X_l + D_l Y_l = I$$

which is the left Bezout identity.

Thus, given the transfer matrix G, making use of the transformation matrices T and T^{-1} we identify the following matrix pairs:

$$(N_r, D_r); (D_l N_l); (Y_r, X_r) \text{ and } (Y_l, X_l)$$

which taken together give complete information about the right and left coprime factors and the coefficient matrices associated with the corresponding Bezout identities.

The advantage of this method is that while T is obtained via a succession of elementary row operations, side by side T^{-1} could also be obtained with little effort via a succession of elementary column operations. This is indicated in the following tabular statement.

	Elementary row operation (corresponding to T)	*Elementary column operation* (corresponding to T^{-1})
1.	$\rho_i \to \rho_j$	$\gamma_i \to \gamma_j$
2.	$\rho_i \to k\rho_i$	$\gamma_i \to k^{-1}\gamma_i$
3.	$\rho_i \to \rho_i + r\rho_j$	$\gamma_j \to \gamma_j - r\gamma_i$

In the above statement the symbol '$\to$' indicates 'replaced by'

To summarize, we have

$$\begin{bmatrix} Y_r & X_r \\ -N_l & D_l \end{bmatrix} \begin{bmatrix} D_r & -X_l \\ N_r & Y_l \end{bmatrix} = \begin{bmatrix} I_q & 0 \\ 0 & I_p \end{bmatrix} \tag{4.5}$$

(4.5) is known as the *generalized Bezout identity*. It is also sometimes called the *doubly coprime fractional representation of G*.

4.5 The Smith–McMillan Form

Earlier we saw how a polynomial matrix could be reduced to Smith form by premultiplying and post multiplying it by unimodular polynomial matrices. We now consider the more general case of matrix valued transfer functions, whose elements are rational functions of s. It turns out that by performing elementary row and column operations, it is possible to reduce these matrices to diagonal form, with the non-zero entries being rational functions of s. This reduced form is known as *Smith-McMillan (SM) form.*

Though not used in practical calculations, it serves as a useful analytical tool for deriving theoretical results.

THEOREM 4.6 (SM form)
Let $\boldsymbol{G} \in \mathbb{R}\ (s)^{p\times m}$ have a normal rank r. Then there exist unimodular matrices $\boldsymbol{L} \in \mathbb{R}\ [s]^{p\times p}$ and $\boldsymbol{R} \in \mathbb{R}\ [s]^{m\times m}$ such that

$$\boldsymbol{G} = \boldsymbol{LMR}$$

The SM form $\boldsymbol{M}$ is given by

$$\boldsymbol{M} = \left[\begin{array}{cccc|c} \varepsilon_1/\psi_1 & & & & \\ & \varepsilon_2/\psi_2 & & & 0 \\ & & \ddots & & \\ & & & \varepsilon_r/\psi_r & \\ \hline & & 0 & & 0 \end{array}\right] \in \mathbb{R}\ (s)^{p\times m} \qquad (4.6)$$

where (ε_i, ψ_i) are pairs of monic polynomials coprime to each other, satisfying the following divisibility properties

$$\psi_{i+1}|\psi_i \qquad i = 1, 2, \ldots r-1$$
$$\varepsilon_i|\varepsilon_{i+1} \qquad i = 1, 2, \ldots, r-1$$

where $\psi_1 = d$ is the monic least common divisor of all the elements of $\boldsymbol{G}$

Proof: Omitted (Refer to Callier and Desoer (1982))

■

The pole and zero polynomials associated with $G(s)$ are given by

$$p(s) = \psi_1(s) \cdot \psi_2(s) \ldots \psi_r(s)$$
$$z(s) = \varepsilon_1(s) \cdot \varepsilon_2(s) \ldots \varepsilon_r(s)$$

The roots of $p(s)$ and $z(s)$ are the poles and zeros of the transfer function $\boldsymbol{G}(s)$. The zeros thus defined are the *transmission zeros* of $\boldsymbol{G}(s)$. Clearly the rank of $\boldsymbol{G}(s)$ drops below it normal rank at the transmission zeros of $\boldsymbol{G}(s)$.

The *McMillan degree* of $\boldsymbol{G}(s)$ is the degree of the polynomial $p(s)$. It represents the order of the minimal state space realization of $\boldsymbol{G}(s)$

4.5.1 Coprime Factorization via SM Form

Given the SM form of $\boldsymbol{G} \in \mathbb{R}\ (s)^{m\times q}$, it is straightforward to arrive at a coprime factorization of $\boldsymbol{G}$. This may be obtained as follows:

Define

$$E = \begin{bmatrix} \text{diag }(\varepsilon_1, \varepsilon_2 \ldots \varepsilon_r) & 0 \\ 0 & 0 \end{bmatrix}$$

$$\psi_r = \begin{bmatrix} \text{diag }(\psi_1, \psi_2 \ldots \psi_r) & 0 \\ 0 & I_{m-r} \end{bmatrix}$$

$$\psi_l = \begin{bmatrix} \text{diag }(\psi_1, \psi_2, \ldots \psi_r) & 0 \\ 0 & I_{p-r} \end{bmatrix}$$

G could be expressed in two different ways as follows

$$\begin{aligned} G = LMR &= (LE)(R^{-1}\psi_r)^{-1} \\ &= N_r D_r^{-1} \end{aligned}$$

or

$$\begin{aligned} G = LMR &= (\psi_l L^{-1})^{-1}(ER) \\ &= D_l^{-1} N_l \end{aligned}$$

Note that even though the non-zero polynomial elements of ψ_r and ψ_l are the same, ψ_l is a $p \times p$ matrix and ψ_r is a m $\times$ m matrix.

We now show that the polynomial matrix pair (N_r, D_r) is coprime. We have

$$\begin{bmatrix} D_r \\ N_R \end{bmatrix} = \begin{bmatrix} R^{-1}\psi_r \\ LE \end{bmatrix} = \begin{bmatrix} R^{-1} & 0 \\ 0 & L \end{bmatrix} \begin{bmatrix} \psi_r \\ E \end{bmatrix}$$

$\begin{bmatrix} D_r \\ N_r \end{bmatrix}$ and $\begin{bmatrix} \psi_r \\ E \end{bmatrix}$ have the same rank because the transformation matrix, relating them is unimodular.

Now consider the rank of $\begin{bmatrix} \psi_r \\ E \end{bmatrix}$

The first r columns of $\begin{bmatrix} \psi_r \\ E \end{bmatrix}$ are linearly independent and the individual columns do not vanish for any $s \in \mathbb{C}$. This follows from the fact that the only non-zero elements of the ith column namely ψ_i and ϵ_i do not vanish simultaneously for any $s \in \mathbb{C}$ as they are prime to each other. The remaining (m – r) columns have only one non-zero entry, namely unity for each column and they are linearly independent.

Hence rank of $\begin{bmatrix} \psi_r \\ E \end{bmatrix}$ is equal to m for all $s \in \mathbb{C}$. The matrix pair $(\boldsymbol{LE}, \boldsymbol{R}^{-1}\psi_r)$ is therefore right coprime. Similarly it can be shown that $(\psi_l \boldsymbol{L}^{-1}, \boldsymbol{ER})$ is left coprime.

For any $\boldsymbol{G} \in \mathbb{R}\ (s)^{p\times q}$ the following facts could be easily established.

1. All the numerator matrices, whether they belong to the right or left coprime factorization of $\boldsymbol{G}$ are equivalent and have the same Smith form.
2. The denominator matrices of any coprime factorization of $\boldsymbol{G}$ (right or left) have the same invariant polynomials.

4.6 Factorization Over a Ring of Stable Matrices

Fractional representation approach for the design of controllers has come in for a great deal of attention in recent times. A comprehensive exposition of the method can be found in the books of Vidyasagar (1985) and Callier and Desoer (1982).

In the design of feedback systems, if there is any single property that stands out as mandatory, it is the stability property. Hence it appears logical to model the plant and the controller as quotients of stable rational functions. To make this concept more explicit, consider an SISO system with transfer function $\boldsymbol{p}\ (s) = \dfrac{a(s)}{b(s)}$ where $a(s)$ are $b(s)$ are coprime polynomials possibly with some of the zeros of $b(s)$ located in the closed right half plane. The plant is therefore unstable. However it is always possible to model the plant as the quotient of two stable rational functions as follows.

$$\boldsymbol{p}(s) = \frac{a(s)}{b(s)} = \frac{a(s)/m(s)}{b(s)/m(s)} = \frac{\boldsymbol{n}(s)}{\boldsymbol{d}(s)}$$

where $m(s)$ is a Hurwitz polynomial (i.e with zeros is the open left half plane) of degree equal to that of $\boldsymbol{b}(s)$. Now the rational functions $\boldsymbol{n}(s)$ and $\boldsymbol{d}(s)$ are not only coprime, they are in additional stable. Hence by applying Euclid's algorithm, we can obtain two stable rational functions $\boldsymbol{x}(s)$ and $\boldsymbol{y}(s)$ show that

$$\boldsymbol{x}(s)\boldsymbol{n}(s) + y(s)\boldsymbol{d}(s) = 1.$$

In fact this is the condition for coprimeness between two stable rational functions $\boldsymbol{n}(s)$ and $\boldsymbol{d}(s)$.

Note that $\boldsymbol{n}(s)$ and $\boldsymbol{d}(s)$ have no common zeros in Re $s \geq 0$ including the point at infinity.

If for example such a common zero, say s_o, existed then

$$\boldsymbol{x}(s_o)\boldsymbol{n}(s_o) + \boldsymbol{y}(s_o)\boldsymbol{d}(s_o) = 0$$

violating the condition of coprimeness.

As is readily seen coprime factorization is not unique. Essentially what is attempted in to obtain a coprime factorization of a scalar transfer function over the ring of stable, proper and real rational functions.

The following examples illustrate some of the concepts involved.

EXAMPLE 4.1

Consider an unstable transfer function

$$\boldsymbol{g}(s) = \frac{(s-1)(s+2)}{(s+1)(s-3)}$$

We are required to find stable transfer functions $\boldsymbol{n}(s)$ and $\boldsymbol{d}(s)$ such that $\boldsymbol{g}(s) = \boldsymbol{n}(s)\boldsymbol{d}^{-1}(s)$ with $\boldsymbol{n}(s)$ and $\boldsymbol{d}(s)$ coprime. This is achieved by dividing the numerator and denominator polynomials of $\boldsymbol{g}(s)$ by a stable polynomial whose degree is equal that of the denominator polynomial. Choose for example $(s+1)^2$ as such a polynomial. We then have

$$\boldsymbol{n}(s) = \frac{(s-1)(s+2)}{(s+1)^2} \text{ and } \boldsymbol{d}(s) = \frac{(s+1)(s-3)}{(s+1)^2} = \frac{s-3}{s+1}$$

and we obtain $\boldsymbol{g}(s)$ as

$$\begin{aligned}\boldsymbol{g}(s) &= \boldsymbol{n}(s)\boldsymbol{d}^{-1}(s)\\ &= \frac{(s-1)(s+2)}{(s+1)^2}\cdot\left(\frac{s-3}{s+1}\right)^{-1}\end{aligned}$$

thus obtaining a coprime factorization of $\boldsymbol{g}(s)$ in terms of stable rational functions.

Another possible factorization out of the many which could be chosen is given by

$$\boldsymbol{n}(s) = \frac{s-1}{s+1} \text{ and } \boldsymbol{d}(s) = \frac{s-3}{s+2} \text{ yielding}$$

$$\boldsymbol{g}(s) = \left(\frac{s-1}{s+1}\right)\cdot\left(\frac{s-3}{s+2}\right)^{-1}$$

EXAMPLE 4.2
Consider a case where

$$\boldsymbol{n}(s) = \frac{s-1}{(s+1)^2} \text{ and } \boldsymbol{d}(s) = \frac{s-2}{(s+3)(s+4)}$$

Both $\boldsymbol{n}(s)$ and $\boldsymbol{d}(s)$ are stable rational functions. However, they are not stable coprime factors of

$$\boldsymbol{g}(s) = \frac{\boldsymbol{n}(s)}{\boldsymbol{d}(s)} = \frac{s-1}{(s+1)^2}\frac{(s+3)(s+4)}{(s-2)}$$

because $\boldsymbol{n}(s)$ and $\boldsymbol{d}(s)$ have a common zero at infinity.

■

The concept of factorizing a scalar transfer function in terms of stable, coprime rational transfer functions can be extended to the more general case where the transfer function is matrix valued.

In this case $\boldsymbol{G} \in \mathbb{R}(s)^{p\times q}$ and it could be factored as

$$\boldsymbol{G} = \boldsymbol{N}_r\boldsymbol{D}_r^{-1} \text{ where } \boldsymbol{N}_r \in \mathcal{RH}_\infty^{p\times q} \text{ and } \boldsymbol{D}_r \in \mathcal{RH}_\infty^{q\times q}$$

or

$$\boldsymbol{G} = \boldsymbol{D}_l^{-1}\boldsymbol{N}_l \text{ where } \boldsymbol{N}_l \in \mathcal{RH}_\infty^{p\times q} \text{ and } \boldsymbol{D}_l \in \mathcal{RH}_\infty^{p\times p}$$

further $(\boldsymbol{N}_r, \boldsymbol{D}_r)$ and $(\boldsymbol{D}_l, \boldsymbol{N}_l)$ and coprime.
Note that the factors are not polynomial matrices but matrix valued rational functions. In this case, coprimeness may be defined as follows

DEFINITION 4.8
The transfer function matrices $\boldsymbol{N}_r$ and $\boldsymbol{D}_r \in \mathcal{RH}_\infty^{p\times p}$ with identical number of columns are said to be *right coprime* if there exists matrices $\boldsymbol{X}_r, \boldsymbol{Y}_r \in \mathcal{RH}_\infty$ such that

$$\boldsymbol{X}_r\boldsymbol{N}_r + \boldsymbol{Y}_r\boldsymbol{D}_r = \boldsymbol{I}$$

similarly $\boldsymbol{D}_l$, $\boldsymbol{N}_l$ having identical number of rows, are said to be *left coprime* if these exist matrices $\boldsymbol{X}_l, \boldsymbol{Y}_l \in \mathcal{RH}_\infty$ such that

$$\boldsymbol{N}_l\boldsymbol{X}_l + \boldsymbol{D}_l\boldsymbol{Y}_l = \boldsymbol{I}$$

DEFINITION 4.9
If the matrix pair $(\boldsymbol{N}_r, \boldsymbol{D}_r)$ is right coprime and further if $\boldsymbol{D}_r$ is nonsingular, then $\boldsymbol{N}_r\boldsymbol{D}_r^{-1}$ represents as a *right coprime factorization.*

Similarly if $(\boldsymbol{D}_l, \boldsymbol{N}_l)$ is left coprime and if $\boldsymbol{D}_l$ is nonsingular, then $\boldsymbol{D}_l^{-1}\boldsymbol{N}_l$ represents a *left coprime factorization.*

EXAMPLE 4.3

This example illustrates the factorization of a matrix valued transfer function in terms of stable rational matrix functions.

$$\text{Let } \boldsymbol{G}(s) = \begin{bmatrix} \dfrac{s-1}{s+2} & \dfrac{s+1}{s-3} \end{bmatrix}$$

By inspection, we may write down the right coprime factorization as

$$\boldsymbol{G}(s) = \begin{bmatrix} \dfrac{s-1}{s+2} & \dfrac{s+1}{s+4} \end{bmatrix} \begin{bmatrix} 1 & 0 \\ 0 & \dfrac{s-3}{s+4} \end{bmatrix}^{-1}$$

Similarly a possible left coprime factorization can be

$$\boldsymbol{G}(s) = \left(\frac{s-s}{s+4}\right)^{-1} \begin{bmatrix} \dfrac{s-1}{s+2} \dfrac{s-3}{s+4} & \dfrac{s+1}{s+4} \end{bmatrix}$$

THEOREM 4.7

Let $(\boldsymbol{N}_1, \boldsymbol{D}_1)$ and $(\boldsymbol{N}_2, \boldsymbol{D}_2)$ be two right coprime factorization pairs belonging to $\mathcal{RH}_\infty$ such that $\boldsymbol{N}_1\boldsymbol{D}_1^{-1} = \boldsymbol{N}_2\boldsymbol{D}_2^{-1}$. Then $\boldsymbol{N}_2 = \boldsymbol{N}_1\boldsymbol{T}$ and $\boldsymbol{D}_2 = \boldsymbol{D}_1\boldsymbol{T}$ where $\boldsymbol{T}$ and $\boldsymbol{T}^{-1}$ both belong to $\mathcal{RH}_\infty$.

Further if $\boldsymbol{G} = \boldsymbol{N}\boldsymbol{D}^{-1}$ with $\boldsymbol{G} \in \mathcal{RH}_\infty$ and $\boldsymbol{N}$ and $\boldsymbol{D}$ are coprime, then $\boldsymbol{D}^{-1} \in \mathcal{RH}_\infty$.

Proof:

Since $\quad \boldsymbol{N}_1\boldsymbol{D}_1^{-1} = \boldsymbol{N}_2\boldsymbol{D}_2^{-1}$

We have $\quad \boldsymbol{N}_1\boldsymbol{D}_1^{-1}\boldsymbol{D}_2 = \boldsymbol{N}_2$

Define $\quad \boldsymbol{D}_1^{-1}\boldsymbol{D}_2 = \boldsymbol{T}$. Hence $\boldsymbol{N}_1\boldsymbol{T} = \boldsymbol{N}_2$

Also $\quad \boldsymbol{D}_2 = \boldsymbol{D}_1\boldsymbol{D}_1^{-1}\boldsymbol{D}_2 = \boldsymbol{D}_1\boldsymbol{T}$

Since $(\boldsymbol{N}_2, \boldsymbol{D}_2)$ is right coprime

$$\boldsymbol{X}_2\boldsymbol{N}_2 + \boldsymbol{Y}_2\boldsymbol{D}_2 = \boldsymbol{I}$$
$$\boldsymbol{X}_2\boldsymbol{N}_1\boldsymbol{T} + \boldsymbol{Y}_2\boldsymbol{D}_1\boldsymbol{T} = \boldsymbol{I}$$
$$(\boldsymbol{X}_2\boldsymbol{N}_1 + \boldsymbol{Y}_2\boldsymbol{D}_1)\boldsymbol{T} = \boldsymbol{I}$$

or $\quad \boldsymbol{T}^{-1} = (\boldsymbol{X}_2\boldsymbol{N}_1 + \boldsymbol{Y}_2\boldsymbol{D}_1)$

Hence $\quad \boldsymbol{T}^{-1} \in \mathcal{RH}_\infty$

Similarly $$X_1 N_1 + Y_1 D_1 = I$$

Hence $$(X_I N_2 + Y_1 D_2) T^{-1} = I$$

or $$T \in \mathcal{RH}_\infty$$

To prove the second part of the theorem we note that if $G \in \mathcal{RH}_\infty$ with $G = ND^{-1}$ (N and D coprime) then $XN + YD = I \Rightarrow XND^{-1} + Y = D^{-1}$ or $XG + Y = D^{-1}$. Hence $D^{-1} \in \mathcal{RH}_\infty$.

■

The above theorem shows that coprime factorization is unique only up to a unit of $\mathcal{RH}_\infty$.

Normally for a stable matrix G factorized as ND^{-1}, N being stable does not mean that D^{-1} should also be stable because of the possibility of pole – zero cancellation. However Theorem 4.7 asserts that provided N and D are coprime factors, D and $N \in \mathcal{RH}_\infty$ necessarily implies that $D^{-1} \in \mathcal{RH}_\infty$ D for any stable matrix G.

4.6.1 The Fractional Representation Approach

In the previous section we saw how a matrix transfer function could be factorized over a ring of stable rational matrices. The underlying concept is much deeper and could be generalized further by applying an axiomatic fractional representation theory as proposed by Desoer *et al.* (1980). In view of its importance as a powerful concept, which could be applied to more general situations, we give a brief account of this theory. We recall that in the stable factorization approach, the plant (whether it be stable or not) is modelled as a quotient of two operators each one of them being individually stable. The same applies to modelling of controllers which are to be designed. Thus the ultimate goal, namely preservation of internal stability in this case dictates the type of modelling structure which one would like to choose. Generalizing this concept we can visualize the plant being modelled as a quotient of two operators, belonging to a prescribed ring H. The problem is to characterize all controllers which will 'place' the feedback system in H. Let G be a ring with an identity and let H be a sub-ring of G with its identity in H. The feedback system and its subsystems will be represented by operators which belong to G. The controller is chosen in such a way that overall system is represented by an operator which belongs to the sub-ring H. We further define two subsets of H as follows.

$$I = \{h \in H : h^{-1} \in G\}$$
$$J = \{h \in H : h^{-1} \in H\}$$

Thus I represents those elements of H which have an inverse in G and J represents those elements of H which have an inverse in H itself. Therefore J represents the set of all units in H. Clearly

$$J \subset I \subset H \subset G$$

Assuming the above structure, an element $\boldsymbol{g} \in G$ has a right fractional representation if there exists a $\boldsymbol{n} \in H$ and a $\boldsymbol{d} \in I$ such that $\boldsymbol{g} = \boldsymbol{n}\boldsymbol{d}^{-1}$. Further, the pair $(\boldsymbol{n}, \boldsymbol{d})$ is right coprime if there exists $\boldsymbol{x}$, $\boldsymbol{y} \in H$ such that

$$\boldsymbol{x}\boldsymbol{n} + \boldsymbol{y}\boldsymbol{d} = I$$

The above type of generalized axiomatic structure $\{G, H, I, J\}$ with the sets suitably defined, can be used to cover problems occuring in both finite dimensional and infinite dimensional spaces. A few examples given below illustrate these points.

EXAMPLE 4.4

Let G be a ring of rational functions with coefficients in $\mathbb{R}$. (denoted by $\mathbb{R}(s)$). Let H be the ring of polynomials with coefficients in $\mathbb{R}$ (denoted by $\mathbb{R}[s]$). With G and H thus specified, the sets I and J are automatically fixed. Elements of H should belong to G and have their inverses in G. All non-zero polynomials qualify for membership in I. Similarly elements of J should belong to H and have their inverses in H. Since the inverse of a polynomial can be a polynomial if and only if the polynomial is a non-zero constant, J represents the set of all polynomials in H which are non-zero constants.

EXAMPLE 4.5

This example is typical of the stable fractional matrix representation approach used to design controllers in MIMO systems. The relevant sets are:

G: The ring of $n \times n$ matrices whose elements are proper rational functions denoted by $\mathbb{R}_p(s)^{n\times n}$

H: The ring of $n \times n$ matrices whose elements are proper rational functions, which are in addition stable, denoted by $\mathcal{RH}_\infty^{n\times n}$ The above choice of G and H points towards the following choice for I and J. Thus we have

I: The set of all invertible matrices $\boldsymbol{M} \in \mathcal{RH}_\infty^{n\times n}$ with the following properties:

i) $\det \boldsymbol{M}(s) \neq 0$ (this ensures invertibility)

ii) $\lim_{s\to\infty} \det \boldsymbol{M}(s) = a$ non-zero constant (This ensures that elements of $\boldsymbol{M}^{-1}(s)$ are proper). We then have

$$\boldsymbol{M}^{-1}(s) = \frac{\text{Adjoint } \boldsymbol{M}(s)}{\det \boldsymbol{M}(s)} \in \mathbb{R}_p\,(s)^{n\times n}$$

J: The set of all invertible matrices $M \in \mathcal{RH}_\infty^{n\times n}$ with the following properties:
i) $\det \boldsymbol{M}(s) \neq 0$
ii) $\lim_{s\to\infty} \det \boldsymbol{M}(s) =$ a non-zero constant
iii) $\det \boldsymbol{M}(s)$ has no zero in $\mathrm{Re}\, s \geq 0$. Condition (iii) ensures that $\boldsymbol{M}^{-1}(s)$ is not only proper but also stable.

■

In the above context, a question may arise as to whether coprime fractional representation is always possible for any choice of sets G and H. We know that this is possible if G represents proper rational functions and H represents stable rational functions. More general cases are not discussed here as they fall beyond the scope of this book

4.7 Coprime Fractional Representation in State Space

Given $\boldsymbol{G}(s) \in \mathcal{RH}_\infty$, its factorization in terms of stable coprime transfer matrices is of interest in the analysis and synthesis of systems. The frequency domain methods, which can possibly be used in this connection are quite cumbersome. However if we know the state space realization of **G**(s) (not necessarily minimal) then there is a direct method of obtaining a stable coprime factorization of **G**(s). The method (refer to Nett *et al.* 1984) generates not only the right and left stable coprime factorization of **G**(s) but also stable coefficient matrices associated with the right and left Bezout identities.

Before proceeding further, we first discuss some preliminaries.

Let
$$\boldsymbol{G} = \boldsymbol{N}_r\boldsymbol{D}_r^{-1} \text{ and}$$
$$\overline{\boldsymbol{X}}_r\boldsymbol{N}_r + \overline{\boldsymbol{Y}}_r\boldsymbol{D}_r = \boldsymbol{I}$$
where $\boldsymbol{N}_r, \boldsymbol{D}_r, \overline{\boldsymbol{X}}_r$ and $\overline{\boldsymbol{Y}}_r$ belong to $\mathcal{RH}_\infty$. Further

Let
$$\boldsymbol{G} = \boldsymbol{D}_l^{-1}\boldsymbol{N}_l \text{ and}$$
$$\boldsymbol{N}_l\overline{\boldsymbol{X}}_l + \boldsymbol{D}_l\overline{\boldsymbol{Y}}_l = \boldsymbol{I}$$
where $\boldsymbol{N}_l, \boldsymbol{D}_l, \overline{\boldsymbol{X}}_l, \overline{\boldsymbol{Y}}_l \in \mathcal{RH}_\infty$.

The above relationships may be expressed in matrix form as

$$\begin{bmatrix} \overline{\boldsymbol{Y}}_r & \overline{\boldsymbol{X}}_r \\ -\boldsymbol{N}_l & \boldsymbol{D}_l \end{bmatrix} \begin{bmatrix} \boldsymbol{D}_r & -\overline{\boldsymbol{X}}_l \\ \boldsymbol{N}_r & \overline{\boldsymbol{Y}}_l \end{bmatrix} = \begin{bmatrix} \boldsymbol{I} & \boldsymbol{R} \\ 0 & \boldsymbol{I} \end{bmatrix}$$

where $\boldsymbol{R} = -\overline{\boldsymbol{Y}}_r\overline{\boldsymbol{X}}_l + \overline{\boldsymbol{X}}_r\overline{\boldsymbol{Y}}_l$.

Post multiply both sides of the above equations by the unimodular-transformation $\begin{bmatrix} I & -R \\ 0 & I \end{bmatrix}$. We have

$$\begin{bmatrix} \overline{\boldsymbol{Y}}_r & -\overline{\boldsymbol{X}}_r \\ -\boldsymbol{N}_l & \boldsymbol{D}_l \end{bmatrix} \begin{bmatrix} \boldsymbol{D}_r & -\boldsymbol{D}_r\boldsymbol{R} & -\overline{\boldsymbol{X}}_l \\ \boldsymbol{N}_r & -\boldsymbol{N}_r\boldsymbol{R} & +\overline{\boldsymbol{Y}}_l \end{bmatrix} = \begin{bmatrix} I & 0 \\ 0 & I \end{bmatrix}$$

Redesignate the block matrices as follows

$$\overline{\boldsymbol{X}}_r = \boldsymbol{X}_r; \overline{\boldsymbol{Y}}_r = \boldsymbol{Y}_r; \boldsymbol{D}_r\boldsymbol{R} + \overline{\boldsymbol{X}}_l = X_l$$
$$-\boldsymbol{N}_r\boldsymbol{R} + \overline{\boldsymbol{Y}}_l = \boldsymbol{Y}_l$$

we then have

$$\begin{bmatrix} \boldsymbol{Y}_r & \boldsymbol{X}_r \\ -\boldsymbol{N}_l & \boldsymbol{D}_l \end{bmatrix} \begin{bmatrix} \boldsymbol{D}_r & -\boldsymbol{X}_l \\ \boldsymbol{N}_r & \boldsymbol{Y}_l \end{bmatrix} = \begin{bmatrix} I & 0 \\ 0 & I \end{bmatrix}$$

which is the same as (4.5). The above method shows how if we have independent right and left characterizations of a matrix transfer function and their corresponding Bezout identity, we can recast them to obtain the generalized Bezout identity. Using the method referred to earlier, it is possible to arrive at state space realization of each of the block matrices in the generalized Bezout identity. These are given in (4.7) and (4.8). The proof is postponed to Chapter 5 as it is tied up with observer based controllers discussed in that chapter. We assume here that the state space realization of $\boldsymbol{G}(s)$ namely $\{A, B, C, D\}$ is known. We further assumed that (A, B) is stabilizable and (C, A) is detectable. F and H are the state feedback gain matrix and the observer gain matrix respectively. They are so chosen that $(A - BF)$ and $(A - HC)$ are stable matrices. This is always possible because of the stablizability and detectability assumption. The state space realization of the block matrices on the left hand side of (4.7) and (4.8) can be directly read off from the state space representation shown on the right hand side of these equations, following the usual convention.

$$\begin{bmatrix} \boldsymbol{D}_r & -\boldsymbol{X}_l \\ \boldsymbol{N}_r & \boldsymbol{Y}_l \end{bmatrix} = \left[\begin{array}{c|cc} A - BF & B & H \\ \hline -F & I & 0 \\ C - DF & D & I \end{array}\right] \tag{4.7}$$

$$\begin{bmatrix} \boldsymbol{Y}_r & \boldsymbol{X}_r \\ -\boldsymbol{N}_l & \boldsymbol{D}_l \end{bmatrix} = \left[\begin{array}{c|cc} A - HC & B - HD & H \\ \hline F & I & 0 \\ -C & -D & I \end{array}\right] \tag{4.8}$$

To give two examples, the state space realization of $\boldsymbol{D}_r$ in (4.7) is given by

$$\boldsymbol{D}_r : \{(A - BF), B, -F, I\}$$

The stable space realization $-\boldsymbol{N}_l$ in (4.8) is given by

$$-\boldsymbol{N}_l : \{(A - HC), (B - HD), -C, -D\}$$

4.8 Summary

A brief summary of the points covered in this chapter is given below.

1. Equivalent matrices are obtained by premultiplying and postmultiplying a given polynomial matrix by unimodular matrices. Equivalent matrices are rank invariant.
2. Two commonly used standard forms are discussed. They are: (i) Hermite form; (ii) Smith form. The former is either upper or lower quasi-triangular and the latter is diagonal. All equivalent matrices have the same Smith form.
3. Coprimeness of two polynomial matrices may be checked by verifying whether the Bezout identity is satisfied. Another test is the rank test where the matrices are arranged either row wise or column wise as the case may be. Then it is verified whether the composite matrix has maximum column or row rank for all values of $s \in \mathbb{C}$.
4. Transfer matrices may be factorized either in terms of polynomial matrices or in terms of matrices whose elements are rational functions. In the latter case we may insist that the factors are stable matrices. Coprimeness in this context means that the 'numerator' and 'denominator' factors have no common zeros in Re $s \geq 0$.
5. Given the stabilizable and detectable state space realization of a transfer matrix, it is possible to obtain the state space realizations of all the eight block matrices associated with the generalized Bezout identity. The method is based on the design of an observor based controller for the given system.

Notes and Additional References

Readers may consult two books – one by Vidyasagar (1985) and the other by Callier and Desoer (1982) to obtain additional information about most of the topics discussed in this chapter. The theoretical underpinnings of the fractional representation approach are dealt by Desoer *et al.* (1980) and

Desoer and Gustafson (1984). Hermite forms, Smith forms and SM forms are discussed in most of the standard books on control theory, e.g. Kailath (1980). Nett, Jacobson and Balas (1984) are credited with proposing a method for obtaining state space realizations of block matrices associated with the generalized Bezout identity.

Exercises

1.a) Determinal the normal rank and local rank of the following polynomial matrices

(i) $$\boldsymbol{H}(\mathrm{s}) = \begin{bmatrix} \mathrm{s}+3 & \mathrm{s}+4 \\ \mathrm{s}^2+4\mathrm{s}+3 & \mathrm{s}^2+5\mathrm{s}+4 \end{bmatrix}$$

(ii) $$\boldsymbol{H}(s) = \begin{bmatrix} s+2 & s+2 \\ s+4 & s+5 \end{bmatrix}$$

b) Determine the poles and zeros of the transfer matrix

$$\boldsymbol{G}(s) = \begin{bmatrix} \dfrac{1}{(s+2)(s+3)} & \dfrac{s+2}{(s+1)(s+3)} & \dfrac{s+3}{(s+1)(s+2)} \\ O & \dfrac{1}{s+3} & \dfrac{1}{s+1} \end{bmatrix}$$

2. Given that

$$\boldsymbol{G}(s) = \begin{bmatrix} 1 & -1 \\ s^2+s-4 & 2s^2-s-8 \\ s^2-4 & 2s^2-8 \end{bmatrix}$$

Obtain its Smith canonical form

3.a) Obtain the Smith–McMillan form of

$$\boldsymbol{P}(s) = \frac{1}{(s+2)(s+3)} \begin{bmatrix} 1 & 0 \\ -2 & 3(s+2)^2 \end{bmatrix}$$

(b) Let the Smith–McMillan form of a transfer matrix be given by

$$\boldsymbol{M}(s) = \begin{bmatrix} \dfrac{s+4}{(s+2)^2(s+3)} & 0 \\ 0 & \dfrac{(s+4)^2}{(s+2)(s+3)} \\ 0 & 0 \end{bmatrix}$$

(i) Obtain the poles and transmission zeros of $\boldsymbol{M}(s)$
(ii) McMillan degree of $\boldsymbol{M}(s)$

4. Demonstrate the transmission blocking properties of the transmission zeros of a transfer matrix.

5. Given $\boldsymbol{g}(s) = \dfrac{1}{(s-1)(s-2)}$ a possible stable coprime factorization of $\boldsymbol{g}(s)$ is given by

$$\boldsymbol{g}(s) = \frac{\boldsymbol{n}(s)}{\boldsymbol{d}(s)} \text{ where } \boldsymbol{n}(s) = \frac{1}{(s+1)^2} \quad \boldsymbol{d}(s) = \frac{(s-1)(s-2)}{(s+1)^2}$$

obtain stable rational functions $\boldsymbol{x}(s)$ and $\boldsymbol{y}(s)$ which satisfy the Bezout identity

$$\boldsymbol{x}(s)\boldsymbol{n}(s) + \boldsymbol{y}(s)\boldsymbol{d}(s) = 1$$

6. Let $\boldsymbol{N}_r \in \mathbb{R}[s]^{p\times q}$ and $\boldsymbol{D}_r \in \mathbb{R}[s]^{q\times q}$ constitute a matrix pair $(\boldsymbol{N}_r, \boldsymbol{D}_r)$ with $\boldsymbol{D}_r$ non-singular. Prove that $(\boldsymbol{N}_r, \boldsymbol{D}_r)$ is right coprime if and only if

$$\text{Rank}\begin{bmatrix} \boldsymbol{D}_r \\ \boldsymbol{N}_r \end{bmatrix} = q \quad \forall s \in \mathbb{C}$$

7. Consider a transfer matrix

$$\boldsymbol{G}(s) = \begin{bmatrix} \dfrac{17s+11}{(s+1)(s-2)} & \dfrac{3}{s-2} \end{bmatrix}$$

A controllable and observable realization of G(s) is given by

$$A = \begin{bmatrix} -1 & 0 \\ 0 & 2 \end{bmatrix}; B = \begin{bmatrix} 2 & 0 \\ 5 & 1 \end{bmatrix}; C = [1 \quad 3]; D = 0$$

Select suitable F and H matrices such that $(A+BF)$ and $(A+HC)$ are stable matrices. Hence obtain the right and left coprime stable factorization of $\boldsymbol{G}(s)$. Also obtain the state space realizations of these factors.

5

Parametrization of Stabilizing Controllers

5.1 Introduction

Parametrization is a time-tested concept used to generate classes of functions endowed with specific properties. In control theory we are interested in knowing the set of all controllers which could stabilize a given system. The contents of this chapter are built around this major topic. In Section 5.2 we discuss both bounded input bounded output (BIBO) stability and internal stability of multivariable systems and derive the conditions under which they are synonymous. The Youla parametrization approach forms the subject matter of Section 5.3. This is an important concept and has materially influenced our approach regarding synthesis of the set of stabilizing controllers. Youla parameterization is based on the coprime factorization of plant transfer matrix. Such a factorization can be performed in state space in a relatively easy manner. This forms the subject matter of Section 5.4. Here, we carry forward the discussions initiated in Chapter 4 and proceed to outline the method adopted for obtaining the required factorization. Section 5.5 deals with the concept of strong stabilization, i.e. stabilization using a controller which, by itself, is stable. A simple test based on the 'parity interlacing property' is explained, which checks whether, the given system is strongly stabilizable or not.

5.2 Well-Posedness and Internal Stability

From Chapter 2 we recall that the principal components of a closed loop configuration are the plant, the sensor and the controller. The plant quite often includes the actuator also. The sensor measures the plant output and the controller generates the plant output. This is schematically represented in Figure 5.1.

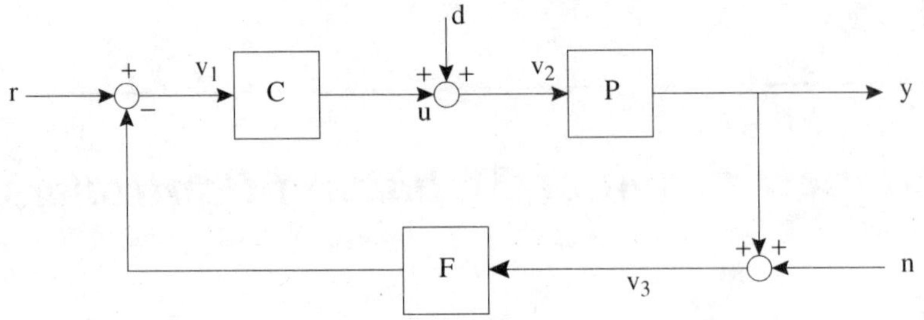

Figure 5.1 The plant controller sensor configuration.

In the figure $\boldsymbol{P}, \boldsymbol{C}, \boldsymbol{F}$ represent the plant, the controller and sensor respectively. We assume that the associated transfer functions are proper i.e., bounded at infinity. The signals marked in the figure represent the following:

r – reference or command input; d – external disturbance
n – sensor noise; v_1, v_2 and v_3 – input signals to controller, plant and sensor respectively
y – plant output, which is also the measured signal

From Figure 5.1 the transfer matrix connecting the exogenous inputs r, d and n to the internal inputs v_1, v_2 and v_3 can be easily deduced. Thus

$$\begin{bmatrix} \boldsymbol{I} & 0 & \boldsymbol{F} \\ -\boldsymbol{C} & \boldsymbol{I} & 0 \\ 0 & -\boldsymbol{P} & \boldsymbol{I} \end{bmatrix} \begin{bmatrix} v_1 \\ v_2 \\ v_3 \end{bmatrix} = \begin{bmatrix} r \\ d \\ n \end{bmatrix} \tag{5.1}$$

From (5.1) we can obtain the transfer function relating the outputs to the inputs as given below:

$$\begin{bmatrix} v_1 \\ v_2 \\ v_3 \end{bmatrix} = \begin{bmatrix} (\boldsymbol{I}+\boldsymbol{FPC})^{-1} & -(\boldsymbol{I}+\boldsymbol{FPC})^{-1}\boldsymbol{FP} & -(\boldsymbol{I}+\boldsymbol{FPC})^{-1}\boldsymbol{F} \\ (\boldsymbol{I}+\boldsymbol{CFP})^{-1}\boldsymbol{C} & (\boldsymbol{I}+\boldsymbol{CFP})^{-1} & -(\boldsymbol{I}+\boldsymbol{CFP})^{-1}\boldsymbol{CF} \\ (\boldsymbol{I}+\boldsymbol{PCF})^{-1}\boldsymbol{PC} & (\boldsymbol{I}+\boldsymbol{PCF})^{-1}\boldsymbol{P} & (\boldsymbol{I}+\boldsymbol{PCF})^{-1} \end{bmatrix} \begin{bmatrix} r \\ d \\ n \end{bmatrix} \tag{5.2}$$

Let the transfer matrix in (5.2) be designated as $\boldsymbol{H}(s)$. We require that all the nine block sub-matrices of $\boldsymbol{H}$ be proper i.e., $\lim_{s\to\infty} \boldsymbol{H}(s)$ be a real finite matrix. Noting that $\boldsymbol{P}, \boldsymbol{C}$ and $\boldsymbol{F}$ are all proper, this can happen if and only if $\det\,[\,\boldsymbol{I}+\boldsymbol{P}(\infty)\boldsymbol{C}(\infty)\boldsymbol{F}(\infty)] \neq 0$. Thus the existence of a proper transfer matrix $\boldsymbol{H}$ is linked with the non singularity of the return difference matrix $[\boldsymbol{I}+\boldsymbol{P}(\infty)\boldsymbol{C}(\infty)\boldsymbol{F}(\infty)]$. A closed loop system as shown in Figure 5.1 is *well posed* if and only if $\det\,[\boldsymbol{I}+\boldsymbol{P}(\infty)\boldsymbol{C}(\infty)\boldsymbol{F}(\infty)] \neq 0$

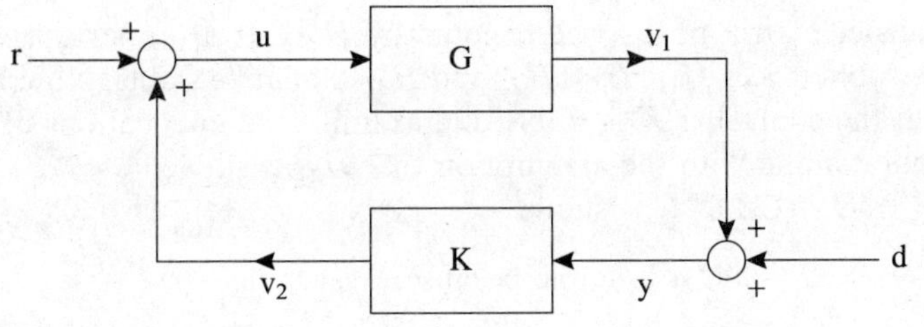

Figure 5.2 A plant controller feedback system.

Internal stability

The problem of internal stability arises in a closed loop system, in the following manner.

Consider the feedback system shown in Figure 5.2 where $\boldsymbol{G}(s)$ is the feed forward transfer matrix and $\boldsymbol{K}(s)$ the feedback transfer matrix. As usual $r(s)$ is the command signal and $d(s)$ the disturbance signal. Note the positive feedback convention used here. This is done to ensure that the formulae developed are in conformity with similar ones derived in the generalized case in later chapters.

From inspection of Figure 5.2 we may write

$$\begin{bmatrix} u(s) \\ y(s) \end{bmatrix} = \begin{bmatrix} I & -\boldsymbol{K} \\ -\boldsymbol{G} & I \end{bmatrix}^{-1} \begin{bmatrix} r(s) \\ d(s) \end{bmatrix} \tag{5.3}$$

and it follows that

$$\begin{bmatrix} u(s) \\ y(s) \end{bmatrix} = \begin{bmatrix} (I - \boldsymbol{KG})^{-1} & (I - \boldsymbol{KG})^{-1}\boldsymbol{K} \\ (I - \boldsymbol{GK})^{-1}\boldsymbol{G} & (I - \boldsymbol{GK})^{-1} \end{bmatrix} \begin{bmatrix} r(s) \\ d(s) \end{bmatrix} \tag{5.4}$$

We now make use of the following well-known matrix identities,

$$(I - \boldsymbol{KG})^{-1}\boldsymbol{K} = \boldsymbol{K}(I - \boldsymbol{GK})^{-1}$$

and

$$\begin{aligned} (I - \boldsymbol{KG})^{-1} &= I + (\boldsymbol{I} - \boldsymbol{KG})^{-1}\boldsymbol{KG} \\ &= I + \boldsymbol{K}(\boldsymbol{I} - \boldsymbol{GK})^{-1}\boldsymbol{G} \end{aligned}$$

(5.4) may now be rewritten as

$$\begin{bmatrix} u(s) \\ y(s) \end{bmatrix} = \begin{bmatrix} I + \boldsymbol{K}(I - \boldsymbol{GK})^{-1}\boldsymbol{G} & \boldsymbol{K}(I - \boldsymbol{GK})^{-1} \\ (I - \boldsymbol{GK})^{-1}\boldsymbol{G} & (I - \boldsymbol{GK})^{-1} \end{bmatrix} \begin{bmatrix} r(s) \\ d(s) \end{bmatrix} \tag{5.5}$$

Let the transfer matrix in (5.5) be denoted by $\boldsymbol{H}(s)$. If $\boldsymbol{H}(s)$ is stable them each of the submatrices $\boldsymbol{H}_{11}$, $\boldsymbol{H}_{12}$, $\boldsymbol{H}_{21}$ and $\boldsymbol{H}_{22}$ should be stable. Now, let us assume that the controller $\boldsymbol{K}$ is stable and examine the implications of such an assumption along with the assumption that $\boldsymbol{H}_{21}$ is stable.
$\boldsymbol{H}_{21}$ stable $\Rightarrow (I - \boldsymbol{GK})^{-1}\boldsymbol{G}$ is stable.

$$\boldsymbol{H}_{11} = I + \boldsymbol{KH}_{21} \text{ is stable because } \boldsymbol{K} \text{ is stable}$$

$$\boldsymbol{H}_{22} = (I - \boldsymbol{GK})^{-1} = I + (I - \boldsymbol{GK})^{-1}\boldsymbol{GK} = (I + \boldsymbol{H}_{21}\boldsymbol{K})$$

$\boldsymbol{H}_{22}$ is stable becomes $I + \boldsymbol{H}_{21}\boldsymbol{K}$ is stable. Finally $\boldsymbol{H}_{12} = \boldsymbol{KH}_{22}$ is stable.

We have thus established the fact that provided the controller is stable, the stability of $\boldsymbol{H}_{21}$ guarantees the stability of $\boldsymbol{H}$.

In general, the fact that the transfer matrix $\boldsymbol{H}_{21}$ connecting $r(s)$ and $y(s)$ is stable is no guarantee that other transfer matrices $\boldsymbol{H}_{11}$, $\boldsymbol{H}_{12}$ and $\boldsymbol{H}_{22}$ are stable. This is precisely the problem posed by internal stability considerations. If $\boldsymbol{K}(s)$ is unstable, then the possibility of unstable pole zero cancellation inside the feedback loop cannot be ruled out. Thus, the transfer matrix $\boldsymbol{H}_{21}$ may exhibit stable characteristics, while the other transfer matrices may be unstable.

EXAMPLE 5.1

Let $\boldsymbol{g}(s) = \dfrac{s}{s+1} \qquad \boldsymbol{k}(s) = \dfrac{-(s+2)}{s}$

clearly, the controller is unstable. We have $(1 - \boldsymbol{gk})^{-1} = \dfrac{s+1}{2s+3}$.

$$\text{Hence } \boldsymbol{h}_{21} = (1 - \boldsymbol{gk})^{-1}\boldsymbol{g} = \frac{s}{2s+3} \text{ which is stable.}$$

$$\text{However } \boldsymbol{h}_{12} = (1 - gk)^{-1}\boldsymbol{k} = \frac{-(s+1)(s+2)}{s(2s+3)} \text{ is unstable.}$$

This example illustrates how the system can be input-output stable without it being internally stable.

■

Internal stability from the state space viewpoint implies that all the eigen values of the A-matrix of the closed loop system are located in Re $s < 0$. Viewed differently, internal stability ensures that there are no right half plane pole zero cancellations between subsystems $\boldsymbol{G}$ and $\boldsymbol{K}$.

A controller which ensures internal stability is known as an *admissible controller.*

The next theorem provides an answer to the question – when exactly is internal stability equivalent to input-output stability?

THEOREM 5.1
Let $\Sigma_1 : (A_1, B_1, C_1, D_1)$ and $\Sigma_2 : (A_2, B_2, C_2, D_2)$ be the realization of the plant and controller respectively in Figure 5.2. Let Σ_1 and Σ_2 be stabilizable and detectable and further let det $(I - \boldsymbol{G}(\infty)\boldsymbol{K}(\infty)) =$ det $(I - \boldsymbol{D}_1\boldsymbol{D}_2) \neq 0$. Under these conditions, the closed loop system shown in Figure 5.1 is internally stable if and only if the transfer matrix $\boldsymbol{H}$ defined by (5.5) belongs to $R\mathcal{H}_\infty$.

REMARKS
Σ_1 and Σ_2 being stabilizable and detectable implies that there are no hidden unstable modes associated with the sub systems. Det $(I - D_1D_2) \neq 0$ ensures that the problem is well posed.

Proof: Referring to Figure 5.2, the state space and output equations of systems Σ_1 and Σ_2 are:

$$\Sigma_1 : \quad \dot{x} = A_1x + B_1u \tag{5.6a}$$

$$v_1 = c_1x + D_1u \tag{5.6b}$$

$$\Sigma_2 : \quad \dot{p} = A_2p + B_2y \tag{5.7a}$$

$$v_2 = C_2p + D_2y \tag{5.7b}$$

Using the above equations we obtain the state space equations of the composite system in terms of the inputs r and d and the outputs u and y. This is done by eliminating v_1 and v_2 between equations (5.6) and (5.7) making use of the relationships $v_1 + d = y$ and $v_2 + r = u$. The state space and input output equations of the composite system are

$$\begin{bmatrix} \dot{x} \\ \dot{p} \end{bmatrix} = \begin{bmatrix} A_1 + B_1M_2D_2C_1 & B_1M_2C_2 \\ B_2M_1C_1 & A_2 + B_2M_1D_1C_2 \end{bmatrix} \begin{bmatrix} x \\ p \end{bmatrix} + \begin{bmatrix} B_1M_2 & B_1M_2D_2 \\ B_2M_1D_1 & B_2M_1 \end{bmatrix} \begin{bmatrix} r \\ d \end{bmatrix} \tag{5.8a}$$

$$\begin{bmatrix} u \\ y \end{bmatrix} = \begin{bmatrix} M_2D_2C_1 & M_2C_2 \\ M_1C_1 & M_1D_1C_2 \end{bmatrix} \begin{bmatrix} x \\ p \end{bmatrix} + \begin{bmatrix} M_2 & M_2D_2 \\ M_1D_1 & M_1 \end{bmatrix} \begin{bmatrix} r \\ d \end{bmatrix} \tag{5.8b}$$

where $M_1 = (I - D_1D_2)^{-1}$ and $M_2 = (I - D_2D_1)^{-1}$. The indicated inverses exist in view of the hypothesis.

To prove the theorem all we need to show is that the realization obtained from (5.8a) and (5.8b) is stabilizable and detectable. We prove detectability using the PBH test for observability (see Appendix *B* for details)

The proof is by contradiction. Let us assume that the composite system is not detectable. Then there exists a λ for which Re $\lambda \geq 0$ satisfying the following equations (as specified by the PBH test on observability) for a non zero vector $\begin{bmatrix} w_1 \\ w_2 \end{bmatrix}$.

$$\begin{bmatrix} A_1 + B_1 M_2 D_2 C_1 - \lambda I & B_1 M_2 C_2 \\ B_2 M_1 C_1 & A_2 + B_2 M_1 D_1 C_2 - \lambda I \\ M_2 D_2 C_1 & M_2 C_2 \\ M_1 C_1 & M_1 D_1 C_2 \end{bmatrix} \begin{bmatrix} w_1 \\ w_2 \end{bmatrix} = 0 \qquad (5.9)$$

Expanding the last and the last but one rows in equation above, we have

$$M_1 C_1 w_1 + M_1 D_1 C_2 w_2 = 0$$
$$M_2 D_2 C_1 w_1 + M_2 C_2 w_2 = 0$$

Since M_1 and M_2 are non-singular, the above equations simplify to

$$C_1 w_1 + D_1 C_2 w_2 = 0$$
$$D_2 C_1 w_1 + C_2 w_2 = 0$$

from which it follows that

$$(I - D_1 D_2) C_1 w_1 = 0 \Rightarrow C_1 w_1 = 0$$

because $I - D_1 D_2$ is non-singular.
Similarly we can show that $C_2 w_2 = 0$.
Substituting $C_1 w_1 = 0$ and $C_2 w_2 = 0$ in the first two equations of (5.9) we get

$$(A_1 - \lambda I) w_1 = 0$$
$$C_1 w_1 = 0$$
$$(A_2 - \lambda I) w_2 = 0$$
$$C_2 w_2 = 0$$

Because (C_1, A_1) and (C_2, A_2) are detectable the above relationships can be true if and only if $w_1 = w_2 = 0$, in view of λ in the above equations having its real part greater than or equal to zero. This violates our original assumption that there exists a non zero vector $\begin{bmatrix} w_1 \\ w_2 \end{bmatrix}$ satisfying (5.9) for Re $\lambda \geq 0$.

Hence the composite system is detectable. Dual arguments can prove that the composite system is also stabilizable.

Stabilizability and detectability of the composite system implies that the system has no uncontrollable and unobservable modes in Re $s \geq 0$. The controllable and observable modes are the poles of the transfer matrix $\boldsymbol{H}$. Hence $\boldsymbol{H} \in \mathcal{RH}_\infty$ implies that the system is internally stable.

The converse of this statement is easily proved because internal stability implies that all the eigen values of the system A-matrix are located in the open left half plane and hence $\boldsymbol{H}$ is stable.

■

Theorem 5.1 is important because under certain mild restrictions, stability of the transfer matrix has been shown to be equivalent to internal stability of the composite system.

5.3 The Youla Parametrization Approach

The transfer matrix $\boldsymbol{H}$ of the feedback system shown in Figure 5.2 is according to (5.3) given by

$$\boldsymbol{H} = \begin{bmatrix} I & -\boldsymbol{K} \\ -\boldsymbol{G} & I \end{bmatrix}^{-1}$$

Theorem 5.1 asserts that the closed loop system is internally stable if and only if $\boldsymbol{H} \in \mathcal{RH}_\infty$. From (5.5) it is clear that $\boldsymbol{H}_{ij}$ $(i, j = 1, 2)$ are non-linear functions of $\boldsymbol{K}$. This is quite inconvenient and makes problem solving difficult. However Youla *et al.* (1976a, 1976b) proposed a parametrization method which delineated the set of all controllers that stabilized the plant. This was later modified by Desoer *et al.* (1980) who pointed out the advantage of working with coprime factors in $\mathcal{RH}_\infty$ instead of the usual coprime factors over a ring of polynomials. One of the merits of the Youla Parameterization approach is that the transfer function sub-matrices are affine functions of the free parameter $\boldsymbol{Q}$ which belongs to $\mathcal{RH}_\infty$. As $\boldsymbol{Q}$ ranges over the $\mathcal{RH}_\infty$ space, the set of all stabilizing controllers are generated.

The Youla parametrization scheme is stated in the following theorem.

THEOREM 5.2 (Youla Parametrization)
Let $\boldsymbol{G} = \boldsymbol{N}_r\boldsymbol{D}_r^{-1} = \boldsymbol{D}_l^{-1}\boldsymbol{N}_l$ be the right and left coprime factorization of $\boldsymbol{G}$ and let

$$\begin{bmatrix} \boldsymbol{V}_r & \boldsymbol{U}_r \\ -\boldsymbol{N}_l & \boldsymbol{D}_l \end{bmatrix} \begin{bmatrix} \boldsymbol{D}_r & -\boldsymbol{U}_l \\ \boldsymbol{N}_r & \boldsymbol{V}_l \end{bmatrix} = \begin{bmatrix} I & O \\ O & I \end{bmatrix}$$

with all of that sub matrices in $\mathcal{RH}_\infty$. Then the set of all controllers that stabilize the plant is given by either

$$\mathcal{K} = \{\boldsymbol{K} = (\boldsymbol{D}_r\boldsymbol{Q} - \boldsymbol{U}_l)(\boldsymbol{N}_r\boldsymbol{Q} + \boldsymbol{V}_l)^{-1} : \boldsymbol{Q} \in \mathcal{RH}_\infty \text{ and det } (\boldsymbol{N}_r\boldsymbol{Q} + \boldsymbol{V}_l) \neq 0\} \tag{5.10}$$

or

$$\mathcal{K} = \{\boldsymbol{K} = (\boldsymbol{V}_r + \boldsymbol{Q}\boldsymbol{N}_l)^{-1}(\boldsymbol{Q}\boldsymbol{D}_l - \boldsymbol{U}_r) : \boldsymbol{Q} \in \mathcal{RH}_\infty \text{ and det } (\boldsymbol{V}_l + \boldsymbol{Q}\boldsymbol{N}_l) \neq 0\} \tag{5.11}$$

Note: (5.10) and (5.11) are derived using the positive feedback convention

Proof: The stabilizing controller $\boldsymbol{K}$ is expressed in two ways

$$\boldsymbol{K} = \boldsymbol{K}_1\boldsymbol{K}_2^{-1} = (\boldsymbol{D}_r\boldsymbol{Q} - \boldsymbol{U}_l)(\boldsymbol{N}_r\boldsymbol{Q} + \boldsymbol{V}_l)^{-1}$$

or

$$\boldsymbol{K} = \boldsymbol{K}_4^{-1}\boldsymbol{K}_3 = (\boldsymbol{V}_r + \boldsymbol{Q}\boldsymbol{N}_l)^{-1}(\boldsymbol{Q}\boldsymbol{D}_l - \boldsymbol{U}_r)$$

In matrix form the above relationships are,

$$\begin{bmatrix} \boldsymbol{K}_1 \\ \boldsymbol{K}_2 \end{bmatrix} = \begin{bmatrix} \boldsymbol{D}_r & -\boldsymbol{U}_l \\ \boldsymbol{N}_r & \boldsymbol{V}_l \end{bmatrix} \begin{bmatrix} \boldsymbol{Q} \\ \boldsymbol{I} \end{bmatrix} \tag{5.12}$$

$$[\boldsymbol{K}_4 \quad -\boldsymbol{K}_3] = [\boldsymbol{I} \quad -\boldsymbol{Q}] \begin{bmatrix} \boldsymbol{V}_r & \boldsymbol{U}_r \\ -\boldsymbol{N}_l & \boldsymbol{D}_l \end{bmatrix} \tag{5.13}$$

We first shown that $\boldsymbol{K} = \boldsymbol{K}_4^{-1}\boldsymbol{K}_3$ stabilizes that system.

We know that for any $\boldsymbol{K}$ which stabilizes the feedback system.

$$\begin{bmatrix} \boldsymbol{I} & -\boldsymbol{K} \\ -\boldsymbol{G} & \boldsymbol{I} \end{bmatrix}^{-1} \in \mathcal{RH}_\infty \tag{5.14}$$

Substituting $\boldsymbol{K} = \boldsymbol{K}_4^{-1}\boldsymbol{K}_3$ and $\boldsymbol{G} = \boldsymbol{D}_l^{-1}\boldsymbol{N}_l$ in (5.14) we get,

$$\begin{bmatrix} \boldsymbol{I} & -\boldsymbol{K} \\ -\boldsymbol{G} & \boldsymbol{I} \end{bmatrix} = \begin{bmatrix} \boldsymbol{I} & -\boldsymbol{K}_4^{-1}\boldsymbol{K}_3 \\ -\boldsymbol{D}_l^{-1}\boldsymbol{N}_l & \boldsymbol{I} \end{bmatrix}$$

$$= \begin{bmatrix} \boldsymbol{K}_4^{-1} & 0 \\ 0 & \boldsymbol{D}_l^{-1} \end{bmatrix} \begin{bmatrix} \boldsymbol{K}_4 & -\boldsymbol{K}_3 \\ -\boldsymbol{N}_l & \boldsymbol{D}_l \end{bmatrix}$$

Substituting for $[\boldsymbol{K}_4 - \boldsymbol{K}_3]$ from (5.13) in the above equation we get,

$$\begin{bmatrix} \boldsymbol{I} & -\boldsymbol{K} \\ -\boldsymbol{G} & \boldsymbol{I} \end{bmatrix} = \begin{bmatrix} \boldsymbol{K}_4^{-1} & 0 \\ 0 & \boldsymbol{D}_l^{-1} \end{bmatrix} \begin{bmatrix} \boldsymbol{I} & -\boldsymbol{Q} \\ 0 & \boldsymbol{I} \end{bmatrix} \begin{bmatrix} \boldsymbol{V}_r & \boldsymbol{U}_r \\ -\boldsymbol{N}_l & \boldsymbol{D}_l \end{bmatrix}$$

and

$$\begin{bmatrix} I & -K \\ -G & I \end{bmatrix}^{-1} = \begin{bmatrix} V_r & U_r \\ -N_l & D_l \end{bmatrix}^{-1} \begin{bmatrix} I & Q \\ 0 & I \end{bmatrix} \begin{bmatrix} K_4 & 0 \\ 0 & D_l \end{bmatrix} \tag{5.15}$$

The right hand side of (5.15) belongs to $\mathcal{RH}_\infty$. Hence $K = K_4^{-1} K_3$ stabilizes the feedback system.

We next show that any controller which stabilizes the feedback system belongs to the set defined in (5.11). If this is proved then the set $\mathcal{K}$ covers all stabilizing controllers without exception.

Choose a stabilizing controller K with left coprime factorization $K = X^{-1} Y$, satisfying the left Bezout identity

$$YB + XA = I \qquad A, B \in \mathcal{RH}_\infty$$

we have

$$\begin{bmatrix} I & -K \\ -G & I \end{bmatrix} = \begin{bmatrix} I & -X^{-1}Y \\ -D_l^{-1}N_l & I \end{bmatrix}$$

$$= \begin{bmatrix} X^{-1} & 0 \\ 0 & D_l^{-1} \end{bmatrix} \begin{bmatrix} X & -Y \\ -N_l & D_l \end{bmatrix}$$

$$\begin{bmatrix} I & -K \\ -G & I \end{bmatrix}^{-1} = \begin{bmatrix} X & -Y \\ -N_l & D_l \end{bmatrix}^{-1} \begin{bmatrix} X & 0 \\ 0 & D_l \end{bmatrix} \tag{5.16}$$

We now show that (5.16) represents a left coprime factorization because Bezout's left identity is satisfied as follows

$$\begin{bmatrix} X & -Y \\ -N_l & D_l \end{bmatrix} \begin{bmatrix} 0 & -U_l \\ -B & 0 \end{bmatrix} + \begin{bmatrix} X & 0 \\ 0 & D_l \end{bmatrix} \begin{bmatrix} A & U_l \\ B & V_l \end{bmatrix} = \begin{bmatrix} I & 0 \\ 0 & I \end{bmatrix}$$

In (5.16) since the left hand side is in $\mathcal{RH}_\infty$ and the right hand side represents a left coprime factorization, according to Theorem 4.7 in Chapter 4, we note that

$$\begin{bmatrix} X & -Y \\ -N_l & D_l \end{bmatrix}^{-1} \in \mathcal{RH}_\infty$$

But

$$\begin{bmatrix} X & -Y \\ -N_l & D_l \end{bmatrix} \begin{bmatrix} D_r & -U_l \\ N_r & V_l \end{bmatrix} = \begin{bmatrix} XD_r - YN_r & -XU_l - YV_l \\ 0 & I \end{bmatrix}$$

In the above equation, the left hand side is in $\mathcal{RH}_\infty$ and is also invertible in $\mathcal{RH}_\infty$ which means that $(XD_r - YN_r)^{-1} \in \mathcal{RH}_\infty$.

We have

$$[X \quad -Y] = [X \quad -Y]\begin{bmatrix} D_r & -U_l \\ N_r & V_l \end{bmatrix}\begin{bmatrix} V_r & U_r \\ -N_l & D_l \end{bmatrix}$$

$$= [XD_r - YN_r \quad -XU_l - YV_l]\begin{bmatrix} V_r & U_r \\ -N_l & D_l \end{bmatrix}$$

and

$$(XD_r - YN_r)^{-1}[X \quad -Y] = [I \quad -Q]\begin{bmatrix} V_r & U_r \\ -N_l & D_l \end{bmatrix}$$

where $Q = (XD_r - YN_r)^{-1}(XU_l + YV_l)$ belongs to $\mathcal{RH}_\infty$.

Hence

$$[K_4 \quad -K_3] = [I \quad -Q]\begin{bmatrix} V_r & U_r \\ -N_l & D_l \end{bmatrix}$$

The stabilizing controller chosen namely $X^{-1}Y$ is thus shown to be equal to $K_4^{-1}K_3$ and therefore belongs to the set of all stabilizing controllers generated by the Youla parameterization formula.

■

REMARKS

1. Q is a free parameter belonging to $\mathcal{RH}_\infty$ with no constraint except that det $(N_rQ + V_l) \neq 0$ or det $(V_r + QN_l) \neq 0$ as the case may be depending on whether we are opting for right or left coprime factorization. This is not really a restriction as it can be shown that 'almost all Q' satisfies this constraint.
2. Theorem 5.2 delineates the set of all controllers in terms of a free parameter Q. There is a one-to-one correspondence between Q and the controllers generated by applying (5.10) or (5.11).

■

We shall now examine how parametrization results in simplification of the transfer matrix.

Consider the transfer matrix H defined in (5.5) as

$$H = \begin{bmatrix} I + K(I - GK)^{-1}G & K(I - GK)^{-1} \\ (I - GK)^{-1}G & (I - GK)^{-1} \end{bmatrix} \tag{5.17}$$

To illustrate the simplification achieved, take for example the $\boldsymbol{H}_{22}$ matrix defined as $(I - \boldsymbol{GK})^{-1}$. This is identified as the sensitivity operator $\boldsymbol{S}$ for the closed loop system. We have

$$\begin{aligned}(I - \boldsymbol{GK})^{-1} &= [I - \boldsymbol{D}_l^{-1}\boldsymbol{N}_l(\boldsymbol{D}_r\boldsymbol{Q} - \boldsymbol{U}_l)(\boldsymbol{N}_r\boldsymbol{Q} + \boldsymbol{V}_l)^{-1}]^{-1} \\ &= \{\boldsymbol{D}_l^{-1}[\boldsymbol{D}_l(\boldsymbol{N}_r\boldsymbol{Q} + \boldsymbol{V}_l) - \boldsymbol{N}_l(\boldsymbol{D}_r\boldsymbol{Q} - \boldsymbol{U}_l)](\boldsymbol{N}_r\boldsymbol{Q} + \boldsymbol{V}_l)^{-1}\}^{-1} \\ &= (\boldsymbol{N}_r\boldsymbol{Q} + \boldsymbol{V}_l)[(\boldsymbol{D}_l\boldsymbol{V}_l + \boldsymbol{N}_l\boldsymbol{U}_l) + (\boldsymbol{D}_l\boldsymbol{N}_r - \boldsymbol{N}_l\boldsymbol{D}_r)\boldsymbol{Q}]^{-1}\boldsymbol{D}_l\end{aligned}$$

But $\boldsymbol{D}_l\boldsymbol{V}_l + \boldsymbol{N}_l\boldsymbol{U}_l = I$ and $\boldsymbol{D}_l\boldsymbol{N}_r - \boldsymbol{N}_l\boldsymbol{D}_r = 0$
Hence $(I - \boldsymbol{GK})^{-1} = (\boldsymbol{N}_r\boldsymbol{Q} + \boldsymbol{V}_l)\boldsymbol{D}_l$

The sensitivity operator is clearly an affine function of $\boldsymbol{Q}$ instead of being as non-linear function of $\boldsymbol{K}$. In a similar manner all the block entries of (5.17) could be expressed in terms of affine functions of $\boldsymbol{Q}$. We have

$$\boldsymbol{H} = \begin{bmatrix} I + (\boldsymbol{D}_r\boldsymbol{Q} - \boldsymbol{U}_l)\boldsymbol{N}_l & (\boldsymbol{D}_r\boldsymbol{Q} - \boldsymbol{U}_l)\boldsymbol{D}_l \\ (\boldsymbol{N}_r\boldsymbol{Q} + \boldsymbol{V}_l)\boldsymbol{N}_l & (\boldsymbol{N}_r\boldsymbol{Q} + \boldsymbol{V}_l)\boldsymbol{D}_l \end{bmatrix} \tag{5.18}$$

In the special case when $\boldsymbol{G} \in R\mathcal{H}_\infty$, the right and left coprime factors of the plant are $(\boldsymbol{G}, I)$ and $(I, \boldsymbol{G})$ respectively. We then have $\boldsymbol{U}_r = \boldsymbol{U}_l = 0$; $\boldsymbol{V}_r = \boldsymbol{V}_l = \boldsymbol{D}_r = \boldsymbol{D}_l = I$ and $\boldsymbol{N}_r = \boldsymbol{N}_l = \boldsymbol{G}$ and the stabilizing controller is given by $\boldsymbol{K} = \boldsymbol{Q}(I + \boldsymbol{GQ})^{-1} = (I + \boldsymbol{QG})^{-1}\boldsymbol{Q}$.

Since every controller generated via either (5.10) or (5.11) guarantees internal stability, the designer is spared the burden of checking for stability at every stage and the class of stabilizing controllers available can be used to achieve various performance objectives.

5.4 Coprime Factorization in State Space Revisited

5.4.1 Coprime Factorization of Plant

Youla parameterization requires the knowledge of not only the right and left coprime stable factors of the plant but also the various coefficient matrices associated with the Bezout identities. The most effective way of obtaining this information is via state space methods. We have alluded to this briefly in Chapter 4. Considering the matter in more detail we explain a constructive procedure to obtain all the eight transfer matrices associated with the doubly coprime factorization of the plant transfer function.

We start with the following premises. We are given the plant transfer function G, with known state space realization (not necessarily minimal) (A, B, C, D) with (A, B) stabilizable and (C, A) detectable.

As a result, there exist real matrices F and H such that $(A+BF)$ and $(A+HC)$ are stable. We note that the state equations of G are.

$$\dot{x} = Ax + Bu \tag{5.19a}$$

$$y = Cx + Du \tag{5.19b}$$

Define $u = Fx + v$ where F is so chosen that $(A+BF)$ is stable. On substitution in (5.19) we get

$$\dot{x} = (A+BF)x + Bv \tag{5.20a}$$

$$y = (C+DF)x + Dv \tag{5.20b}$$

The transfer matrix from u to y is

$$\boldsymbol{G} = C(sI - A)^{-1}B + D$$

To obtain the right coprime factorization proceed as follows.

First obtain the transfer function from an intermediate variable v to u, i.e. $u = \boldsymbol{D}_r v$. Then obtain the transfer function from v to y i.e $y = \boldsymbol{N}_r v$.

Hence $y = \boldsymbol{N}_r \boldsymbol{D}_r^{-1} u$

To obtain $\boldsymbol{D}_r : v \rightarrow u$ we note that

$$\dot{x} = (A+BF)x + Bv$$

$$u = Fx + v$$

Hence $\boldsymbol{D}_r : \{(A+BF), B, F, I\}$ (5.21)

To obtain $\boldsymbol{N}_r : v \rightarrow y$

$$\dot{x} = (A+BF)x + Bv$$

$$y = (C+DF)x + Dv$$

Hence $\boldsymbol{N}_r : \{(A+BF), B, (C+DF), D\}$ (5.22)

Thus $\boldsymbol{G} : u \rightarrow y$ is given by

$$\boldsymbol{G} = \boldsymbol{N}_r \boldsymbol{D}_r^{-1}$$

which is a right factorization of $\boldsymbol{G}$ (we will show subsequently that it is indeed a right coprime factorization).

To obtain the left coprime factorization of $\boldsymbol{G}$, first choose H so that $(A+HC)$ is stable.

The observer equation for the system with realization (A, B, C, D) is given by

$$\begin{aligned}\dot{\xi} &= A\xi + Bu + H(C\xi + Du - y)\\ &= (A+HC)\xi - Hy + (B+HD)u\end{aligned}$$

Now define $\eta = (C\xi + Du - y)$.
Let η be expressed in terms of y and u as

$$\eta = \boldsymbol{D}_l y \text{ and } \eta = \boldsymbol{N}_l u$$

and we obtain $\boldsymbol{G} = \boldsymbol{D}_l^{-1}\boldsymbol{N}_l$.

To obtain $\boldsymbol{D}_l : y \rightarrow \eta$ make use of the equations

$$\dot{\xi} = (A + HC)\xi - Hy + (B + HD)u$$
$$\eta = C\xi + Du - y$$

we have

$$\boldsymbol{D}_l : \{(A + HC), -H, C, -I\} \tag{5.23}$$

From the above set of equations we can also obtain

$$\boldsymbol{N}_l : u \rightarrow \eta$$
$$\boldsymbol{N}_l : \{(A + HC), (B + HD), C, D\} \tag{5.24}$$

Note that the matrix pairs $(\boldsymbol{N}_r, \boldsymbol{D}_r)$ and $(\boldsymbol{D}_l, \boldsymbol{N}_l)$ we have constructed are stable and only their coprimeness need to be proved.

5.4.2 Coprime Factorization of Controller

The next step is to synthesize a stabilizing controller and thereafter obtain its coprime factorization. A direct way of achieving this is to opt for an observer based controller. It is well known that the closed loop system, with such a controller in the feedback path has eigenvalues which are obtained from the union of the eigenvalues of $(A + BF)$ and $(A + HC)$. Since both these matrices are stable, the composite system is internally stable. It should be borne in mind that such a stabilizing controller obtained for specific choices of F and H is only one among the infinitely many stabilizing controllers generated by applying the Youla parametrization procedure.

The structure of the observer-based controller is schematically represented in Figure 5.3.

The observer-based controller is governed by the following observer and controller equations:

$$\left.\begin{aligned} \dot{\xi} &= A\xi + Bu + H(\hat{y} - y) \\ \hat{y} &= C\xi + Du \end{aligned}\right\} \text{observer equations}$$

$$u = F\xi \qquad \text{(controller equation)}$$

Define $\mu = y - \hat{y} = y - C\xi - Du$

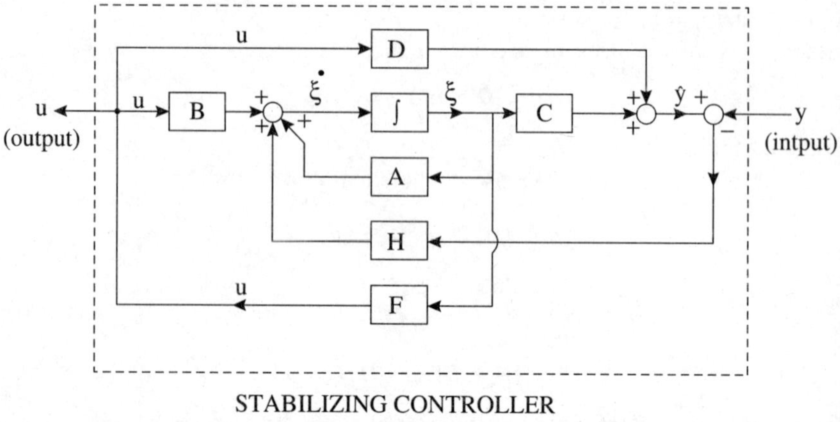

Figure 5.3 Observer-based stabilizing controller.

The observer-based controller has a transfer function $\boldsymbol{K}$ which maps y on to u.

To obtain right coprime factorization of $\boldsymbol{K}$ we make use of an intermediate variable μ and express y and u interms of μ. We have

$$y = \boldsymbol{R}_r\mu \quad ; \quad u = \boldsymbol{S}_r\mu \text{ so that}$$
$$u = \boldsymbol{S}_r\boldsymbol{R}_r^{-1}y = \boldsymbol{K}y$$

To obtain $\boldsymbol{R}_r : \mu \to y$ make use of the following equations

$$\begin{aligned}\dot{\xi} &= A\xi + Bu - H\mu \\ &= (A + BF)\xi - H\mu \\ y &= C\xi + Du + \mu \\ &= (C + DF)\xi + \mu\end{aligned}$$

We have

$$\boldsymbol{R}_r = \{(A + BF), -H, (C + DF), I\} \tag{5.25}$$

To obtain

$$\boldsymbol{S}_r : \mu \to u$$

Make use of the following equations

$$\dot{\xi} = (A + BF)\xi - H\mu$$
$$u = F\xi$$

we have

$$\boldsymbol{S}_r : \{(A + BF), -H, F, 0\} \tag{5.26}$$

The left coprime factorization of the controller $\boldsymbol{K}$ is obtained as follows.
Use an intermediate variable v so that

$$v = \boldsymbol{R}_l u \text{ and } v = \boldsymbol{S}_l y$$
$$\text{so that } u = \boldsymbol{R}_l^{-1}\boldsymbol{S}_l y = \boldsymbol{K}y$$

To obtain

$$\boldsymbol{R}_l : u \to v$$

Define a signal $v = -F\xi + u$ and use the following equations

$$\dot{\xi} = A\xi + Bu + H(C\xi + Du - y)$$
$$= (A + HC)\xi + (B + HD)u - Hy$$
$$v = -F\xi + u$$

We have

$$\boldsymbol{R}_l : \{(A + HC), (B + HD), -F, I\} \tag{5.27}$$

To obtain

$$\boldsymbol{S}_l : y \to v$$

We make use of the same set of equations as was used for obtaining $\boldsymbol{R}_l$
We have

$$\boldsymbol{S}_l : \{(A + HC), -H, -F, 0\} \tag{5.28}$$

The overall transfer function of the observer-based controller can be obtained from:

$$\xi = (A + BF + HC + HDF)\xi - Hy$$
$$u = F\xi$$

and the observer-based controller transfer function is given by

$$\boldsymbol{K} = -F[sI - (A + BF + HC + HDF)]^{-1}H \tag{5.29}$$

5.4.3 State Space Realization

The state space realization of $\boldsymbol{D}_r, \boldsymbol{N}_r, \boldsymbol{R}_r$ and $\boldsymbol{S}_r$ obtained from (5.21), (5.22), (5.25) and (5.26) may be compactly represented as follows

$$\begin{bmatrix} \boldsymbol{D}_r & \boldsymbol{S}_r \\ \boldsymbol{N}_r & \boldsymbol{R}_r \end{bmatrix} = \left[\begin{array}{c|c|c} A+BF & B & -H \\ \hline F & I & 0 \\ \hline C+DF & D & I \end{array}\right] \tag{5.30}$$

Similarly, the state space realization of $\boldsymbol{D}_l, \boldsymbol{N}_l, \boldsymbol{R}_l$ and $\boldsymbol{S}_l$ obtained from (5.23), (5.24), (5.27) and (5.28) may be represented as

$$\begin{bmatrix} \boldsymbol{R}_l & -\boldsymbol{S}_l \\ -\boldsymbol{N}_l & \boldsymbol{D}_l \end{bmatrix} = \left[\begin{array}{c|c|c} A+BC & -(B+HD) & H \\ \hline F & I & 0 \\ \hline C & -D & I \end{array}\right] \tag{5.31}$$

We shall now prove that

$$\begin{bmatrix} \boldsymbol{R}_l & -\boldsymbol{S}_l \\ -\boldsymbol{N}_l & \boldsymbol{D}_l \end{bmatrix} \begin{bmatrix} \boldsymbol{D}_r & \boldsymbol{S}_r \\ \boldsymbol{N}_r & \boldsymbol{R}_r \end{bmatrix} = \begin{bmatrix} I & 0 \\ 0 & I \end{bmatrix} \tag{5.32}$$

We prove this by demonstrating that the inverse matrix of

$$\begin{bmatrix} \boldsymbol{D}_r & \boldsymbol{S}_r \\ \boldsymbol{N}_r & \boldsymbol{R}_r \end{bmatrix}$$

has a realization identical to the realization of

$$\begin{bmatrix} \boldsymbol{R}_l & -\boldsymbol{S}_l \\ -\boldsymbol{N}_l & \boldsymbol{D}_l \end{bmatrix}$$

as given in (5.31).

Recall that (see Appendix B) if (A, B, C, D) is the realization of a transfer matrix, its inverse matrix has a realization $(A - BD^{-1}C, -BD^{-1}, D^{-1}C, D^{-1})$. From (5.30) we know the realization of

$$\begin{bmatrix} \boldsymbol{D}_r & \boldsymbol{S}_r \\ \boldsymbol{N}_r & \boldsymbol{R}_r \end{bmatrix}$$

and applying the above formula we can obtain the realization of the inverse function. It can be verified that this coincides with the realization as given in (5.31), thus proving (5.31).

This result is stated in a theorem as follows.

THEOREM 5.3
Let $\boldsymbol{G}$ represent the plant transfer function and $\boldsymbol{K}$ the transfer function of its stabilizing controller. Let $\boldsymbol{G} = \boldsymbol{N}_r\boldsymbol{D}_r^{-1} = \boldsymbol{D}_l^{-1}\boldsymbol{N}_l$ and $\boldsymbol{K} = \boldsymbol{S}_r\boldsymbol{R}_r^{-1} = \boldsymbol{R}_l^{-1}\boldsymbol{S}_l$ be that right and left stable factors of $\boldsymbol{G}$ and $\boldsymbol{K}$. Then these factors could be so chosen that (5.32) is satisfied. Further the matrix pairs $(\boldsymbol{N}_r, \boldsymbol{D}_r), (\boldsymbol{D}_l, \boldsymbol{N}_l)(\boldsymbol{S}_r, \boldsymbol{R}_r)$ and $(\boldsymbol{R}_l, \boldsymbol{S}_l)$ are coprime.

Proof: The constructive procedure adopted to obtain the matrix pairs $(\boldsymbol{N}_r, \boldsymbol{D}_r), (\boldsymbol{D}_l, \boldsymbol{N}_l), (\boldsymbol{S}_r, \boldsymbol{R}_r)$ and $(\boldsymbol{R}_l, \boldsymbol{S}_l)$ ensures that all the matrices are stable. Coprimeness of the respective matrix pairs follow from (5.32) For example, we have from (5.32)

$$\boldsymbol{R}_l\boldsymbol{D}_r + (-\boldsymbol{S}_l)\boldsymbol{N}_r = \boldsymbol{I} \tag{5.33}$$

Since $\boldsymbol{R}_l, \boldsymbol{D}_r, \boldsymbol{S}_l$ and $\boldsymbol{N}_r \in \mathcal{RH}_\infty$ the above equation is the right Bezout identity for the matrix pair $(\boldsymbol{N}_r, \boldsymbol{D}_r)$ which implies coprimeness of the pair. The coprimeness of the other matrix pairs could be similarly established

■

REMARKS
The left coprime factorization of $\boldsymbol{K}$ is given by $\boldsymbol{K} = \boldsymbol{R}_l^{-1}\boldsymbol{S}_l$. From (5.33) we note that $\boldsymbol{R}_l$ and $-\boldsymbol{N}_l$ happen to be the coefficient matrices associated with $\boldsymbol{D}_r$ and $\boldsymbol{N}_r$ in the right Bezout identity. This shows that once the right coprime factors of the plant are known as well as the associated Bezout identity, the stabilizing controller transfer function in left coprime factorized form is immediately deducible. In other words if $\boldsymbol{G} = \boldsymbol{N}\boldsymbol{D}^{-1}$ and $\boldsymbol{X}\boldsymbol{N} + \boldsymbol{Y}\boldsymbol{D} = \boldsymbol{I}$, then provided $\boldsymbol{Y}$ is non-singular, then $\boldsymbol{K} = -\boldsymbol{Y}^{-1}\boldsymbol{X}$ is a stabilizing controller for $\boldsymbol{G}$

5.5 Strong Stabilization

In the previous section we delineated the set of all controllers which internally stabilizes a given plant. In this context, we would like to know whether among the class of internally stabilizing controllers, there are any controllers which are stable. We shall presently see how for certain plants, it is just not possible to have stable stabilizing controllers. The reasons why designers prefer stable stabilizing controllers to unstable ones are as follows.

1. If for some reasons (e.g. attenuator or sensor failure) the feedback loop is broken, the open loop stability is still maintained if the plant and the stabilizer are individually stable.
2. It is well known that if an unstable plant is stabilized by feedback compensation, then the zeros of the plant in the right half plane continue to be zeros of the closed loop transfer function no matter what type of controller is used. If an unstable controller is used, it can be shown that it will introduce additional zeros in Re $s \geq 0$ over and above those contributed by the plant. But as discussed in Chapter 6, right half plane zeros impose limitations on controller design and therefore their presence is undesirable.

DEFINITION 5.1 (Strong Stabilization)
A plant is said to be strongly stabilizable if there exists a stable controller which internally stabilizes the plant.

Youla *et al* (1974) obtained necessary and sufficient conditions for a multivariable plant to be strongly stabilizable. We first discuss that SISO version of this result.

THEOREM 5.3 (Strong Stabilization – Scalar Case)
Given a proper scalar transfer function $\boldsymbol{p} \in \mathbb{R}(s)$ let $\sigma_1, \sigma_2, \ldots \sigma_p$ be the distinct real zeros of $\boldsymbol{p}$ in Re $s \geq 0$ (including the zero at infinity) arranged in descending order. Let the number of real poles of $\boldsymbol{p}$ in the interval (σ_i, σ_{i+1}) be given by ν_i where the poles are counted according to their multiplicity. Let η denote the number of odd integers in the sequence $\{\nu_1, \nu_2, \ldots \nu_{p-1}\}$. Then every stabilizing controller of $\boldsymbol{p}$ has at least η poles in Re $s \geq 0$. Moreover, the lower bound is exact in the sense that there exists at least one stabilizing controller with η unstable poles

Proof: The proof is omitted.

REMARKS
The above theorem may be paraphrased as follows. A plant $\boldsymbol{P}$ is strongly stabilizable if and only if the number of real poles of $\boldsymbol{p}$ (multiplicity included) between every pair of real zeros of $\boldsymbol{p}$ located in Re $s \geq 0$ is even. The above property is known as *parity interlacing property*.

EXAMPLE 5.1

Consider the following plant transfer functions

i) $p(s) = \dfrac{(s+1)(s-2)(s-3)}{s^2(s+2)(s-4)}$

ii) $p(s) = \dfrac{s(s-1)^2(s-3)}{(s+1)(s-2)^2(s-4)}$

In (i) the number of real poles and zeros located in Re $s \geq 0$ is given below

poles	**zeros**
$s = 0$ (repeated twice)	$s = 2$
$s = 4$	$s = 3$
	$s = \infty$

Between the zeros 2 and 3, there are no poles. Hence $\nu_2 = 0$. Between zeros 3 and ∞, there is one pole at $s = 4$. Hence $\nu_1 = 1$. The number of odd integers in the sequence (ν_1, ν_2) is 1. Hence the plant is not strongly stabilizable and every controller which stabilizes it will at least have one pole in Re $s \geq 0$.

The pole–zero pattern for the transfer function given in (ii) is as follows:

poles	**zeros**
$s = 2$ (repeated twice)	$s = 0$
$s = 4$	$s = 1$ (repeated twice)
	$s = 3$

Between zeros 0 and 1, there are no poles. Hence $\nu_2 = 0$. Between zeros 1 and 3, there are 2 poles. Hence $\nu_2 = 2$. Hence $(\nu_1, \nu_2) = (0, 2)$ and the plant is strongly stabilizable.

Note that when multiple zeros are present they are considered as one distinct zero. Obviously there cannot be poles between coincident zeros.

■

Before generalizing the result to cover the multiple input multiple output (MIMO) case, the following points are clarified. Let $\boldsymbol{P} \in \mathbb{R}(s)^{p \times m}$. The real zeros of P in Re $s \geq 0$ is taken to mean in the present context the *blocking zeros* of $\boldsymbol{P}$. These are zeros of $\boldsymbol{P}$ in Re $s \geq 0$ for which $\boldsymbol{P}(s) = 0$.

The multiplicity of each pole of $\boldsymbol{P}(s)$ is reckoned by its McMillan degree.

If $\boldsymbol{P}$ is strictly proper, than $\boldsymbol{P}(\infty) = 0$ and hence $s = \infty$ is a blocking zero of $\boldsymbol{P}$.

In SISO systems, every zero is a blocking zero and hence no distinction need be made between transmission zeros and blocking zeros of the transfer function.

With these clarifications, we now state that MIMO version of Theorem 5.3.

THEOREM 5.4 (Strong Stabilization – MIMO Case)
Let $\boldsymbol{P}$ be a matrix valued transfer function. Let $(\sigma_1, \sigma_2, \ldots \sigma_p)$ be the real blocking zeros of $\boldsymbol{P}$ in Re $s \geq 0$ (including ∞ if $\boldsymbol{P}$ is strictly proper) arranged in descending order. Let ν_i be the number of poles of $\boldsymbol{P}$ in the interval (σ_i, σ_{i+1}) counted according to their McMillan degrees and let η denote the number of odd integers in the sequence $\{\nu_1, \nu_2, \ldots \nu_{p-1}\}$. Then every controller which internally stabilizes $\boldsymbol{P}$ has at least η poles in Re $s \geq 0$. Moreover the lower bound is exact in the sense that there exists at least one stabilizing controller with exactly η real poles in Re $s \geq 0$.

Proof: Proof is omitted. Proofs for Theorems 5.3 and 5.4 can be found in Vidyasager (1985).

REMARKS
In actual practice, MIMO systems rarely possess blocking zeros, except the blocking zero at infinity for strictly proper plants. If there is only one blocking zero, the question of real poles being present between real zeros (in Re $s \geq 0$) does not arise. Hence invariably $\boldsymbol{P}$ is strongly stabilizable.

5.6 Summary

We summarize below the salient points covered in this chapter:

1. Well posedness implies that $\lim_{s \to \infty} \boldsymbol{H}(s)$ is finite where $\boldsymbol{H}(s)$ is the transfer matrix for the closed loop system. This is possible only if det $(I - \boldsymbol{G}(\infty)\boldsymbol{K}(\infty))$ is not zero.
2. Provided the plant and the controller are stabilizable and detectable, the realization of the closed loop system matrix is also stabilizable and detectable.
3. An offshoot of the above result is that stability of the input–output system guarantees internal stability of the system.
4. It is possible to delineate the set of all stabilizing controllers for a given plant using the Youla Parametrization approach. Further, by merely knowing one such stabilizing controller for the plant, we can generate the set of all stabilizing controllers for the plant.
5. The free parameter $\boldsymbol{Q}$ in the Youla Parameterization formula, is affinely related to important operators such as sensitivity and

complementary sensitivity functions. As a result, studies in $\mathcal{H}_2$ and $\mathcal{H}_\infty$ optimization are considerably simplified.

6. Given that the state space realization (A, B, C, D) for the plant, under mild restrictions, it is possible to generate the left and right stable coprime factors of the plant as well as the unknown stabilizing controller. This is a constructive procedure, which makes use of the observer-based controller theory.
7. An easy way of finding out whether a given plant can be stabilized by a stable controller or not is by checking for parity interlacing property.

Notes and Additional References

The Youla parameterization approach has dominated controller design for over a decade. Original results are found in Youla *et al.* (1976a, 1976b) and a modified version of the same in Desoer *et al.* (1980). The results are rigorously proved in Vidyasager (1985) and Green and Limebeer (1995) for the multivariable case. State space methods for stable coprime factorization are described in Nett *et al.* (1984). The procedure adopted in this chapter closely follows Maciejowski (1989). For results pertaining to strong stabilization the reader may refer to Youla *et al.* (1974) and Vidyasager (1985).

Exercises

1. Let the transfer matrix connecting inputs r and d with outputs u and y in Figure 5.2 be denoted by $\boldsymbol{H}$. In equation 5.5 the submatrices $\boldsymbol{H}_{11}$, $\boldsymbol{H}_{12}$, $\boldsymbol{H}_{21}$ and $\boldsymbol{H}_{22}$ of $\boldsymbol{H}$ are expressed in terms of $(I - \boldsymbol{GK})^{-1}$. Obtain an alternative expression for $\boldsymbol{H}$ where all the submatrices are expressed in terms of $(I - \boldsymbol{KG})^{-1}$.
2. Using Figure 5.2 obtain a transfer matrix $\boldsymbol{W}$ connecting r and d with v_1 and v_2. Prove that $\boldsymbol{W} \in \mathcal{RH}_\infty$ if and only if $\boldsymbol{H} \in \mathcal{RH}_\infty$.
3. Let $(\boldsymbol{N}_r, \boldsymbol{D}_r)$ and $(\boldsymbol{D}_l, \boldsymbol{N}_l)$ be the right and left coprime factorization of plant $\boldsymbol{G}$. Similarly let $(\boldsymbol{S}_r, \boldsymbol{R}_r)$ and $(\boldsymbol{R}_l, \boldsymbol{S}_l)$ be the right and left coprime factorization of the controller $\boldsymbol{K}$. Then show that the following conditions are equivalent.

   (i) $\boldsymbol{H}$ $(\boldsymbol{G}, \boldsymbol{K})$ representing the transfer matrix of the closed loop system in Fig. 5.2 belongs to $\mathcal{RH}_\infty$.
   (ii) $\boldsymbol{S}_l\boldsymbol{N}_r - \boldsymbol{R}_l\boldsymbol{D}_r$ and its inverse belong to $\mathcal{RH}_\infty$.
   (iii) $\boldsymbol{N}_l\boldsymbol{S}_r - \boldsymbol{D}_l\boldsymbol{R}_r$ and its inverse belong to $\mathcal{RH}_\infty$.

4. The expressions for Youla Parameterization as given in (5.10) and (5.11) were obtained using the positive feedback convention. Show that if the negative feedback convention is used then the following expressions could be derived.

$$\begin{aligned} \boldsymbol{K} &= (\boldsymbol{U}_l + \boldsymbol{D}_r\boldsymbol{Q})(\boldsymbol{V}_l - \boldsymbol{N}_r\boldsymbol{Q})^{-1} \\ &= (\boldsymbol{V}_r - \boldsymbol{Q}\boldsymbol{N}_l)^{-1}(\boldsymbol{U}_r + \boldsymbol{Q}\boldsymbol{D}_l) \end{aligned}$$

5. Design a feedback controller for an unstable plant $\boldsymbol{p}(s) = \dfrac{s+1}{s^2-4}$ which will simultaneously place the poles of the feedback system in the region Re $(s) < -1$ and cause the system to asymptotically track a unit step input.

Hint: Select the coprime factors of $\boldsymbol{p}(s)$ in such a way that all the poles are located in Re $(s) < -1$. The Bezout identify coefficients will also exhibit this property. Controller $\boldsymbol{c}(s)$ is chosen via Youla parameterization and the free parameter $\boldsymbol{q}$ is so chosen that all its poles are located in Re $(s) < -1$.
(The problem is taken from Desoer *et al.* 1980)

6.a) Test whether the plant

$$\boldsymbol{p}(s) = \frac{(s-1)^2(s^2-s-1)}{(s-2)^2(s+1)^3}$$

is strongly stabilizable or not using the 'parity interlacing property'.

b) If $\boldsymbol{p}(s)$ given in (a) is strongly stabilizable than construct a stable stabilizing controller for the plant.

7. Recall that the internal stability of a closed loop system (using positive feedback convention) is equivalent to

$$\begin{bmatrix} \boldsymbol{I} & -\boldsymbol{K} \\ -\boldsymbol{G} & \boldsymbol{I} \end{bmatrix}^{-1} \in \mathcal{RH}_\infty$$

Let $\boldsymbol{G} = \boldsymbol{N}_r\boldsymbol{D}_r^{-1} = \boldsymbol{D}_l^{-1}\boldsymbol{N}_l$ and
Let $\boldsymbol{K} = \boldsymbol{S}_r\boldsymbol{R}_r^{-1} = \boldsymbol{R}_l^{-1}\boldsymbol{S}_l$
be the right and left coprime representation of $\boldsymbol{G}$ and $\boldsymbol{K}$. Prove that the feedback system is stable if and only if

$$\text{both} \begin{bmatrix} \boldsymbol{D}_r & \boldsymbol{S}_r \\ \boldsymbol{N}_r & \boldsymbol{R}_r \end{bmatrix}^{-1} \text{and} \begin{bmatrix} \boldsymbol{R}_l & -\boldsymbol{S}_l \\ -\boldsymbol{N}_l & \boldsymbol{D}_l \end{bmatrix}^{-1}$$

belong to $\mathcal{RH}_\infty$.

8. An observer-based controller is used to stabilize a given plant. Write down the state space equations governing the closed loop system. Hence show that out of the $2n$ eigen values of the closed loop system (n for the plant and n for the controller) half the eigen values may be independently adjusted by the control gain matrix F and the remaining half may be adjusted independently by the observer gain matrix H. (Note that this is a manifestation of the separation principle).

6

Sensitivity Minimization and Robust Stabilization

6.1 Introduction

This chapter introduces the study of $\mathcal{H}_\infty$ optimization problems in MIMO systems. The two problems discussed are – sensitivity minimization and robust stabilization. To keep the discussion simple, initially we deal with systems with the number of inputs equal to or more than the number of outputs. Later on in Section 6.3 this condition is relaxed and the more general $\mathcal{H}_\infty$ optimization problem leading to 2-block and 4-block optimization studies are presented. A complete solution to the 4-block problem using the spectral factorization approach is discussed. The sensitivity optimization studies of non-minimum phase transfer functions lead to a phenomenon referred to in literature as the 'water bed effect'. This is explained in Section 6.5. and is followed by a discussion on the design limitations imposed by the right half plane zeros. Section 6.6 explains the concept of robust stabilization for different types of model uncertainties. Sufficient conditions for robust stabilization are derived using the small gain theorem. Section 6.7 shows how the robust stabilization problem may be posed as a NP interpolation problem.

6.2 Sensitivity Minimization

6.2.1 The Design Philosophy

In a typical feedback design problem, we are given a plant, a description of the type of inputs we are likely to come across, and also the specifications regarding acceptable performance standards. We are then required to design a controller, which stabilizes the closed loop system and meets prescribed specifications.

In conventional design, the inputs are known. For example they may be steps, ramps, exponentials, etc. in the deterministic case or specified values of mean and covariance of the input signals in the stochastic case. Under such conditions, the *LQG* theory provides a useful design tool. The requirements imposed in the case of robust design are much more stringent. Indeed, the stability and performance requirements have to be satisfied. In addition, there is a demand for the satisfaction of these requirements over a specified range of uncertainty. Further, the inputs and disturbances are no longer specified but are assumed to belong to a class of signals. We are then required to minimize the energy content or power of the output signal for the worst possible input signals belonging to the stipulated class. This gives rise to a typical mini–max problem, where the effect on the output due to the worst possible exogenous signal (measured by its $\mathcal{L}_2$ norm) is sought to be offset by an appropriate choice of controller. As shown in Chapter 4, the maximum ratio of the $\mathcal{L}_2$ norm of the output to the $\mathcal{L}_2$ norm of the input turns out to be the $\mathcal{H}_\infty$ norm of the transfer matrix. The problem thus reduces to the minimization of the $\mathcal{H}_\infty$ norm of a weighted transfer function. This is in sharp contrast to the Wiener-Hopf approach, where the disturbance class is a singleton namely white noise of unit intensity in the stochastic formulation. The $\mathcal{H}_\infty$ norm minimization approach in control theory was pioneered by Zames (1981) and later modified and refined in a series of publications by Zames and a number of other researchers.

6.2.2 Problem Formulation

We are concerned about the effect of the disturbances d on the output y of a feedback system shown in Figure 6.1. $\boldsymbol{W}_1$ and $\boldsymbol{W}_2$ are square rational matrices known as weighting matrices. $\boldsymbol{W}_1$ is chosen to reflect the variables to be attenuated at the output and $\boldsymbol{W}_2$ is chosen to represent the class of exogenous signals. The significance of their choice is explained in a subsequent section.

In Figure 6.1 plant $\boldsymbol{G}$ is a $p \times q$ matrix and controller $\boldsymbol{K}$ is a $q \times p$ matrix. Both are matrix valued proper real functions in the Laplace variable s.

The transfer matrix from v to y is denoted by

$$\boldsymbol{X} = \boldsymbol{W}_1 \boldsymbol{S} \boldsymbol{W}_2$$

where $\boldsymbol{S}$ is the sensitivity operator defined as

$$\boldsymbol{S} = (I - \boldsymbol{GK})^{-1}$$

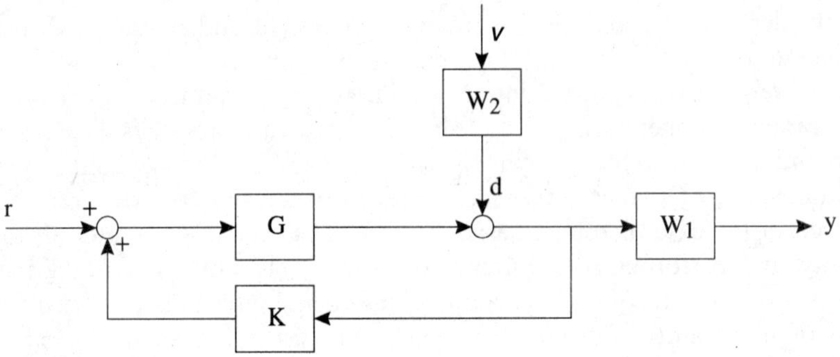

Figure 6.1 Feedback control system.

The class of disturbances is denoted by $\mathcal{D}$ which is described as follows

$$\mathcal{D} = \{d : d = \boldsymbol{W}_2 v, \text{ energy of } v \leq 1\}$$

The problem posed is the minimization of the $\mathcal{H}_\infty$ norm of the weighted sensitivity function.

$$\boldsymbol{X} = \boldsymbol{W}_1(\boldsymbol{I} - \boldsymbol{GK})^{-1}\boldsymbol{W}_2$$

over all real rational proper $\boldsymbol{K}$s which achieve internal stability. To make this problem mathematically tractable, the following assumptions are made.

1. The weighting matrices $\boldsymbol{W}_1$ and $\boldsymbol{W}_2$ and the plant transfer function $\boldsymbol{G}$ are real rational matrices.
2. $\boldsymbol{G}$ is strictly proper. It has no poles on the imaginary axis and has maximum row rank over all values of ω.
3. $\boldsymbol{W}_1$ and $\boldsymbol{W}_2$ are non-singular, proper and stable and they have stable inverses.

Given $\boldsymbol{G}$ a $p \times q$ matrix, the rank assumption in item 2 implies that $p \leq m$. The number of outputs is thus less than or equal to the number of inputs. This is the so called 'fat plant case'. The zeros of $\boldsymbol{G}$ on the imaginary axis are precisely the points where the row rank of $\boldsymbol{G}$ is lowered. Since this is ruled out by assumption 2, $\boldsymbol{G}$ has neither poles nor zeros on the imaginary axis. The assumption that $\boldsymbol{G}$ be strictly proper ensures that $\lim_{s\to\infty}(\boldsymbol{I} - \boldsymbol{GK}) = \boldsymbol{I}$. Hence, $\det(\boldsymbol{I} - \boldsymbol{GK}) \neq 0$ and therefore $(\boldsymbol{I} - \boldsymbol{GK})$ is invertible.

The first step in the minimization of

$$\boldsymbol{X} = \boldsymbol{W}_1(\boldsymbol{I} - \boldsymbol{GK})^{-1}\boldsymbol{W}_2 \tag{6.1}$$

is the replacement of $\boldsymbol{K}$ in this expression by its Youla parametrized version. Accordingly, we have $\boldsymbol{K} = (\boldsymbol{D}_r\boldsymbol{Q} - \boldsymbol{U}_l)(\boldsymbol{N}_r\boldsymbol{Q} + \boldsymbol{V}_l)^{-1}$ (refer to equation (5.10)).

Similarly G in (6.1) is replaced by its left coprime factorization $D_l^{-1}N_l$. Making these substitutions, (6.1) is reduced to

$$X = W_1(N_rQ + V_l)D_lW_2 \tag{6.2}$$

We now seek the minimization of the $\mathcal{H}_\infty$ norm of X as the free parameter $Q \in \mathcal{RH}_\infty$ is varied. Let

$$\begin{aligned}\mu &= \mathrm{Inf}\{\|W_1(N_rQ + V_l)D_lW_2\|_\infty : Q \in \mathcal{RH}_\infty\} \\ &= \mathrm{Inf}\{\|T_1 + T_2QT_3\|_\infty : Q \in \mathcal{RH}_\infty\}\end{aligned} \tag{6.3}$$

where

$$\begin{aligned}T_1 &= W_1V_lD_lW_2 \\ T_2 &= W_1N_r \\ T_3 &= D_lW_2\end{aligned}$$

It can be shown that the infimum in (6.3) is not actually achieved because of the constraint that Q be proper. If we temporarily relax this constraint and permit Q to be improper, then the infimum is actually achieved under mild sufficient conditions. These conditions state that the ranks of $T_2(j\omega)$ and $T_3(j\omega)$ be constant for $0 \le \omega \le \infty$. It is important to note that μ cannot be reduced further merely by Q belonging to a wider class of functions denoted by $\mathcal{H}_\infty$ instead of being confined to $\mathcal{RH}_\infty$. That is:

$$\min\{\|T_1 + T_2QT_3\|_\infty : Q \in \mathcal{RH}_\infty\} = \min\{\|T_1 + T_2QT_3\|_\infty : Q \in H_\infty\} \tag{6.4}$$

In general there exist a family of Qs achieving the infimum.

6.2.3 The Model Matching Problem

The above formulation of the $\mathcal{H}_\infty$ optimization problem is amenable to a nice geometric interpretation. As Q varies over $\mathcal{H}_\infty$, the product T_2QT_3 generates a closed subspace and μ represents the minimum distance (using $\mathcal{H}_\infty$ norm) from T_1 to the subspace $T_2\mathcal{H}_\infty T_3$. This is defined as follows

$$T_2\mathcal{H}_\infty T_3 = \{T_2QT_3 : Q \in \mathcal{H}_\infty\} \tag{6.5}$$

We may write

$$\mu = \mathrm{dist}\,(-T_1, T_2\mathcal{H}_\infty T_3) \tag{6.6}$$

The rank condition on T_2 and T_3 ensures that the subspace defined in (6.5) is closed.

This in turn guarantees that there exists a matrix valued function in $T_2\mathcal{H}_\infty T_3$ closest to the specified matrix T_1.

Minimizing $\|T_1 + T_2QT_3\|_\infty$ over all $Q \in \mathcal{H}_\infty$ is known as the *Model matching problem.*

In this formulation we choose Q in such a manner that the matrix T_2QT_3 matches the model $-T_1$ as closely as possible. It is relevant to note that in many cases there exists no Q such that exact model matching viz. $T_2QT_3 = -T_1$ is possible. This is because we insist on a Q which is stable and proper and the Q which achieves exact model matching (in cases where such a Q exists) may not be stable and proper. Hence we have to be content with approximate matching. The significance of the model matching format arises out of the fact that many $\mathcal{H}_\infty$ optimization problems in control theory may ultimately be recast in this format, which has a simple structure. A solution to the $\mathcal{H}_\infty$ optimization of the model matching problem therefore provides solutions to a whole class of problems.

To continue with the weighted sensitivity minimization problem for the 'fat plant' case we note that T_2 is a $p \times q$ matrix and T_3 a $p \times p$ matrix. Now perform the inner and outer factorization of T_2 and T_3. We have

$$T_2 = (T_2)_i(T_2)_o$$
$$T_3 = (T_3)_{co}(T_3)_{ci}$$

where the subscripts 'i', 'o', 'co' and 'ci' stand for inner, outer, co-outer and co-inner respectively. Since $p \leq q$ (for the fact plant case) $(T_2)_i$ is a $p \times p$ matrix; $(T_2)_o$ is a $p \times q$ matrix; $(T_3)_{co}$ is a $p \times p$ matrix and $(T_3)_{ci}$ is a $p \times p$ matrix. We have

$$T_1 + T_2QT_3 = T_1 + (T_2)_i(T_2)_oQ(T_3)_{co}(T_3)_{ci}$$

We may now replace $(T_2)_oQ(T_3)_{co}$ by a free parameter which for notational convenience is redesignated as Q. This replacement is valid because $(T_2)_o$ is right invertible and $(T_3)_{co}$ is left invertible in $\mathcal{RH}_\infty$. The assumption that $T_2(j\omega)$ and $T_3(j\omega)$ have no zeros on the imaginary axis, is made use of here as it is because of this property that both $(T_2)_o$ and $(T_3)_{co}$ are respectively right and left invertible in $\mathcal{RH}_\infty$. Again for notational convenience denote $(T_2)_i$ by U_i and $(T_3)_{ci}$ by U_{ci}. The model matching problem is thus reduced to

$$\min\|X\|_\infty = \min\|T_1 + U_iQU_{ci}\|_\infty$$

for $Q \in \mathcal{H}_\infty$. We now invoke a result which states that the ∞-norm of a matrix is preserved if it is either premultiplied by an inner matrix or postmultiplied by the adjoint of an inner matrix. In the present context since both U_i and U_{ci} are square, the distinction between inner and co-inner disappears.

We note that

$$U_i U_i^* = U_i^* U_i = I$$

and

$$U_{ci}^* U_{ci} = U_{ci} U_{ci}^* = I$$

Premultiplying X by U_i^* and postmultiplying it U_{ci}^*, we have

$$\begin{aligned}\|X\|_\infty &= \|U_i^*(T_1 + U_i Q U_{ci}) U_{ci}^*\|_\infty \\ &= \|U_i^* T_1 U_{ci}^* + Q\|_\infty \\ &= \|R^* + Q\|_\infty\end{aligned}$$

where $R^* = U_i^* T_1 U_{ci}^*$. Limebeer and Hung (1987) have shown that $R = U_{ci} T_1^* U_i$ has only stable poles. We further note that the ∞-norm of a matrix valued function and its adjoint are identical. We may therefore write

$$\|X\|_\infty = \|R + Q^*\|_\infty \tag{6.7}$$

and

$$\mu = \inf_{Q \in \mathcal{H}_\infty} \|R + Q^*\|_\infty \tag{6.8}$$

where Q^* has only unstable poles and R has only stable poles.

The model matching problem thus reduces to approximating a known stable transfer matrix R by an unstable transfer matrix $-Q^*$. This corresponds to the well known zeroth order Hankel norm approximation problem which is discussed in Chapter 7.

Recall that while simplifying the model matching problem, in order to reduce it to the Hankel norm approximation problem we have made use of the fact that the relevant inner and co-inner matrices are square. This does not hold when $p > q$, i.e. when the plant has more outputs than inputs (tall plant case) Again representing $T_2 = (T_2)_i (T_2)_o$ we note that $(T_2)_i$ is no longer square (it has dimension $p \times q$) and the foregoing method needs modification. This is explained in Section 6.5. The next section discusses the rationale behind the choice of weighting functions used in optimization studies.

6.2.4 Choice of Weighting Functions

Whether it be the $\mathcal{H}_2$ or $\mathcal{H}_\infty$ optimization problem, choice of the weighting functions is crucial for obtaining meaningful results. We discuss this choice in the sequel with particular reference to $\mathcal{H}_\infty$ optimization problems. Straightforward $\mathcal{H}_\infty$ minimization of for example the sensitivity operator $(I - GK)^{-1}$ without using weighting functions does not make much engineering sense. This is so because in $\mathcal{H}_\infty$ optimization, we are seeking a uniform bound over all frequencies. However, notwithstanding our description of the input signal space as a unit 'ball' in $\mathcal{L}_2$ space, we do have from practical considerations some idea of the distribution of the energy spectrum of the input signal. For example we may have *a priori* knowledge that the energy spectrum of the input signal is concentrated over a frequency range $[\omega_1, \omega_2]$ and the signal practically vanishes outside this range. If such be the case, the $\mathcal{H}_\infty$ norm of the sensitivity function should have the lowest possible value in this range and outside this range its value may not matter much, as the disturbance signals almost vanish there. By a proper choice of the weighting function $\boldsymbol{w}$, this information could be incorporated into optimization studies.

To illustrate, take for example a scalar weighting function $\boldsymbol{w}$ with $|\boldsymbol{w}(j\omega)| = 1$ for $\omega \in [\omega_1, \omega_2]$ and $|\boldsymbol{w}(j\omega)| = \epsilon$, elsewhere. Here ϵ denotes an arbitrarily small positive number. With such a choice of $\boldsymbol{w}$, $\|\boldsymbol{ws}\|_\infty$ attains the maximum value of $|(\boldsymbol{s}(j\omega)|$ in the frequency range $[\omega_1, \omega_2]$. The role of $\boldsymbol{w}$ here is that of a band-pass filter.

The action of the weighting function may also be viewed from another angle. To continue with the scalar case, suppose the disturbance signal d in Figure 6.1 is constrained to belong to a class

$$\{d : d = \boldsymbol{w}v \text{ for some } v \in \mathcal{H}_2, \|v\|_2 \leq 1\}$$

where we assume that $\boldsymbol{w}$ and $\boldsymbol{w}^{-1}$ are in $\mathcal{H}_\infty$.

Since $v = \boldsymbol{w}^{-1}d$, the disturbance signals are constrained to satisfy the inequality

$$\frac{1}{2\pi} \int_{-\infty}^{\infty} |\boldsymbol{w}(j\omega)|^{-2} |d(j\omega)|^2 \, d\omega \leq 1$$

which can be interpreted as a constraint on the weighted energy of d. Note that the energy density spectrum $|d(j\omega)|^2$ is weighted by a factor $|\boldsymbol{w}(j\omega)|^{-2}$. If for example $|\boldsymbol{w}(j\omega)|$ is relatively large on some frequency band and relatively small off it, then we can generate a class of signals with their energy concentrated in this band. Thus by a proper choice of $\boldsymbol{w}$ we can generate a class of narrow band signals, whose spectra are confined to the chosen frequency band.

To summarize, the weights are so chosen that it reflects the band of frequencies over which the input signal is concentrated. In SISO systems, expressing performance criteria in terms of the desired slope of the Bode plot is well known. This immediately translates into appropriate choice of weights. However in MIMO systems, the choice of a suitable weighting function is a difficult task. One way of doing it is through dynamic simulation studies.

6.3 1-Block, 2-Block and 4-Block Problems

Consider a model matching problem

$$X = T_1 + T_2 Q T_3$$

where T_2 has more rows than column and T_3 has more columns than rows. We may initially proceed on the same lines as we did in the fat plant case and obtain

$$X = T_1 + (T_2)_i Q (T_3)_{ci}$$

where $T_1, (T_2)_i, (T_3)_{ci}$ and Q are in $\mathcal{RH}_\infty$.

Using the geometric concept of minimal distance between the fixed T_1 and the closed subspace $(T_2)_i \mathcal{H}_\infty (T_3)_{ci}$ we may write

$$\mu = \text{ dist } (-T_1, U_i \mathcal{H}_\infty U_{ci})$$

where U_i denotes $(T_2)_i$ and U_{ci} denotes $(T_3)_{ci}$. Note that $U_i^\sim U_i = I$ (since U_i is inner). Also recall that $U_i^\sim = U_i^T(-s)$.

Now define the following transformations:

$$E = \begin{bmatrix} U_i^\sim \\ I - U_i U_i^\sim \end{bmatrix}$$

and

$$L = \begin{bmatrix} U_{ci} \\ I - U_{ci}^\sim U_{ci} \end{bmatrix}$$

It is easily verified that $E^\sim E = I$ and $L^\sim L = I$ and hence both are inners.

As a result the $\mathcal{L}_\infty$ norm of a matrix is unchanged if it is premultiplied by E and postmultiplied by $L^\sim$. Hence

$$\begin{aligned} \|X\|_\infty &= \|E(T_1 + U_i \mathcal{H}_\infty U_{ci}) L^\sim\|_\infty \\ &= \|E T_1 L^\sim + E U_i \mathcal{H}_\infty U_{ci} L^\sim\|_\infty \end{aligned}$$

and we have

$$\mu = \text{ dist } (-E T_1 L^\sim, E U_i \mathcal{H}_\infty U_{ci} L^\sim)$$

It may be verified that

$$E U_i = \begin{bmatrix} I \\ 0 \end{bmatrix} \text{ and } U_{ci} L^\sim = [I \quad 0]$$

Define $-E T_1 L^\sim = R$.

Then

$$\mu = \text{ dist } \left(R, \begin{bmatrix} I \\ 0 \end{bmatrix} \mathcal{H}_\infty [I \quad 0] \right) \tag{6.9}$$

for $Q \in \mathcal{H}_\infty$ we have,

$$\begin{bmatrix} I \\ 0 \end{bmatrix} Q [I \quad 0] = \begin{bmatrix} Q & 0 \\ 0 & 0 \end{bmatrix}$$

and

$$\mu = \operatorname*{Inf}_{Q \in \mathcal{H}_\infty} \left\| \begin{bmatrix} R_{11} - Q & R_{12} \\ R_{21} & R_{22} \end{bmatrix} \right\|_\infty \tag{6.10}$$

(6.10) represents a typical 4-block problem where the 4-block matrices R_{11}, R_{12}, R_{21}, and R_{22} are in $\mathcal{L}_\infty$ and Q is in $\mathcal{H}_\infty$. In fact, the generalized regulator problem in Chapter 9 is equivalent to a 4-block problem.
The following interesting special cases can occur, which simplifies the 4-block problem.

1. T_2 has full column rank and T_3 has full row rank.
2. T_2 has full column rank and T_3 full column rank.
3. T_2 has full row rank and T_3 has full row rank.
4. T_2 has full row rank and T_3 has full column rank.

Case 1: This relates to the most general situation and results in the 4-block problem.

Case 2: Here U_{ci} is square and non-singular and

$$U_{ci}^{\sim} = U_{ci}^{-1} \text{ Hence}$$
$$L^{\sim} = \begin{bmatrix} U_{ci}^{-1} & 0 \end{bmatrix}$$
$$R = E(-T_1)L^{\sim}$$
$$= \begin{bmatrix} R_{11} & 0 \\ R_{21} & 0 \end{bmatrix}$$

and

$$\mu = \operatorname*{Inf}_{Q \in \mathcal{H}_\infty} \left\| \begin{bmatrix} R_{11} - Q & 0 \\ R_{21} & 0 \end{bmatrix} \right\|_\infty$$

This is the 2-block problem.

Case 3: U_i is square and non-singular and

$$U_i^{-1} = U_i^{\sim}$$
$$E = \begin{bmatrix} U_i^{\sim} \\ 0 \end{bmatrix}$$

and

$$\mu = \operatorname*{Inf}_{Q \in \mathcal{H}_\infty} \left\| \begin{bmatrix} R_{11} - Q & R_{12} \\ 0 & 0 \end{bmatrix} \right\|_\infty$$

which is another version of the 2-block problem.

Case 4: Both U_i and U_{ci} are square and consequently $U_i^{-1} = U_i^{\sim}$ and $U_{ci}^{-1} = U_{ci}^{\sim}$ and we obtain

$$\mu = \operatorname*{Inf}_{Q \in \mathcal{H}_\infty} \left\| \begin{bmatrix} R_{11} - Q & 0 \\ 0 & 0 \end{bmatrix} \right\|_\infty$$

This is the simplest case. We have already come across this case in sensitivity minimization studies of fat plants.

6.3.1 The Equivalent 1-Block Problem

The main computational bottleneck in $\mathcal{H}_\infty$ optimization studies is that the solution cannot be obtained in a finite number of steps. However, in practice, exact solution of $\mathcal{H}_\infty$ optimization problem is not required. What is required is a sub-optimal solution where we seek to obtain the set of all sub-optimal controllers which make $\|X\|_\infty < \gamma$ where $\gamma > \mu$, the optimal solution. Bearing this in mind we now examine a method for obtaining a sub-optimal solution to the 4-block problem.

Briefly what is done is to reduce the 4-block problem into its equivalent 1-block problem, Once this has been achieved, we can apply standard methods to obtain a solution to the 1-block problem. The method outlined though conceptually elegant is computationally quite demanding. Moreover, the state space method suggested by Doyle *et al.* (1989) has rendered the earlier approaches somewhat obsolete from the computational view point. Nevertheless, from a theoretical angle the method is significant. Before proceeding further, we discuss some side issues connected with the problem.

Spectral factorization

Consider a transfer matrix $G \in \mathcal{RH}_\infty$ and let $\|G\|_\infty < \gamma$. Then $\gamma^2 I - G^*(j\omega)G(j\omega) > 0$.

Let $\Phi(j\omega) = \gamma^2 I - G^*(j\omega)G(j\omega)$

Then there exists a transfer matrix $W \in \mathcal{RH}_\infty$ such that $W^{-1} \in \mathcal{RH}_\infty$ and

$$\Phi(j\omega) = W^*(j\omega)W(j\omega) \quad \forall\omega$$

W is called the *spectral factor of* Φ.

Note that for any $\Phi(j\omega)$ such that $\Phi(j\omega) = \Phi^*(j\omega)$ and $\Phi(j\omega) > 0$, for all ω, spectral factorization is possible.

We now explain a practical procedure for obtaining $\boldsymbol{W}$. Let (A, B, C, D) be a realization of $\boldsymbol{G}$. Assume that A is asymptotically stable. It is required to construct the spectral factor of $(\gamma^2 I - \boldsymbol{G}^* \boldsymbol{G})$.

Let T be a non-singular matrix such that

$$T'T = \gamma^2 I - D'D$$

Define $L = -(T')^{-1}(D'C + B'X)$ where X is the solution of Riccati eqn. (3.42) in Chapter 3.

Now define $\boldsymbol{W} = T + L(sI - A)^{-1}B$. Since A is stable $\boldsymbol{W}$ is stable. $\boldsymbol{W}^{-1}$ has a realization $(A - BT^{-1}L, BT^{-1}, -T^{-1}L, T^{-1})$. The A -matrix of $\boldsymbol{W}^{-1}$ is stable because

$$\begin{aligned} A - BT^{-1}L &= A + B(T'T)^{-1}(D'C + B'X) \\ &= A + BS^{-1}(D'C + B'X) \end{aligned}$$

where $S = T'T$ according to the notation used in Chapter 3.

Since X satisfies the Riccati eqn. (3.42) $\hat{A} + RX$ is stable where $\hat{A} = A + BS^{-1}D'C$ and $R = BS^{-1}B'$

Therefore

$$A + BS^{-1}D'C + BS^{-1}B'X \text{ is stable.}$$

i.e.

$$A + BS^{-1}(D'C + B'X) \text{ is stable,}$$

which implies that $A - BT^{-1}L$ is stable.

Hence both $\boldsymbol{W}$ and $\boldsymbol{W}^{-1}$ are in $\mathcal{RH}_\infty$. For the indicated choice of $\boldsymbol{W}$, it is easy to verify that $\boldsymbol{W}^* \boldsymbol{W} = \gamma^2 \boldsymbol{I} - \boldsymbol{G}^* \boldsymbol{G}$, thus proving the result.

The major computation involved in spectral factorization relates to the solution of the Riccati equation (3.42). We also require the Cholesky decomposition of $(\gamma^2 I - D'D)$ for the assumed value of γ. This decomposition is expressed as $T'T$. Thereafter the calculation of the spectral factor W is straightforward.

6.3.2 2-Block Problem

With these preliminaries out of the way, we now examine the 2-block problem, expressed in column format. Specifically consider

$$\mu_0 = \operatorname*{Inf}_{\boldsymbol{Q} \in \mathcal{H}_\infty} \left\| \begin{bmatrix} \boldsymbol{H} - \boldsymbol{Q} \\ \boldsymbol{T} \end{bmatrix} \right\|_\infty < \gamma$$

where $\boldsymbol{H}$ and $\boldsymbol{T} \in \mathcal{L}_\infty$ and $\boldsymbol{Q} \in \mathcal{H}_\infty$.

We have $\mu_0 < \gamma$

iff $\gamma^2 I - [(H-Q)^*(H-Q) + T^*T]$ is positive definite.
iff $[\gamma^2 I - T^*T] - (H-Q)^*(H-Q)$ is positive definite.
iff $M^*M - (H-Q)^*(H-Q)$ is positive definite where M is the spectral factor of $\gamma^2 I - T^*T$.
iff $I - [M^{*^{-1}}(H-Q)^*(H-Q)M^{-1}]$ is positive definite.
iff $\|(H-Q)M^{-1}\|_\infty < 1$.
iff $\|(H_u + H_s - QM^{-1})\|_\infty < 1$ where H_u and H_s are projections of HM^{-1} on $\mathcal{H}_\infty^\perp$ and $\mathcal{H}_\infty$ respectively.
iff $\|H_u - (QM^{-1} - H_s)\|_\infty < 1$.
iff $\|H_u - Z\|_\infty < 1$ where $Z = QM^{-1} - H_s$ and $Z \in \mathcal{H}_\infty$.

The above result is summarized in Theorem 6.1.

THEOREM 6.1
$\mu_0 < \gamma$ if and only if $\|H_u - Z\|_\infty < 1$ for some $Z \in \mathcal{H}_\infty$ where H_u is the unstable projection of HM^{-1} and M satisfies $M^*M = \gamma^2 I - T^*T$. Further, $M \in \mathcal{H}_\infty$ and $M^{-1} \in \mathcal{H}_\infty$. ■

Since the $\mathcal{H}_\infty$ norms of an operator and its adjoint are the same,

$$\|H_u - Z\|_\infty = \|H_u^* - Z^*\|_\infty$$

where $H_u^* \in \mathcal{H}_\infty$ and $Z^* \in \mathcal{H}_\infty^\perp$.

But according to Nehari's theorem (see Chapter 7) for a fixed $H_u^* \in \mathcal{H}_\infty$ and for a variable $Z^* \in \mathcal{H}_\infty^\perp$
$\|H_u^* - Z^*\|_\infty$ attains a minimum value when

$$\operatorname*{Inf}_{Z^* \in \mathcal{H}_\infty^\perp} \|H_u^* - Z^*\|_\infty = \text{Maximum Hankel singular value of } H_u^*$$

Hence by evaluating the maximum Hankel singular value of the stable matrix H_u^* we can check whether $\|H_u - Z\|_\infty$ is less than unity or otherwise.

Computational steps involved
The following steps are involved in verifying whether the chosen γ is greater than the optimal value μ_0.

1. Selection of γ.
2. Spectral factorization of $(\gamma^2 I - T^*T)$.
3. Unstable and stable projections of HM^{-1} namely H_u and H_s.
4. Computation of maximum Hankel singular value $\sigma_{\max}$ of H_u^*.
5. Updating of γ. If $\sigma_{\max} < 1$ decrease γ if $\sigma_{\max} > 1$ increase γ.

To appreciate the amount of computation involved we list out the major operations to be performed in each cycle of the algorithm.

i) Spectral factorization which involves the solution of one algebraic Riccati equation and also Cholesky decomposition of a matrix.
ii) Unstable and stable projections which involve partial fraction decomposition of $\boldsymbol{HM}^{-1}$
iii) Hankel singular value calculation which involves state space realization of $\boldsymbol{H}_u^*$; solution of two matrix Lyapunov equations and eigen value calculations.

All these calculations have to be repeated for each iteration and it may take several iterations before an acceptable value of γ, close enough to μ_o is obtained.

If the 2-block problem has a row format we have

$$\operatorname*{Inf}_{Q \in \mathcal{H}_\infty} \|\boldsymbol{H} - \boldsymbol{Q} \quad \boldsymbol{T}\|_\infty < \gamma$$

following identical arguments as in the previous case we now have

$$(\gamma^2 \boldsymbol{I} - \boldsymbol{TT}^*) = \boldsymbol{LL}^*$$

where LL^* is the co-spectral factorization of $(\gamma^2 \boldsymbol{I} - \boldsymbol{TT}^*)$
We now arrive at a parallel result which states that $\mu_0 < \gamma$

$$\|\boldsymbol{L}^{-1}(\boldsymbol{H} - \boldsymbol{Q})\|_\infty < 1$$

Rest of the steps involved are identical to those described earlier and so, they are not repeated here.

6.3.3 4-Block Problem

We now outline a method to solve the 4-block problem. Specifically the problem is to check whether

$$\operatorname*{Inf}_{Q \in \mathcal{H}_\infty} \left\| \begin{bmatrix} \boldsymbol{A} - \boldsymbol{Q} & \boldsymbol{B} \\ \boldsymbol{C} & \boldsymbol{D} \end{bmatrix} \right\|_\infty < \gamma \tag{6.11}$$

The problem is solved in two stages

$$\text{Let} \begin{bmatrix} \boldsymbol{A} - \boldsymbol{Q} & \boldsymbol{B} \\ \boldsymbol{C} & \boldsymbol{D} \end{bmatrix} = \begin{bmatrix} \boldsymbol{X} \\ \boldsymbol{Y} \end{bmatrix} \tag{6.12}$$

As already proved, inequality (6.11) is satisfied if and only if

$$\|\boldsymbol{XM}^{-1}\|_\infty < 1$$

where $\gamma^2 \boldsymbol{I} - \boldsymbol{Y}^* \boldsymbol{Y} = \boldsymbol{M}^* \boldsymbol{M}$.

Substituting for X we obtain

$$\|[A-Q \quad B]M^{-1}\|_{\infty} < 1$$
$$\|[AM^{-1} - QM^{-1} \quad BM^{-1}]\|_{\infty} < 1$$
$$\|[A_1 - R \quad B_1]\|_{\infty} < 1 \tag{6.13}$$

where $AM^{-1} = A_1$; $QM^{-1} = R$ and $BM^{-1} = B_1$.

Note that A_1 and $B_1 \in \mathcal{L}_{\infty}$ and $R \in \mathcal{H}_{\infty}$.

Let $I - B_1 B_1^* = LL^*$ (co-spectral factorization).

(6.13) is true if and only if

$$\|L^{-1}(A_1 - R)\|_{\infty} < 1$$

Let $L^{-1}A_1 = A_2$ and $L^{-1}R = S$ where $S \in \mathcal{H}_{\infty}$

Then we have

$$\|A_2 - S\|_{\infty} < 1$$

Hence $\gamma > \mu_0$ if and only if

$$\operatorname*{Inf}_{S \in \mathcal{H}_{\infty}} \|A_2 - S\|_{\infty} < 1$$

The above procedure may be summarized as follows.

1. Choose γ as the greater of the two values

$$\|[C \quad D]\|_{\infty} \text{ and } \left\| \begin{bmatrix} B \\ D \end{bmatrix} \right\|_{\infty}$$

   This follows from the fact that the ∞-norm of a matrix must always be equal to or greater than the ∞-norm of its sub matrices.
2. Obtain a spectral factorization of $(\gamma^2 I - Y^* Y)$ yielding M where $Y = [C \quad D]$.
3. Obtain the co-spectral factorization of $(I - BM^{-1}M^{*^{-1}}B^*) = LL^*$.
   If $\|BM^{-1}\|_{\infty} \geq 1$ increase γ and go to step 2 otherwise go to step 4.
4. Compute $\mu = \min_{S \in \mathcal{H}_{\infty}} \|L^{-1}AM^{-1} - S\|_{\infty}$ by evaluating the maximum Hankel singular value of $L^{-1}\, A\, M^{-1}$
   If $\mu > 1$ increase γ and go to step 2.
   If $\mu < 1$ decrease γ and go to step 2.
   If $\mu = 1$ go to step 5.
5. Set $Q = LSM$ where S achieves the minimum in step 4.

The Q thus obtained, minimizes the $\mathcal{H}_{\infty}$ norm of the 4-block problem.

The above procedure provides a theoretical solution to the 4-block problem though its practical implementation is somewhat difficult, because of the computational complexity.

6.4 Sensitivity Trade-Offs for Multivariable Plants

In an earlier section, we saw how the $\mathcal{H}_\infty$ norm of the weighted sensitivity matrix could be minimized over all stable transfer matrices. In these calculations, the weighting matrix is supposed to reflect some of the performance requirements. The resulting controllers are thus optimal with respect to these chosen weighting matrices. However, in feedback design, there are situations where suitable weighting matrices could not be easily selected because of competing requirements. If for example, the disturbance signal $\boldsymbol{d}$ is confined to a band $[0, \omega_1]$, then to prevent excessive amplification at the output $\bar{\sigma}[\boldsymbol{S}]$ has to be kept small over $[0, \omega_1]$. On the other hand $\bar{\sigma}[\boldsymbol{S}]$ has also to be kept relatively small over the rest of the frequency range as dictated by the global bound prescribed.

Let $\|\boldsymbol{S}\|_{[0,\omega_1]} = \max\limits_{0 \leq \omega \leq \omega_1} \bar{\sigma}[\boldsymbol{S}(j\omega)]$.

In order to achieve disturbance rejection and good tracking, the designer may want to know the smallest value of $\|\boldsymbol{S}\|_{[0,\omega_1]}$ (namely ϵ) that can be achieved over the operating band $[0, \omega_1]$ while satisfying the global bound $\|\boldsymbol{S}\|_\infty < \delta$ for all ω. The upper bound of the sensitivity norm profile is shown in Figure 6.2.

Two questions arise in this context. First, given ϵ over $[0, \omega_1]$ what is the smallest global bound δ that one can achieve? Second, given δ, what is the smallest value of ϵ over $[0, \omega_1]$ realizable? Note that the global bound prescribed has to be greater than unity. For non-minimum phase plants, it can be shown that a trade-off situation exists, in the sense that any reduction of ϵ can only be achieved at the expense of a higher value for δ.

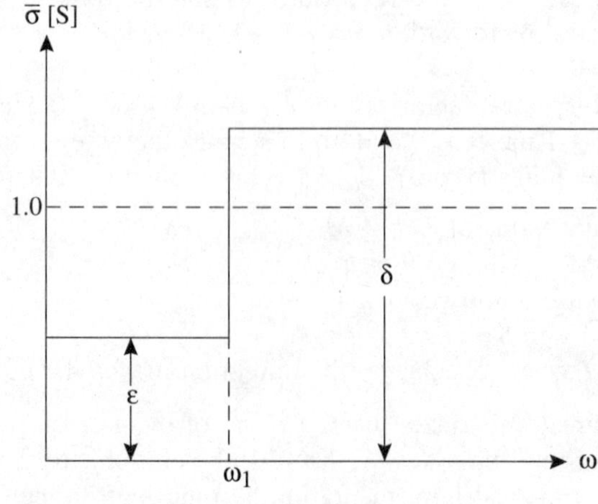

Figure 6.2 Sensitivity norm profile.

We are dealing here with both the right half plane (Re $s \geq 0$) and the unit disk $\overline{U}$, ($|z| \leq 1$). We recapitulate some of the relevant properties in these two spaces. They are related by the mapping $s \to z = \dfrac{1-s}{1+s}$ which maps the right half plane on to the unit disk. A rational function of s is stable, if it is analytic on Re $s \geq 0$. A rational function of z is stable if it is analytic on $|z| \leq 1$.

The Hardy space $\mathcal{H}_\infty$ for the open unit disk U is the Banach space of complex valued functions of z which are analytic and bounded on U.

The Hardy space of matrices $\mathcal{H}_\infty^{p\times m}$ consists of all p × m matrices with entries from $\mathcal{H}_\infty$.

The norm of an element $\boldsymbol{M} \in \mathcal{H}_\infty^{p\times m}$ is given by

$$\|\boldsymbol{M}\|_\infty = \sup_{z\in U} \overline{\sigma}[\boldsymbol{M}(z)]$$

Recall that $\mathcal{RH}_\infty^{p\times m}$ denotes a real rational subspace of $\mathcal{H}_\infty^{p\times m}$.We have

$$\|\boldsymbol{S}\|_{[\theta_1,\theta_2]} = \max_{\theta_1\leq\theta\leq\theta_2} \overline{\sigma}[\boldsymbol{S}(e^{j\theta})]$$

Inner and outer matrices in the z-domain are defined as follows:

$\boldsymbol{M} \in \mathcal{RH}_\infty^{p\times m}$ is *inner* if $[\boldsymbol{M}(e^{j\theta})]^*[\boldsymbol{M}(e^{j\theta})] = I$ for all θ.
$\boldsymbol{M} \in \mathcal{RH}_\infty^{p\times m}$ is *outer* if rank of $\boldsymbol{M}(z) = p$ for all z on U.

6.4.1 Non-minimum Phase Plants

The sensitivity trade-off problem can be stated as the problem of finding the infimum of $\|\boldsymbol{S}\|_{[0,\omega_1]}$ over all stabilizing controllers such that $\|\boldsymbol{S}\|_\infty < \delta$ where $[0,\omega_1]$ is a given operating frequently band and $\delta > 1$ is the global bound.

In the case of minimum phase plants (plants whose transfer functions have no zeros in the closed right half plane) Young and Francis (1985) have proved the following result.

Let the plant have its number of inputs equal to or more than the number of outputs and let the plant transfer function be a minimum phase function. Then for any $\epsilon > 0$ however small it may be and $0 < \omega_1 < \infty$, and $\delta > 1$, there exists a stabilizing controller such that $\|\boldsymbol{S}\|_{[0,\omega_1]} < \epsilon$ and $\|\boldsymbol{S}\|_\infty < \delta$.

Clearly in this case there is no possibility for trade-off as the infimum of $\|\boldsymbol{S}\|_{[0.\omega_1]}$ for any choice of $\delta > 1$ and for any finite ω_1 is zero. Therefore we will concentrate our attention on non-minimum phase plants.

For convenience, we will discuss the problem of finding the infimum of $\|\boldsymbol{S}\|_{[0,\omega_1]}$ in the z-domain. Accordingly all transfer matrices such as $\boldsymbol{G}$, $\boldsymbol{K}$ and $\boldsymbol{S}$ are transformed to the z domain by substituting $s = \dfrac{1-z}{1+z}$

As the imaginary axis is mapped on to a circle of unit radius, the interval $[-j\omega_1, j\omega_1]$ is mapped on to an arc $\{e^{j\theta} : -\theta_1 \leq \theta \leq \theta_2\}$ The point $s = \infty$ is mapped to $z = -1$.

The following assumptions are made regarding the plant matrix $\boldsymbol{G}(z)$ (in the z -domain):

A1: $\boldsymbol{G}$ is a $p \times m$ matrix whose entries are real rational functions in z.
A2: $\boldsymbol{G}(-1) = 0$.
A3: $\boldsymbol{G}$ has no poles on the unit circle.
A4: Rank $\boldsymbol{G}(e^{j\theta})$ equals p for all θ except for $\theta = \pi$.

A2 implies that $\boldsymbol{G}(s)$ is strictly proper and A4 implies that $p \leq q$ and that $\boldsymbol{G}(s)$ has no zeros on the imaginary axis.

According to the above assumptions $\boldsymbol{G}(s)$ has neither poles nor zeros on the imaginary axis

Consider the sensitivity matrix $\boldsymbol{S} = (I - \boldsymbol{GK})^{-1}$. Substitute for G, its left coprime factorization $\boldsymbol{D}_l^{-1}\boldsymbol{N}_l$ and for $\boldsymbol{K}$, its Youla parametrized version $(\boldsymbol{D}_r\boldsymbol{Q} - \boldsymbol{U}_l)(\boldsymbol{N}_r\boldsymbol{Q} + \boldsymbol{V}_l)^{-1}$ and obtain

$$\boldsymbol{S} = (\boldsymbol{N}_r\boldsymbol{Q} + \boldsymbol{V}_l)\boldsymbol{D}_l$$

Let the inner–outer factorization of $\boldsymbol{N}_r$ and $\boldsymbol{D}_l$ be

$$\boldsymbol{N}_r = (\boldsymbol{N}_r)_i(\boldsymbol{N}_r)_0 \text{ and } \boldsymbol{D}_l = (\boldsymbol{D}_l)_0(\boldsymbol{D}_l)_i$$

Note that $(\boldsymbol{N}_r)_i$ and $(\boldsymbol{D}_l)_i$ are both non-singular square matrices of size $p \times p$. Hence they are unitary. Further Adj $(\boldsymbol{N}_r)_i = \det\,(\boldsymbol{N}_r)_i(\boldsymbol{N}_r)_i^{-1}$ is also inner. We have

$$\begin{aligned}\boldsymbol{S} &= [(\boldsymbol{N}_r)_i(\boldsymbol{N}_r)_o\boldsymbol{Q} + \boldsymbol{V}_l](\boldsymbol{D}_l)_o(\boldsymbol{D}_l)_i \\ &= (\boldsymbol{N}_r)_i(\boldsymbol{N}_r)_o\boldsymbol{Q}(\boldsymbol{D}_l)_o(\boldsymbol{D}_l)_i + \boldsymbol{V}_l(\boldsymbol{D}_l)_o(\boldsymbol{D}_l)_i\end{aligned}$$

Premultiplying $\boldsymbol{S}$ by Adj $(\boldsymbol{N}_r)_i$ and postmultiplying it by $(\boldsymbol{D}_l)_i^{-1}$, we obtain

$$\|\boldsymbol{S}\|_\infty = \|\det\,(\boldsymbol{N}_r)_i(\boldsymbol{N}_r)_o\boldsymbol{Q}(\boldsymbol{D}_l)_o + \text{ Adj }(\boldsymbol{N}_r)_i\boldsymbol{V}_l(\boldsymbol{D}_l)_o\|_\infty$$

where we have used the property that multiplication by inners does not affect the norm.

Suppose we require a sensitivity profile as in Figure 6.3(b). To ensure this we assume a weighing function $\hat{u}_\epsilon$ on the boundary (also called the boundary modulus function) whose profile should have (see Figure 6.3(a)) inverse characteristics with respect to the sensitivity profile on the unit circle. If we fix the boundary profile $\hat{u}_\epsilon$ as in Figure 6.3(a). then it is possible to generate an *outer* function u_ϵ, defined on the open disk U, which will yield exactly $\hat{u}_\epsilon$ on the boundary. This is achieved by applying the Poisson integral formula. This formula states that the value of a bounded analytic function at a point in

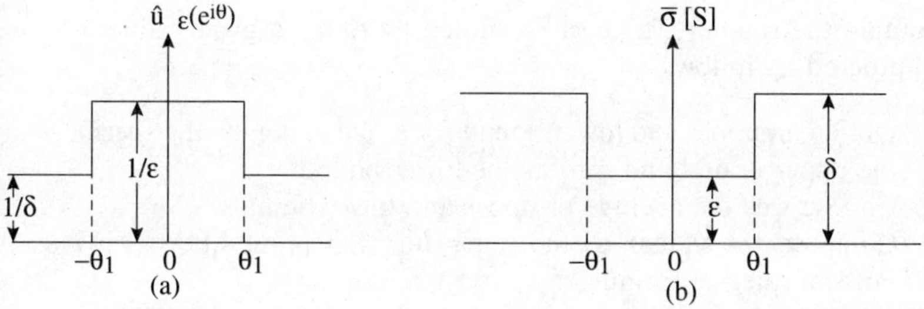

Figure 6.3 Weighting function and sensitivity profile.

the right half plane (unit disk) is completely determined by the coordinates of the point together with the value of the function on the imaginary axis (on unit circle) (Refer to Rudin 1974) The function u_ϵ thus generated in chosen as the weighting function. We now have

$$\|u_\epsilon \boldsymbol{S}\|_\infty = \|u_\epsilon \det\,(\boldsymbol{N}_r)_i(\boldsymbol{N}_r)_o \boldsymbol{Q}(\boldsymbol{D}_l)_o + u_\epsilon \mathrm{Adj}(\boldsymbol{N}_r)_i \boldsymbol{V}_l(\boldsymbol{D}_l)_o\|_\infty$$

Let $\boldsymbol{X} = u_\epsilon$ Adj $(\boldsymbol{N}_r)_i \boldsymbol{V}_l(\boldsymbol{D}_l)_o) + u_\epsilon \det\,(\boldsymbol{N}_r)_i(\boldsymbol{N}_r)_o \boldsymbol{Q}(\boldsymbol{D}_l)_o$
and $\boldsymbol{F} = u_\epsilon$ Adj $(\boldsymbol{N}_r)_i \boldsymbol{V}_l(\boldsymbol{D}_l)_o$

Then $\boldsymbol{X} = \boldsymbol{F} + u_\epsilon$ det $(\boldsymbol{N}_r)_i(\boldsymbol{N}_r)_o \boldsymbol{Q}(\boldsymbol{D}_l)_o$

Since det $(\boldsymbol{N}_r)_i$ is an inner, all its zeros are located in the open unit disk. Representing these zeros by $\lambda_1, \lambda_2 \ldots \lambda_n$, we have $|\lambda_i| < 1$ for $i = 1, 2, \ldots n$.

It follows that

$$X(\lambda_i) = F(\lambda_i) \qquad i = 1, 2, \ldots, n$$

Thus, we have reduced the weighted sensitivity minimization problem to an interpolation problem where we seek a bounded matrix valued function $\boldsymbol{X}(z)$, analytic in the open unit disk and satisfying the interpolation conditions $\boldsymbol{X}(\lambda_i) = \boldsymbol{F}(\lambda_i) \quad i = 1, 2, \ldots n$. This is readily identified as the classical Nevanlinna–Pick (NP) interpolation problem. This problem is discussed in detail in Chapter 8. For the present, we merely note that the solvability of NP problem depends upon the positive definiteness of a matrix N, known as that Pick matrix. Since the Pick matrix is a function of the chosen weighting function profile we denote it by $N(u_\epsilon)$. For a given $\delta > 1$ we are seeking to find the smallest value of ϵ which will make $N(u_\epsilon)$ positive definite. For evaluating $\boldsymbol{F}(\lambda_i)$ we have to first obtain $u_\epsilon(\lambda_i)$ via the Poisson integral relation. Denoting $\boldsymbol{F}(\lambda_k)$ by F_k, the Pick matrix $N = [N_{ij}]$ is given by

$$N_{ij} = \frac{1}{1 - \bar{\lambda}_i \lambda_j} \quad [I - F_i^* F_j]$$

To obtain the smallest value of ϵ denoted by ϵ_o for a given value of δ, we may proceed as follows:

1. Obtain an upper and lower bound for ϵ. We may take the specified δ as the upper bound and zero as the lower bound
2. Choose ϵ as the average of upper and lower bounds.
3. Compute the values of the outer function u_ϵ at $\lambda_1, \lambda_2, \ldots \lambda_n$ using Poisson integral formula.
4. Form the Pick matrix $N(u_\epsilon)$ and check for its positive definiteness.
5. If $N(u_\epsilon)$ is not positive definite, increase ϵ and go to 3.
6. If $N(u_\epsilon)$ is positive definite and the ϵ obtained is of sufficient accuracy then stop. Or else lower ϵ and go to 3.

REMARKS

1. When there is no weighting, we get the smallest value of δ. Hence the δ specified when weighting is employed should always be greater than $\|\boldsymbol{S}\|_\infty$. If this is not so, then $N(u_\epsilon)$ will never be positive definite for any $\epsilon - \delta$ profile chosen.
2. The values of the weighting function u_ϵ at $\lambda_1, \lambda_2, \ldots \lambda_n$ can be determined from the following formula

$$u(\lambda) = c \exp\left[\frac{1}{2\pi}\int_{-\pi}^{\pi} \frac{e^{j\theta} + \lambda}{e^{j\theta} - \lambda} \log \hat{u}(e^{j\theta}) \mathrm{d}\theta\right]$$

where $\hat{u}$ is called the boundary modulus function. λ is in the open unit disk and c is a constant which may be taken to be unity.

The inverse relationships between ϵ and δ for different frequency ranges are shown in Figure 6.4(a) and Figure 6.4(b).

There exists a stabilizing controller, giving rise to a sensitivity function, whose norm lies below the $\epsilon - \delta$ profile shown in Figure 6.3(b) if and only if the (ϵ, δ) pair lies inside the shaded region in Figure 6.4(a) for a given $[0, \omega]$ interval. From Figure 6.4(b) we note that if different frequency intervals $[0, \omega_1]$ and $[0, \omega_2]$ are chosen with $\omega_2, > \omega_1$, for a fixed δ, the larger the interval chosen, the larger will be the value of ϵ.

To summarize, if we try to depress ϵ in a given frequency interval, it results in a higher value of δ outside the interval and vice versa. This situation is analogous to that of a water bed, where pushing down the water level in one region, results in the elevation of water level in a contiguous region. As the sensitivity function of non-minimum phase systems behaves in a similar fashion, the term 'water bed effect' to denote this phenomena seems appropriate.

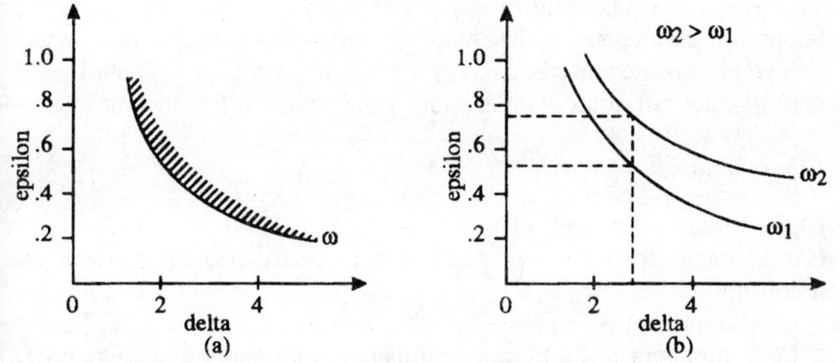

Figure 6.4 Epsilon-delta relationship.

6.5 Design Limitations Due to Right Half Plane Zeros

This section briefly discusses the design limitations imposed due to the presence of right half plane zeros. In this context, we have specifically chosen the sensitivity function for study because it is a key parameter influencing disturbance rejection, and good tracking behaviour.

The effect of the right half plane zeros is best explained in the scalar case as it clearly illustrates the principles involved in a simple manner. Hence consider a scalar loop transfer function $\boldsymbol{gk}$ with $\boldsymbol{g}$ assumed to be strictly proper. This implies that $\lim_{s\to\infty}(1-\boldsymbol{gk})^{-1}=1$. Let $\boldsymbol{gk}$ be denoted by $\boldsymbol{l}$. Then the sensitivity function $\boldsymbol{s}=(1-\boldsymbol{l})^{-1}$. Denote by z and p a zero and a pole of the plant whose transfer function is $\boldsymbol{g}$. We note that $\boldsymbol{s}(z)=1$ and $\boldsymbol{s}(p)=0$. Thus the poles of the plant are the zeros of the sensitivity function.

Let $\boldsymbol{w}$ represent the weighting function. Then $\|\boldsymbol{ws}\|_\infty$ denotes the $\mathcal{H}_\infty$ norm of the weighted sensitivity function. Suppose the loop transfer function $\boldsymbol{l}$ has a zero z in $\mathrm{Re}z \geq 0$. We will show later that

$$\|\boldsymbol{ws}\|_\infty \geq |\boldsymbol{w}(z)|$$

Assuming this to be true, it means that the $\mathcal{H}_\infty$ norm of the weighted sensitivity function is equal to or greater than the modulus of the weighting function evaluated at any one of the zeros of the plant in the open right half plane. This implies that if we require $\|ws\|_\infty \leq 1$, then a necessary condition for this is $|\boldsymbol{w}(z)| \leq 1$. Thus, the plant zeros impose a lower bound on the $\mathcal{H}_\infty$ norm of the weighted sensitivity function. Before we prove this result, we state an important theorem, repeatedly used in robust control theory.

THEOREM 6.2 (Maximum Modulus Theorem)
Suppose $\mathcal{D}$ be a region in the complex plane. Assume that $\mathcal{D}$ represents a non empty open connected set. If f represents an analytic function in $\mathcal{D}$, then $|f|$ does not attain its maximum value at any interior point of $\mathcal{D}$.

Proof: Omitted.

REMARKS
For a simple application of maximum modulus theorem consider the following.

Let $\mathcal{D}$ represent the open right half plane.

Let f represent a stable transfer function and therefore analytic in $\mathcal{D}$. Then modulus of f namely $|f|$ does not attain a maximum anywhere in the open right-half plane. The maximum is attained only on the boundary which is the imaginary axis in this case. Hence we can immediately write:

$$\|f\|_\infty = \sup_{\text{Re } s>0} |f(s)|$$

In general if G is a matrix valued analytic function over Re $s > 0$, the mapping $s \longrightarrow \overline{\sigma}[G(s)]$ need not be analytic and hence strictly speaking the maximum modulus principle need not apply to $\overline{\sigma}[G(\cdot)]$. However it is true that:

$$\|G\|_\infty = \sup_\omega \overline{\sigma}[G(j\omega)] = \sup_{\text{Re } s>0} \overline{\sigma}[G(s)]$$

because of other reasons.

■

We now revert back to the proof of the inequality

$$\|ws\|_\infty \geq |w(z)|$$

We have

$$|w(z)s(z)| = |w(z)| \text{ since } s(z) = 1$$

By applying the maximum modulus theorem

$$\begin{aligned}\|ws\|_\infty &\geq |w(z)s(z)| \\ &= |w(z)|\end{aligned}$$

Because of the above inequality, $\|ws\|_\infty$ must be equal to or greater than max $\{w(z_1), w(z_2), \ldots w(z_r)\}$ where $z_1, z_2, \ldots z_r$ are the r zeros of w in the open right half plane. This constraint is further aggravated when the plant right half plane poles are close to the plant right half plane zeros. This is demonstrated as follows.

As before let p and z be respectively a pole and a zero of the transfer function $\boldsymbol{g}$ in the open right-half plane. For simplicity we will assume that both are real positive. Assume further that the loop transfer function $\boldsymbol{l}$ has no other poles or zeros in the closed right half plane.

Factoring $\boldsymbol{s}$ into its inner and outer we have $\boldsymbol{s} = \boldsymbol{s}_i \boldsymbol{s}_0$. Since the poles of $\boldsymbol{l}$ are the zeros of $\boldsymbol{s}$, we have

$$\boldsymbol{s}_i(s) = \frac{s-p}{s+p}$$

$$\boldsymbol{s}_i(z) = \frac{z-p}{z+p}$$

$$\boldsymbol{s}(z) = \boldsymbol{s}_i(z).\boldsymbol{s}_0(z)$$

But $\boldsymbol{s}(z) = 1$
Hence

$$\boldsymbol{s}_o(z) = \frac{1}{\boldsymbol{s}_i(z)} = \frac{z+p}{z-p}$$

We have

$$\|\boldsymbol{ws}\|_\infty = \|\boldsymbol{ws}_i\boldsymbol{s}_0\|_\infty = \|\boldsymbol{ws}_0\|_\infty \text{(because } \boldsymbol{s}_i \text{ is inner)}$$

But, because of maximum modulus theorem

$$\|\boldsymbol{ws}_0\|_\infty \geq |\boldsymbol{w}(z)\boldsymbol{s}_0(z)|$$
$$= \boldsymbol{w}(z)\left|\frac{z+p}{z-p}\right|$$

Therefore

$$\|\boldsymbol{ws}\|_\infty \geq |\boldsymbol{w}(z)|\left|\frac{z+p}{z-p}\right|$$

If z and p are close to each other, $\boldsymbol{w}(z)$ gets amplified.

We therefore surmise that the plant zeros either alone or in conjunction with neighboring poles (all of them in the open right half plane) set, the lower bound on the achievable value of $\|\boldsymbol{ws}\|_\infty$

Recall that the sensitivity minimization problem is to find a stabilizing controller for which the infinity norm of the weighted sensitivity function attains its minimum value. Thus

$$\min_{\boldsymbol{k}} \|\boldsymbol{ws}\|_\infty = \min_{\boldsymbol{k}} \sup_{\omega} |\boldsymbol{w}(j\omega)\boldsymbol{s}(j\omega)| \tag{6.14}$$

The optimal solution to (6.14) was obtained by Zames and Francis (1983). Their solution states that for a stable minimum phase weighting function provided by the designer, the optimal *unweighted* sensitivity function $\boldsymbol{s}^*(s)$ has the form

$$\boldsymbol{s}^*(s) = \alpha^* \boldsymbol{b}_g(s) \boldsymbol{b}_{k^*}(s) \boldsymbol{w}^{-1}(s) \tag{6.15}$$

where:

$\boldsymbol{b}_g(s)$ is the Blaschke product of unstable plant poles

$\boldsymbol{b}_{k^*}(s)$ is the Blaschke product of the unstable poles of the optimal compensator

α^* is a constant

Since Blaschke products are all pass functions of unit magnitude, it follows from (6.15) that

$$|\boldsymbol{s}^*(j\omega)| = \alpha^* |\boldsymbol{w}^{-1}(j\omega)|$$

Thus the shape of the optimal unweighted sensitivity function is determined by the shape of the weighting function, i.e. its relative magnitude at different frequencies. However the level of sensitivity reduction depends upon α^* which in turn depends upon the zero locations in the right half plane and the chosen weighting function. These points are discussed in Freudenberg and Looze (1986a, 1986b). Simply scaling the weighting function leaves the unweighted optimal sensitivity profile unchanged. Hence in order to change the relative magnitudes of $|\boldsymbol{s}^*(j\omega)|$ at different frequencies, $\boldsymbol{w}(j\omega)$ has to be suitably altered. The important point here is that the right half plane zeros of the plant play an important role in determining the $|\boldsymbol{s}^*(j\omega)|$ profile over the frequency range. This problem is much more complicated in the multivariable case, where the design limitations imposed by right half plane poles and zeros are not yet clearly understood.

6.6 Plant Uncertainty and Robustness

In this section, we are concerned with models of multivariable systems featuring unstructured uncertainty. As already explained in Chapter 2, the two common representations of uncertainty are the additive and multiplicative uncertainty. They are schematically represented in Figure 6.5. For additive perturbations we have:

$$\tilde{\boldsymbol{G}} = \boldsymbol{G} + \Delta_a$$

where $\boldsymbol{G}$ represents the nominal plant and Δ_a represents additive perturbation.

For multiplicative perturbations we have

$$\tilde{\boldsymbol{G}} = (1 + \Delta_m)\boldsymbol{G}$$

As before $\boldsymbol{G}$ represents nominal plant and Δ_m multiplicative perturbation.

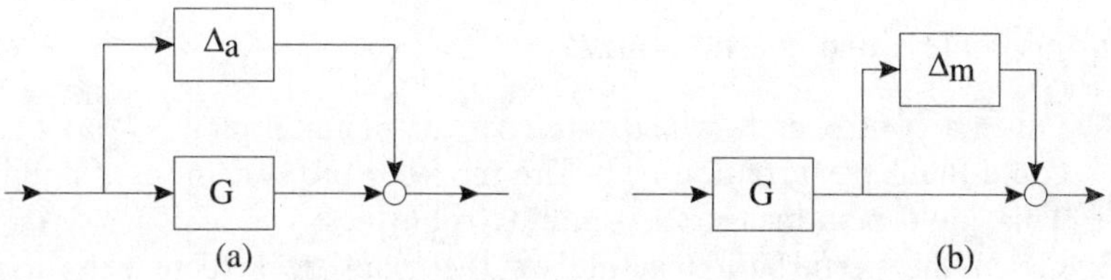

Figure 6.5 Model error representation.

The size of the perturbation in measured by its infinity norm namely $\|\boldsymbol{\Delta}\|_\infty$. We can make the size frequency dependent by specifying $\bar{\sigma}[\boldsymbol{\Delta}(j\omega)] < \boldsymbol{l}(\omega) \quad \forall\omega \geq 0$ where $\boldsymbol{l}(\cdot)$ is a positive scalar function. We note that $\|\boldsymbol{\Delta}_a\|_\infty$ represents absolute magnitude of model error whereas $\|\Delta_m\|_\infty$ represents relative magnitude of model error. The latter is more realistic in most practical applications.

6.6.1 Robust Stability

An important issue in robust control theory is the one concerning robust stability. The concept of robustness which permeates the entire theory may be explained as follows. To begin with we define the set of all perturbed transfer functions for both additive and multiplicative perturbations. Thus for additive perturbation we have:

$$\mathcal{P}_a = \{\tilde{\boldsymbol{G}} : \tilde{\boldsymbol{G}} = \boldsymbol{G} + \boldsymbol{\Delta}_a ; \bar{\sigma}[\boldsymbol{\Delta}_a(j\omega)] < \boldsymbol{l}_a(\omega) \quad \forall\omega\} \tag{6.16}$$

and for multiplicative perturbation:

$$\mathcal{P}_m = \{\tilde{\boldsymbol{G}} : \tilde{\boldsymbol{G}} = (\boldsymbol{I} + \boldsymbol{\Delta}_m)\boldsymbol{G} ; \bar{\sigma}[\boldsymbol{\Delta}_m(j\omega)] < \boldsymbol{l}_m(\omega) \quad \forall\omega\} \tag{6.17}$$

where $\boldsymbol{l}_a(\cdot)$ and $\boldsymbol{l}_m(\cdot)$ are positive scalar functions which define the uncertainty profile at different frequencies.

When we discuss robustness we must be clear about three things:

(i) Plant model set $\mathcal{P}$.
(ii) Feedback controller $\boldsymbol{K}$.
(iii) A characteristic pertaining to the system which is to remain invariant under perturbations.

Plant model set is defined by either (6.16) or (6.17). The property whose invariance under perturbation is the design objective is internal stability. In addition we may consider it desirable to have performance robustness. However stability robustness is mandatory.

In this context two questions arise:

1. Given a controller $\boldsymbol{K}$ which stabilizes the nominal plant $\boldsymbol{G}$, how large could the $\boldsymbol{\Delta}$ perturbation be before the feedback system loses stability? This could be a measure of stability robustness.
2. Given the perturbation acting on the plant does there exist a real rational controller $\boldsymbol{K}$ which can stabilize the feedback system and if so how to synthesize such a controller?

Matters discussed in the sequel have close bearing on these two questions. The main tools used for this purpose are the multivariable Nyquist stability theorem and the small gain theorem.

6.6.2 Generalized Nyquist Stability Criteria

Consider a plant, $\boldsymbol{G}$ with feedback controller $\boldsymbol{K}$. According to Theorem 5.1 in Chapter 5, the closed loop system is internally stable if and only if the transfer matrix:

$$\boldsymbol{H} = \begin{bmatrix} I & -\boldsymbol{K} \\ -\boldsymbol{G} & I \end{bmatrix}^{-1}$$

is stable. (Note that we use positive feedback convention here).

Let $\boldsymbol{G} = \boldsymbol{N}\boldsymbol{D}^{-1}$ and $\boldsymbol{K} = \boldsymbol{U}\boldsymbol{V}^{-1}$ be that right coprime *polynomial factors* of the plant and the controller. Substituting these factors for $\boldsymbol{G}$ and $\boldsymbol{K}$ we get:

$$\boldsymbol{H} = \begin{bmatrix} I & -\boldsymbol{U}\boldsymbol{V}^{-1} \\ -\boldsymbol{N}\boldsymbol{D}^{-1} & I \end{bmatrix}^{-1} = \begin{bmatrix} \boldsymbol{D} & 0 \\ 0 & \boldsymbol{V} \end{bmatrix} \begin{bmatrix} \boldsymbol{D} & -\boldsymbol{U} \\ -\boldsymbol{N} & \boldsymbol{V} \end{bmatrix}^{-1} \qquad (6.18)$$

Because of the coprimeness of (N, D) and (U, V) it is easy to show that

$$\begin{bmatrix} \boldsymbol{D} & 0 \\ 0 & \boldsymbol{V} \end{bmatrix} \text{ and } \begin{bmatrix} \boldsymbol{D} & -\boldsymbol{U} \\ -\boldsymbol{N} & \boldsymbol{V} \end{bmatrix} \text{ are right coprime polynomial matrices.}$$

Hence, they do not have any common zeros in the right half plane.

Further from (6.18) we note that the poles of $\boldsymbol{H}$ are the zeros of the polynomial ϕ where

$$\begin{aligned} \phi &= \det \begin{bmatrix} \boldsymbol{D} & -\boldsymbol{U} \\ -\boldsymbol{N} & \boldsymbol{V} \end{bmatrix} \\ &= \det \boldsymbol{D} \det (\boldsymbol{V} - \boldsymbol{N}\,\boldsymbol{D}^{-1}\boldsymbol{U}) \text{ (by Schur's formula for determinant)} \\ &= \det \boldsymbol{D} \det (I - \boldsymbol{N}\,\boldsymbol{D}^{-1}\boldsymbol{U}\boldsymbol{V}^{-1}) \det \boldsymbol{V} \\ &= \det \boldsymbol{D} \det \boldsymbol{V} \det (I - \boldsymbol{G}\boldsymbol{K}) \end{aligned} \qquad (6.19)$$

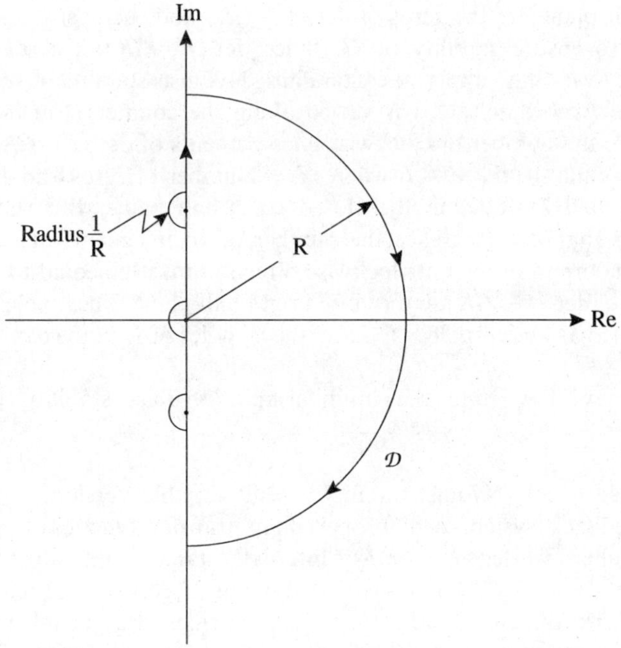

Figure 6.6 Nyquist contour $\mathcal{D}$.

Consider a Nyquist $\mathcal{D}$ contour as shown in Figure 6.6. $\mathcal{D}$ is a closed contour of radius R with semicircular indentation of radius $1/R$ into the left half plane, wherever $\boldsymbol{G}$ and $\boldsymbol{K}$ have poles on the imaginary axis. Radius R is chosen large enough to enclose all the closed right half plane poles of $\boldsymbol{G}$, $\boldsymbol{K}$ and the closed loop transfer matrix $\boldsymbol{H}$. The closed region defined by $\mathcal{D}$ shall exclude all the left half plane poles of $\boldsymbol{G}$, $\boldsymbol{K}$ and $\boldsymbol{H}$. We assume a clockwise orientation for $\mathcal{D}$.

Since $\begin{bmatrix} \boldsymbol{D} & 0 \\ 0 & \boldsymbol{V} \end{bmatrix}$ and $\begin{bmatrix} \boldsymbol{D} & -\boldsymbol{U} \\ -\boldsymbol{N} & \boldsymbol{V} \end{bmatrix}$ are coprime and since the zeros of ϕ are the same as the zeros of $\begin{bmatrix} \boldsymbol{D} & -\boldsymbol{U} \\ -\boldsymbol{N} & \boldsymbol{V} \end{bmatrix}$ we conclude that det $\boldsymbol{D}$ det $\boldsymbol{V}$ have no common zeros with ϕ in the closed right half plane. Consequently as seen from (6.19) every zero of ϕ in the closed right half plane should be a zero of det $(\boldsymbol{I} - \boldsymbol{GK})$.

Further since ϕ is a polynomial, every pole of det $(\boldsymbol{I} - \boldsymbol{GK})$ should cancel with the zeros of det $\boldsymbol{D}$ det $\boldsymbol{V}$ as can be deduced from (6.19). To recapitulate, if $\boldsymbol{H}$ is to be stable, it should not have poles in the closed right half plane. Since the poles of $\boldsymbol{H}$ are the zeros of ϕ this implies that ϕ has no zeros in the closed right half plane. But as already shown, the zeros of ϕ in the closed right half plane are the zeros of det $(I - GK)$ and such zeros should not be present to ensure stability of H. Since det $(I - GK)$ is a scalar rational function, we may apply the standard Nyquist stability criteria to this function. Accordingly as s is varied along the counter $\mathcal{D}$ in the clockwise direction, the number of clockwise encirclements of det $(I - GK)$ about the origin is equal to the $Z - P$ where $Z =$ Number of zeros and $P =$ number poles of det $(I - GK)$ in the closed right half plane. But stability of H demands that $Z = 0$. Hence the number of encirclements of det $(I - GK)$ about the origin in the anticlockwise direction must be equal to the number of poles of det $(I - GK)$ in the closed right half plane. These poles are readily determined as the n_1, poles of G and the n_2 poles of K in the closed right half plane.

We may now state the multivariable Nyquist stability theorem as follows.

THEOREM 6.3 (Nyquist stability – Multivariable Version)
A feedback system with proper plant transfer function G and proper controller transfer function K is internally stable if and only if the Nyquist plot of det $(I - GK)$ as s is varied along the $\mathcal{D}$ contour in Figure 6.6 in the clockwise direction makes $(n_1 + n_2)$ anticlockwise encirclements of the origin without crossing it. Here n_1, and n_2 are the number of poles of G and K respectively in the closed right half plane.

6.6.3 Robustness under Perturbations

Consider Figure 6.7 where the plant is subjected to additive perturbation Δ.

Assume that $\bar{\sigma}[\Delta(j\omega)] < |l(j\omega)|$ for every ω or equivalently $\|l^{-1}\Delta\|_\infty < 1$ where l is a scalar valued function in $\mathcal{RH}_\infty$. Further assume that $\tilde{G} = G + \Delta$ has the same number of unstable poles as G (however these unstable poles need not be identical).

We first obtain the equivalent transfer function of the unperturbed system $(\Delta = 0)$ viewed from the input terminal 'a' and the output terminal 'b' marked in the figure, ignoring the external input r. This is readily obtained as $K(I - GK)^{-1}$. The perturbation transfer matrix Δ is now connected across terminals a and b as shown in Figure 6.7.

It is assumed that the nominal feedback system is internally stable. Hence $K(I - GK)^{-1} \in \mathcal{RH}_\infty$. Then according to the 'small gain theorem' (explained in the next section) the closed loop in Figure 6.8 is stable provided that $\mathcal{H}_\infty$ norm of the loop transfer function is less than unity. Applying this theorem, we obtain

$$\|\Delta K(I - GK)^{-1}\|_\infty < 1 \tag{6.20}$$

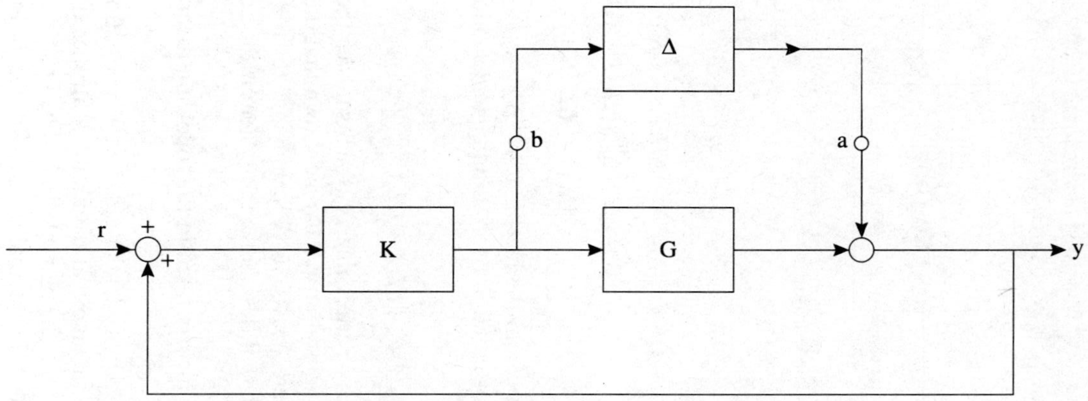

Figure 6.7 Feedback of plant with additive perturbations.

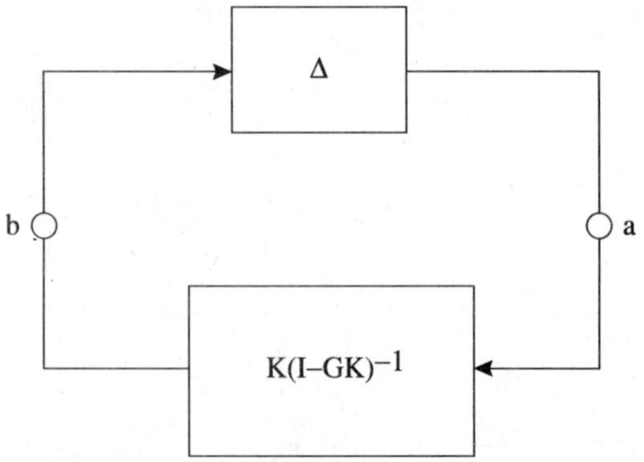

Figure 6.8 Equivalent closed loop.

But

$$\begin{aligned}\|\Delta K(I-GK)^{-1}\|_\infty &= \|\boldsymbol{l}^{-1}\Delta \boldsymbol{l}K(I-GK)^{-1}\|_\infty \\ &\leq \|\boldsymbol{l}^{-1}\Delta\|_\infty \|\boldsymbol{l}K(I-GK)^{-1}\|_\infty \\ &< \|\boldsymbol{l}K(I-GK)^{-1}\|_\infty\end{aligned}$$

The last step follows from the fact that $\|\boldsymbol{l}^{-1}\Delta\|_\infty < 1$.

Hence a sufficient condition for stability under additive perturbation is that

$$\|\boldsymbol{l}K(I-GK)^{-1}\|_\infty < 1 \tag{6.21}$$

(6.20) shows that any permissible Δ which satisfies the inequality specified will guarantee internal stability. It can be shown that the inequality is tight in the sense that there exists a **Δ** belonging to the permissible set of perturbations which will destabilize the system if the condition (6.20) is not strictly met.

These results are incorporated in the following theorem.

THEOREM 6.4 (Stability Under Additive Perturbation)
Let ***G*** and ***K*** be the nominal plant and controller transfer functions of a feedback system with the plant perturbed by an additive error **Δ**. Then the closed loop system is internally stable if the following conditions are satisfied

1. The nominal closed loop system is internally stable.
2. Δ is a rational function such that ***G*** and ***G***+**Δ** have the same number of closed right half plane poles.

3. Δ satisfies the bound

$$\bar{\sigma}[\Delta(s)] < \frac{1}{\bar{\sigma}[K(I - GK)^{-1}(s)]} \quad \forall s \in \mathcal{D}$$

Further, there exists a rational transfer matrix Δ satisfying condition 2 and

$$\bar{\sigma}[\Delta(j\omega)] \leq \frac{1}{\bar{\sigma}[K(I - GK)^{-1}(j\omega)]} \quad \forall \omega$$

which destabilizes the system.

Proof: By hypothesis, the nominal closed loop system is stable. Hence the mapping function det $(I - GK(s))$ makes $n_1 + n_2$ anticlockwise encirclements of the origin as s is varied along the Nyquist contour $\mathcal{D}$. The perturbed system also has $(n_1 + n_2)$ poles in the closed right half plane. The corresponding mapping function for the perturbed closed loop system is det $(I - (G + \Delta)K(s))$. If, as s is varied along the $\mathcal{D}$ contour, the perturbed mapping function also makes $(n_1 + n_2)$ anticlockwise encirclements of the origin, then the perturbed system is stable. To prove this we have to show that as the unperturbed mapping function is continuously warped to coincide ultimately with the perturbed mapping function, at no stage, for any value of s on $\mathcal{D}$ does the mapping function cross the origin. Indeed, if such a thing were to happen, then the number of encirclements will definitely get altered. The above statement can be couched in mathematical terms as

$$\det\,[I - (G + \varepsilon\Delta)K(s)] \neq 0 \quad \forall s \in \mathcal{D} \text{ and } \forall \epsilon \in [0, 1]$$

Note that $\varepsilon = 0$ corresponds to the unperturbed system and $\varepsilon = 1$ corresponds to the perturbed system.

$$\det\,[(I - GK) - \varepsilon\Delta K] \neq 0 \quad \forall s \in \mathcal{D}, \quad \forall \varepsilon \in [0, 1]$$

$$\Rightarrow \det\,[(I - \varepsilon\Delta K(I - GK)^{-1})(I - GK)] \neq 0 \quad \forall s \in \mathcal{D}, \quad \forall \varepsilon \in [0, 1]$$

$$\Rightarrow \det\,[I - \varepsilon\Delta K(I - GK)^{-1}] \neq 0 \quad \forall s \in \mathcal{D}, \quad \forall \varepsilon \in [0, 1]$$

In arriving at the above result, we have used the fact that det $(I - GK) \neq 0 \quad \forall s \in \mathcal{D}$ because the unperturbed closed loop system is stable.

We have to show now that

$$\det\,[I - \varepsilon\Delta K(I - GK)^{-1}] \neq 0 \quad \forall s \in \mathcal{D}, \quad \forall \varepsilon \in [0, 1] \tag{6.22}$$

At this stage we make use of a property in singular value analysis namely that for a square matrix Q its minimum singular value $\underline{\sigma}(Q)$ indicates how close it is to being made singular by a square perturbation matrix R. To make Q singular via perturbation matrix R, it can be shown that the maximum singular value of R, i.e. $\overline{\sigma}(R)$ should at least be equal to or more than $\underline{\sigma}(Q)$. In the present context, I represents $\boldsymbol{Q}$ with $\underline{\sigma}(I) = 1$ and $\varepsilon\boldsymbol{\Delta K}(I - \boldsymbol{GK})^{-1}$ represents R for $s \in \mathcal{D}$ and $\varepsilon \in [0, 1]$.

Hence (6.22) is satisfied if

$$\overline{\sigma}[\varepsilon\boldsymbol{\Delta K}(I - \boldsymbol{GK})^{-1}] < 1 \quad \forall s \in \mathcal{D}, \quad \forall \varepsilon \in [0, 1]$$

But according to condition 3 of the theorem we have

$$\overline{\sigma}[\varepsilon\Delta K(I - \boldsymbol{GK})^{-1}] < 1 \quad \forall s \in \mathcal{D}, \quad \forall \varepsilon \in [0, 1]$$

and hence (6.22) is satisfied.

The establishment of the property that there exists a Δ, which destabilizes the closed loop system, if the inequality condition 3 is not tight is fairly lengthy and is omitted. Interested readers may refer to Green and Limebeer (1995).

■

Results pertaining to stability of closed loop systems involving multiplicative perturbation of the plant can be derived in a similar manner. Figure 6.9 gives details of the interconnection.

The transfer function of the unperturbed system from terminal a to terminal b with zero input is given by

$$\boldsymbol{GK}(I - \boldsymbol{GK})^{-1}$$

The small gain theorem may now be applied to the closed loop comprising the transfer matrices $\boldsymbol{GK}(I - \boldsymbol{GK})^{-1}$ and $\boldsymbol{\Delta}$. We conclude that the closed loop system is internally stable provided

$$\|\boldsymbol{\Delta GK}(I - \boldsymbol{GK})^{-1}\|_{\infty} < 1$$

The relevant theorem for multiplicative perturbation may be stated as follows:

THEOREM 6.5 (Stability Under Multiplicative Perturbation)
Let G and K be the nominal plant and controller transfer functions of a feedback system with the plant perturbed by a multiplicative error Δ.

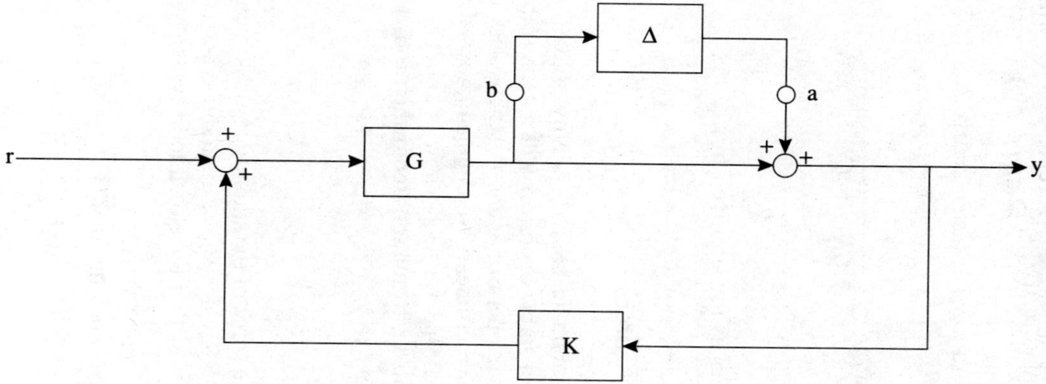

Figure 6.9 Feedback with multiplicative perturbation of plant.

Then the closed loop system is internally stable if the following conditions are satisfied.

1. The nominal closed loop system is internally stable.
2. Δ is a rational transfer function such that G and $(I+\Delta)G$ have the same number of closed right half plane poles.
3. Δ satisfies the bound $\bar{\sigma}[\mathbf{\Delta}(s)] < \dfrac{1}{\bar{\sigma}[\mathbf{GK}(\mathbf{I}-\mathbf{GK})^{-1}(s)]} \quad \forall s \in \mathcal{D}$

Further, there exists a rational transfer matrix $\mathbf{\Delta}$ satisfying condition 2 such that

$$\bar{\sigma}[\mathbf{\Delta}(j\omega) \leq \frac{1}{\bar{\sigma}[\mathbf{GK}(\mathbf{I}-\mathbf{GK})^{-1}(j\omega)]}$$

which destabilizes the system.

Proof: On the same lines as for additive error.

6.6.4 The Small Gain Theorem

This is a very useful theorem in robust stability studies. It states that a feedback loop composed of stable operators will remain stable so long as the product of their individual operator gains is strictly less than unity. The operators may be linear or non-linear. No detailed knowledge of the systems in the loop is required. All that we need know is about the stability of the systems in the loop and also information about their individual gains.

6.6.5 Stability Margin

We begin by recapitulating some of the stability results obtained in the earlier sections. Consider Figure 6.8 which shows an additive perturbation $\mathbf{\Delta}$ for the plant. If $\mathbf{\Delta} = 0$, then we know that the closed loop system denoted by $(\mathbf{G}, \mathbf{K})$ is stable if and only if

(a) $\mathbf{S}, \mathbf{KS}, \mathbf{SG}, (\mathbf{I}-\mathbf{KSG})$ belong to $\mathcal{RH}_\infty$
(b) det $[\mathbf{I}-\mathbf{G}(\infty)\mathbf{K}(\infty)] \neq 0$

where $\mathbf{S} = (\mathbf{I}-\mathbf{GK})^{-1}$

Now, let us consider an additive perturbation $\mathbf{\Delta}$ and impose an upper bound on its infinity norm, namely $\|\mathbf{\Delta}\|_\infty < \varepsilon$.

Denote the perturbed feedback system by $(\mathbf{G}, \mathbf{K}, \varepsilon)$. We have the following definition for robust stability.

DEFINITION 6.1
A feedback system $(\boldsymbol{G}, \boldsymbol{K}, \varepsilon)$ is robustly stable if and only if $(\boldsymbol{G}+\Delta, \boldsymbol{K})$ is internally stable for all $\Delta \in \mathcal{RL}_\infty$ such that:

(a) $\boldsymbol{G}$ and $\boldsymbol{G}+\Delta$ have the same number of poles in the closed right half plane and;

(b) $\|\Delta\|_\infty < \varepsilon$.

Theorem 6.4 gives the conditions under which the feedback system $(\boldsymbol{G}, \boldsymbol{K}, \varepsilon)$ is robustly stable. These are:

(a) that $(\boldsymbol{G}, \boldsymbol{K})$ be internally stable;

(b) $\|\boldsymbol{K}(\boldsymbol{I}-\boldsymbol{G}\boldsymbol{K})^{-1}\|_\infty < \dfrac{1}{\varepsilon}$.

However, for a given ε, the existence of a $\boldsymbol{K}$ that satisfies (a) and (b) above cannot be taken for granted. The maximum value of ε that may be prescribed can be obtained from the condition:

$$\varepsilon\|\boldsymbol{K}(\boldsymbol{I}-\boldsymbol{G}\boldsymbol{K})^{-1}\|_\infty < 1$$

where we are seeking a $\boldsymbol{K}$ from among all stabilizing $\boldsymbol{K}$s which make $\|\boldsymbol{K}(\boldsymbol{I}-\boldsymbol{G}\boldsymbol{K})^{-1}\|_\infty$ a minimum. In other words

$$\operatorname*{Inf}_{\boldsymbol{K}} \|\boldsymbol{K}(\boldsymbol{I}-\boldsymbol{G}\boldsymbol{K})^{-1}\|_\infty < \frac{1}{\varepsilon}$$

is the condition to be satisfied to ensure that $(\boldsymbol{G}, \boldsymbol{K}, \varepsilon)$ be robustly stable. The robust stability problem is therefore nothing but an $\mathcal{H}_\infty$ optimization problem, and the techniques we have discussed in the case of sensitivity minimization problems are relevant in this context also.

Let $\gamma_{\text{opt}} = \operatorname*{Inf}_{\boldsymbol{K}} \|\boldsymbol{K}(\boldsymbol{I}-\boldsymbol{G}\boldsymbol{K})^{-1}\|_\infty$

Since $\gamma_{\text{opt}} \cdot \epsilon = 1$, the smaller the value of γ_{opt} the larger ϵ can be. The value $\epsilon_{\text{sup}} = \dfrac{1}{\gamma_{\text{opt}}}$ may be interpreted as the stability margin associated with the system. In the light of the above discussions, we may redefine the plant model sets $\mathcal{P}_a$ and $\mathcal{P}_m$ defined in (6.16) and (6.17) as follows.

$$\mathcal{P}_a = \{(\boldsymbol{G}+\Delta_a) : \partial(\boldsymbol{G}) = \partial(\boldsymbol{G}+\Delta_a), \|\Delta_a\|_\infty < \epsilon\} \quad (6.23)$$

$$\mathcal{P}_m = \{(\boldsymbol{I}+\Delta_m)\boldsymbol{G} : \partial(G) = \partial[(\boldsymbol{I}+\Delta_m)\boldsymbol{G}], \|\Delta_m\|_\infty < \epsilon\} \quad (6.24)$$

where $\partial(\cdot)$ denotes the number of poles in the closed right half plane.

Summarizing, in robust control, we seek to find a $\boldsymbol{K}$ (if one exists) that achieves internal stability for every $\tilde{\boldsymbol{G}}$ (the perturbed model transfer matrix) belonging to either $\mathcal{P}_a$ or $\mathcal{P}_m$. If the prescribed ϵ is greater than ϵ_{sup} then no such $\boldsymbol{K}$ exists. So the problem is to obtain γ_{opt} as well the controller $\boldsymbol{K}$ which achieves it. This is discussed in the next section.

6.7 Robust Stabilizing Controllers

In a mathematical sense, the robust stabilization problem is no different from the sensitivity minimization problem, discussed earlier in this chapter. However, we make use of this problem to show how the NP interpolaton theory could be applied to it in order to obtain the optimal stabilizing controller.

Recall from (6.21) that the condition for robust stabilization in the case of additive perturbation can be expressed as

$$\|\boldsymbol{l}_a\boldsymbol{K}(I - \boldsymbol{GK})^{-1}\|_\infty < 1$$

Simplifying the above expression by substituting $\boldsymbol{D}_l^{-1}\boldsymbol{N}_l$ for $\boldsymbol{G}$ and the Youla parametrized expression for $\boldsymbol{K}$, we obtain

$$\boldsymbol{l}_a\boldsymbol{K}(I - \boldsymbol{GK})^{-1} = \boldsymbol{l}_a(\boldsymbol{D}_r\boldsymbol{Q} - \boldsymbol{U}_l)\boldsymbol{D}_l$$

and the robust stabilization condition is equivalent to

$$\|\boldsymbol{l}_a\boldsymbol{K}(I - \boldsymbol{GK})^{-1}\|_\infty = \|\boldsymbol{l}_a(\boldsymbol{D}_r\boldsymbol{Q} - \boldsymbol{U}_l)\boldsymbol{D}_l\|_\infty < 1$$

The above inequality is achieved only if

$$\operatorname*{Inf}_{Q\in\mathcal{RH}_\infty} \|\boldsymbol{l}_a(\boldsymbol{D}_r\boldsymbol{Q} - \boldsymbol{U}_l)\boldsymbol{D}_l\|_\infty < 1 \tag{6.25}$$

As before we assume that $\boldsymbol{G}$ has at least as many inputs as outputs and has no poles on the imaginary axis. For notational convenience we replace $\boldsymbol{l}_a$ by $\boldsymbol{r}$. Now perform the inner-outer factorization $\boldsymbol{D}_r, \boldsymbol{D}_l$ and $\boldsymbol{r}$ we have

$$\boldsymbol{D}_r = (\boldsymbol{D}_r)_i(\boldsymbol{D}_r)_0, \quad \boldsymbol{D}_l = (\boldsymbol{D}_l)_0(\boldsymbol{D}_l)_i, \quad \boldsymbol{r} = \boldsymbol{r}_i \cdot \boldsymbol{r}_0$$

$\boldsymbol{D}_r$ and $\boldsymbol{D}_l$ being square non-singular matrices $(\boldsymbol{D}_r)_i$ and $(\boldsymbol{D}_l)_i$ are square and unitary.

Let det $(\boldsymbol{D}_r)_i = \det(\boldsymbol{D}_l)_i = \delta$

δ being inner, belongs to $\mathcal{RH}_\infty$ and its zeros are the poles of $\boldsymbol{G}$ in the open right half plane. Let these poles be simple and denote them by $(\lambda_1, \lambda_2, \ldots, \lambda_n)$. We have

$$\|\boldsymbol{r}(\boldsymbol{D}_r\boldsymbol{Q} - \boldsymbol{U}_l)\boldsymbol{D}_l\|_\infty = \|\boldsymbol{r}_i\boldsymbol{r}_0[(\boldsymbol{D}_r)_i(\boldsymbol{D}_r)_0\boldsymbol{Q}(\boldsymbol{D}_l)_0(\boldsymbol{D}_l)_i - \boldsymbol{U}_l(\boldsymbol{D}_l)_0(\boldsymbol{D}_l)_i]\|_\infty$$

Making use of the norm preserving properties of $(\boldsymbol{D}_l)_i^{-1}$ and $\boldsymbol{r}_i$ we have

$$\|\boldsymbol{r}(\boldsymbol{D}_r\boldsymbol{Q} - \boldsymbol{U}_l)\boldsymbol{D}_l\|_\infty = \|(\boldsymbol{D}_r)_i(\boldsymbol{D}_r)_0\boldsymbol{Q}(\boldsymbol{D}_l)_0\boldsymbol{r}_0 - \boldsymbol{U}_l(\boldsymbol{D}_l)_0\boldsymbol{r}_0\|_\infty$$

Now $(\boldsymbol{D}_r)_0\boldsymbol{Q}(\boldsymbol{D}_l)_0\boldsymbol{r}_0$ may be replaced by $\boldsymbol{Q}$ because $(\boldsymbol{D}_r)_0$ has a right inverse, $(D_l)_0$ a left inverse and $\boldsymbol{r}_0$ has an inverse in RH_∞. Hence we may treat $\boldsymbol{Q}$ as a free parameter. The expression to be optimized is now given by

$$\|(\boldsymbol{D}_r)_i\boldsymbol{Q} - \boldsymbol{U}_l(\boldsymbol{D}_l)_0\boldsymbol{r}_0\|_\infty$$

Premultiplying the above expression by $(\boldsymbol{D}_r)_i^{-1}\det(\boldsymbol{D}_r)_i$ which is an inner, the infinity norm remains unchanged and we obtain

$$\|\delta\boldsymbol{Q} - \text{Adj}\,(\boldsymbol{D}_r)_i\boldsymbol{U}_l(\boldsymbol{D}_l)_0\boldsymbol{r}_0\|_\infty$$

Let $\boldsymbol{F} = \text{Adj}\,(\boldsymbol{D}_r)_i\boldsymbol{U}_l(\boldsymbol{D}_l)_0\boldsymbol{r}_0$
Then the condition for internal stability is

$$\gamma_{\text{Inf}} = \min_{Q\in\mathcal{RH}_\infty} \|\delta\boldsymbol{Q} - \boldsymbol{F}\|_\infty < 1 \tag{6.26}$$

Since $\mathcal{RH}_\infty$ is only a normed space and not a Banach space, this infinium may not be actually achieved if $\boldsymbol{Q}$ is confined to $\mathcal{RH}_\infty$.

To circumvent this difficulty, we modify the problem and permit $\boldsymbol{Q}$ to belong to $\mathcal{H}_\infty$ space. We then have

$$\gamma_{\text{opt}} = \min_{Q\in\mathcal{H}_\infty} \|\delta\boldsymbol{Q} - \boldsymbol{F}\|_\infty \tag{6.27}$$

where γ_{opt} is actually achieved for some $\boldsymbol{Q} \in \mathcal{H}_\infty$. Further it turns out that provided $\boldsymbol{G}$ is strictly proper $\gamma_{\text{Inf}} = \gamma_{\text{opt}}$ (see Francis *et al.* 1984). The difference between (6.26) and (6.27) is that in the former $\boldsymbol{Q} \in \mathcal{RH}_\infty$, and in the latter $\boldsymbol{Q} \in \mathcal{H}_\infty$. The optimal $\boldsymbol{Q}_0$ obtained by solving (6.27) is stable but not proper in general. Hence the associated $\boldsymbol{K}$ is also not proper. This is unacceptable. We therefore generate a sequence $\{\boldsymbol{Q}_n\}$ of $\mathcal{RH}_\infty$ functions which approach $\boldsymbol{Q}_0$ in the limit as $n \to \infty$. We proceed as follows.

If $\boldsymbol{Q}_0$ is not proper, choose an integer β such that $s^{-\beta}\boldsymbol{Q}_0$ is proper. This condition determines β. Now define

$$\boldsymbol{Q}_n = \boldsymbol{Q}_0 J_n$$

where

$$J_n = \left(\frac{n}{s+n}\right)^\beta$$

where n is an integer.
Clearly $\boldsymbol{Q}_n$ is both stable and proper. Hence $\boldsymbol{K}_n$ obtained by using the Youla parameterization formula

$$\boldsymbol{K}_n = (\boldsymbol{U}_l + \boldsymbol{D}_r\boldsymbol{Q}_n)(\boldsymbol{V}_l - \boldsymbol{N}_r\boldsymbol{Q}_n)^{-1} \text{ is proper.}$$

By choosing sufficiently large value of n, we may obtain a γ_n as close to γ_{opt} as desired. The value of γ_n obtained by substituting for $\boldsymbol{Q}_n$ is

$$\gamma_n = \|\delta\boldsymbol{Q}_0 J_n - \boldsymbol{F}\|_\infty$$

■

While solving the $\mathcal{H}_\infty$ optimization problem

$$\inf_{Q\in\mathcal{H}_\infty} \|\delta\boldsymbol{Q} - \boldsymbol{F}\|_\infty$$

the two quantities to be determined are 1. The infimum value and 2. The $\boldsymbol{Q}$ which achieves this infimum.

We assume that the infimum has been determined by some method and it has been verified to be less than unity. This ensures that there exists a $\boldsymbol{Q}$ and therefore a $\boldsymbol{K}$ which ensures robust stability. We are however concerned here about obtaining $\boldsymbol{Q}$ using that NP interpolation theory.

Let $\phi = \delta\, \boldsymbol{Q} - \boldsymbol{F}$.

Both ϕ and $\boldsymbol{F}$ belong to $\mathcal{RH}_\infty$. Hence $(\phi + \boldsymbol{F}) \in \mathcal{RH}_\infty$. But this does not mean that $\boldsymbol{Q}$ should belong to $\mathcal{RH}_\infty$ because there could be pole zero cancellations. This is especially so because δ is inner, belongs to $\mathcal{RH}_\infty$ and has its distinct zeros $\lambda_1, \lambda_2, \ldots, \lambda_n$ located in Re $s > 0$. Since we require that $\boldsymbol{Q}$ be stable, this stability constraint may be translated into an interpolation constraint as follows.

If $\phi(\lambda_i) = -\boldsymbol{F}(\lambda_i) \quad i = 1, 2, \ldots n$, this means that no zero of δ has been cancelled against a pole of $\boldsymbol{Q}$ in which case $\phi(\lambda_i)$ would not have been equal to $-\boldsymbol{F}(\lambda_i)$. Thus, the interpolation constraint ensures that the stability constraint is satisfied. If $\boldsymbol{F}$ and $\boldsymbol{Q} \in \mathcal{RH}_\infty$ then it is readily seen that $\phi \in \mathcal{RH}_\infty$.

To summarize, we are seeking a function $\phi \in \mathcal{RH}_\infty$ such that $\|\phi\|_\infty < 1$ and ϕ satisfies the interpolation constraints

$$\phi(\lambda_i) = -\boldsymbol{F}(\lambda_i) \qquad i = 1, 2, \ldots, n$$

We recognize that this is precisely the NP problem. If for example for the prescribed uncertainty bound $\|\Delta\|_\infty < \epsilon$, the value of ϵ is beyond the permissible limit, then NP problem will not have a solution. This will be reflected in the Pick matrix, which according to the non-solvability condition will cease to be positive definite. Once ϕ is obtained and knowing $\boldsymbol{F}$ the determination of $\boldsymbol{Q}$ is straightforward.

In the MIMO case, the $\boldsymbol{Q}$ obtained as a solution of the $\mathcal{H}_\infty$ optimization problem is not unique and consequently $\boldsymbol{K}$ is also not unique. This non-uniqueness actually helps in choosing a particular $\boldsymbol{K}$ from among the class of solutions in order to satisfy certain other performance criteria.

6.8 Summary

The main points discussed in the chapter may be summarized as follows:

1. Disturbances and noise are significant in certain known frequency ranges. The system sensitivity should be suitably adjusted to cover these ranges.

2. This adjustment of sensitivity may be accomplished by a choice of appropriate weighting functions, whose frequency spectra are akin to those of the disturbance signals.
3. Optimization problems such as sensitivity minimization and robust stabilization may in the final analysis be reduced to a model matching problem. The function to be optimized is given by $(\boldsymbol{T}_1 + \boldsymbol{T}_2\boldsymbol{Q}\boldsymbol{T}_3)$ where $\boldsymbol{T}_1, \boldsymbol{T}_2$ and $\boldsymbol{T}_3 \in \mathcal{RH}_\infty$. These are known and $\boldsymbol{Q} \in \mathcal{RH}_\infty$ is the unknown function.
4. The $\mathcal{H}_\infty$ optimization problem is posed as the determination of $\operatorname*{Inf}_{Q\in\mathcal{RH}_\infty} \|\boldsymbol{T}_1 + \boldsymbol{T}_2\boldsymbol{Q}\boldsymbol{T}_3\|_\infty$. The infimum in this problem is achieved only if $\boldsymbol{T}_2(j\omega)$ and $\boldsymbol{T}_3(j\omega)$ have constant rank for all ω and also if $\boldsymbol{Q}$ is allowed to range over $\mathcal{H}_\infty$, instead of its subspace $\mathcal{RH}_\infty$.
5. The modified problem is posed as

$$\operatorname*{Inf}_{Q\in\mathcal{H}_\infty} \|\boldsymbol{T}_1 + \boldsymbol{T}_2\boldsymbol{Q}\boldsymbol{T}_3\|_\infty$$

   The infimum is achieved in this case and the value of the infimum is the same as that for the original problem.
6. Let $\boldsymbol{Q}_0 \in \mathcal{H}_\infty$ be the solution of the modified problem. Then under mild restrictions, there exist a sequence of functions $\{\boldsymbol{Q}_0 J_n\}$ belonging to $\mathcal{RH}_\infty$ such that as $n \to \infty$, $\|\boldsymbol{T}_1 + \boldsymbol{T}_2\boldsymbol{Q}_0 J_n \boldsymbol{T}_3\|_\infty$ tends to the same infimum as in the case of the modified problem.
7. For the MIMO case, there exists no unique $\boldsymbol{Q}$ but a family of $\boldsymbol{Q}$s which achieve the infimum. This is in contrast to the SISO case where the optimal $\boldsymbol{Q}$ is a unique all pass function.
8. Depending upon whether $\boldsymbol{T}_2$ and $\boldsymbol{T}_3$ have full column/row rank, the model matching problem may be categorized as 1-block, 2-block or 4-block problems.
9. The 4-block and 2-block problems have to be first reduced to an equivalent 1-block problem, via spectral factorization before a solution can be attempted. However, this approach has now been superseded by recent developments.
10. Sensitivity minimization over a specified finite frequency interval for non-minimum phase plants involves a trade-off. The smaller the value of the norm of the sensitivity functions, within the interval, the larger it will be outside this interval.
11. In robust stability studies we are interested in findings: (a) For a given controller which stabilizes the nominal plant, how big could the perturbation be before the closed loop system becomes internally unstable? (b) Given the perturbation bound, does there exist a $\boldsymbol{K}$ would which guarantees internal stability for all permissible plant perturbations within this bound?

12. With the help of the small gain theorem, it is possible to establish sufficient conditions for internal stability for specified plant perturbation $\mathbf{\Delta}$. The conditions are

$$\|\Delta_a\|_\infty \|\boldsymbol{K}(I - \boldsymbol{GK})^{-1}\|_\infty < 1 \text{ (for additive perturbation)}$$
$$\|\Delta_m\|_\infty \|\boldsymbol{GK}(I - \boldsymbol{GK})^{-1}\|_\infty < 1 \text{ (for multiplicative perturbation)}$$

13. One way of solving the $\mathcal{H}_\infty$ optimization problem is through the application of NP interpolation theory. This is demonstrated in the case of a robust stabilization problem by replacing the stability constraint by an interpolation constraint.

Notes and Additional References

The application of $\mathcal{H}_\infty$ optimization to control theory problems was pioneered by Zames (1981). Zames and Francis (1983) carried this study further by discussing sensitivity to disturbances and robustness under plant perturbations. This was followed up by a detailed study of SISO systems by Francis and Zames (1984). They applied the theory of Sarason (1967) to obtain the optimal weighted sensitivity function. Sensitivity minimization in the case of MIMO systems was solved by Francis *et al.* (1984) applying the theory of Ball and Helton (1983). To solve this problem, they adopted the operator-theoretic approach. During the same period, Chang and Pearson (1984) following the NP interpolation approach solved a similar problem. Sensitivity trade-offs in the case of mulvariable plants were discussed by Young and Francis (1985).

The limitations imposed by right half plane plant zeros are clearly brought out by Freudenberg and Looze in two papers (1986(a) and 1986(b)). Robust stabilization has been discussed by a number of authors – Kimura (1984); Vidyasagar and Kimura (1986); Glover (1986).

The reduction of 4-block and 2-block problems to an equivalent 1-block problem is discussed in Vidyasagar (1985). Multivariable Nyquist criteria and its applications in robust stabilization studies are fully covered in Maciejowski (1989) and Green and Limebeer (1995). These books also serve as reference texts for much of the material covered in this chapter. For those who are interested in robust control of SISO systems there is no better book than the one written by Doyle *et al.* (1992).

Exercises

1.a) In a scalar interpolation problem, it is required to obtain a $\phi \in \mathcal{RH}_\infty$ and satisfying $\|\phi\|_\infty \leq 1$ as well as the following interpolation conditions

$$\phi(\alpha_i) = \beta_i, |\beta_i| \leq 1 \quad i = 1, 2, \ldots n$$

with all α_is in the open right-half plane. If one of the $|\beta_i|$s happen to be unity then show that $\phi(s) \equiv 1$ and hence a constant. The interpolation condition can then be satisfied if and only if the remaining $|\beta_i|$s are also equal to unity.

b) Show through an example how the above reasoning breaks down in the matrix case. Suggest a matrix version of maximum modulus principle.

2. Let the plant transfer function be

$$\boldsymbol{g}(s) = \frac{5(s-3)}{(s+1)^2}$$

It is required to design a stabilizing controller, which produces a global sensitivity bound of $\|\boldsymbol{s}\|_\infty = 1.5$. What is the lowest value of $\max_{0 \leq \omega \leq 0.2} |\boldsymbol{s}(j\omega)|$ that can possibly be achieved in this case?

3. Show that the smallest global sensitivity bound δ which is achievable over all stabilizing controllers is obtained when the weighting envelope is a constant over all frequencies.

4.a) Given $\boldsymbol{g}(s) = \dfrac{9 - s^2}{s^4 + 6s^2 + 25}$ obtain a spectral factorization of $\boldsymbol{g}(s)$. If the numerator polynomial is $s^2 - 9$ instead of $9 - s^2$, then spectral factorization is not possible. Why?

b) Given a matrix transfer function

$$\boldsymbol{\phi} = \boldsymbol{W}^{\sim}\boldsymbol{W}$$

where $\boldsymbol{\phi} \in \mathcal{RH}_\infty$ and $\boldsymbol{W}$ and $\boldsymbol{W}^{-1}$ belong to $\mathcal{RH}_\infty$ prove that $\boldsymbol{\phi}^{\sim} = \boldsymbol{\phi}$ and $\boldsymbol{\phi}(j\omega) > 0$.

5. Prove Theorem 6.5 pertaining to robust stability under multiplicative perturbation.

6. Given the plant transfer functions and the weighting functions

$$\boldsymbol{g}(s) = \frac{s-1}{(s+1)(s-0.5)}$$

$$\boldsymbol{w} = \frac{s+0.1}{s+1}$$

design a controller $\boldsymbol{k}$ (not necessarily proper) such that $\|\boldsymbol{wgk}(1-\boldsymbol{gk})^{-1}\|_{\infty}$ is a minimum. Hence obtain the maximum value of ϵ pertaining to the multiplicative disturbance $\|\delta\|_{\infty} \leq \epsilon$ that the closed loop system can tolerate without losing stability. Denoting this value by ϵ_{sup}, assume an arbitrary $\epsilon < \epsilon_{\text{sup}}$ and design a *proper controller* which ensures robust stability for the chosen ϵ
(Hint: Parameterize the controller using the Youla formula. Obtain a $\boldsymbol{Q}_0$ which minimizes the weighted $\mathcal{H}_{\infty}$ norm. If it is improper, set $\boldsymbol{Q} = J_n\boldsymbol{Q}_0$ where $J_n = \left(\dfrac{n}{s+n}\right)^{\beta}$. β is determined by $\boldsymbol{Q}_0$ and n is arbitrarily chosen, depending upon the approximation desired).

7. Explain how stability margin information is provided by the Nyquist plot. Prove that the distance from the critical point $(-1, 0)$ to the nearest point on the Nyquist plot of $\boldsymbol{gk}$ is given by $\dfrac{1}{\|\boldsymbol{s}\|_{\infty}}$ where $\boldsymbol{s}$ is the sensitivity operator. Hence show that if the perturbed plant $\tilde{\boldsymbol{g}}$ has the same number of unstable poles as the nominal plant $\boldsymbol{g}$, and the inequality

$$|\tilde{\boldsymbol{g}}(jw)\boldsymbol{k}(jw) - \boldsymbol{g}(jw)\boldsymbol{k}(jw)| < \frac{1}{\|\boldsymbol{s}\|_{\infty}}$$

is satisfied then internal stability is preserved. Show that this is a conservative condition for establishing internal stability.

8. Using small gain theorem obtain stability conditions for the following perturbed models:
   i) $\tilde{\boldsymbol{G}} = \boldsymbol{G}(\boldsymbol{I} + \boldsymbol{\Delta}\,\boldsymbol{G})^{-1}$
   ii) $\tilde{\boldsymbol{G}} = \boldsymbol{G}(\boldsymbol{I} + \boldsymbol{\Delta})^{-1}$

9. In the case of SISO systems, obtain:
   i) The condition for simultaneously achieving nominal sensitivity performance and robust stability;
   ii) The robust performance condition, i.e. the condition which simultaneously ensures robust stability and also robust sensitivity performance.

7

Balanced Realization and Hankel Norm Approximation

7.1 Introduction

This chapter essentially covers two topics, which are not only important in their own right but play a significant part in the solution of $\mathcal{H}_\infty$ optimization problems. Although seemingly uncorrelated they exhibit deep connections as revealed in the rest of this chapter. In Section 7.2 the underlying physical reasoning for adopting a particular reference frame, in the case of balanced systems is explained. An important property exhibited by balanced realizations is that if we bifurcate the system into subsystems, then each of these subsystems is observable and controllable in addition to being asymptotically stable. This is a useful property which can be used in system approximation. Continuing with the theme of approximation, in Section 7.3 we discuss $\mathcal{H}_2$ approximation in Hilbert space. An attractive feature of this method is the geometric insight provided by such approximations. We next pass on to the more difficult and demanding problem of $\mathcal{H}_\infty$ approximation. The Hankel operator plays a major role in $\mathcal{H}_\infty$ optimization studies from an operator-theoretic viewpoint. This is explained in Section 7.4. This section also deals with Hankel norm approximation which follows directly from the Hankel norm concept. Section 7.5 provides the meeting ground for the twin concepts of balanced realization and Hankel norm approximation. Together they help resolve the $\mathcal{H}_\infty$ optimization problem in the state space frame work. An algorithm which makes this possible, proposed by Glover (1984) is discussed in Section 7.5. This chapter concludes with a summary.

7.2 Balanced Realization

It is a well known fact that the transfer function represents only the controllable and observable part of a dynamic system. Even among controllable and observable modes, the question often arises – how controllable or observable are these modes? For example if a particular mode is weakly controllable in the sense that its effect on the state is only marginal, then it does not play a significant role in regulating the input–output relationships. The same remarks apply to weakly observable modes also. Therefore, it is intuitively clear that a good approximation of the input-output relationship may be obtained if we retain only the most controllable and the most observable modes of the system. This is the underlying principle behind model reduction based on the balanced state space representation of the model.

In this method, the controllability and observability gramians are used to measure controllability and observability in certain directions in state space. The gramians are not invariant under state space transformations. It turns out that there exists a particular coordinate reference frame in which both the gramians are diagonal and equal. The system representation in this particular reference frame is called balanced representation.

DEFINITION 7.1 (Balanced realization)
A realization (A, B, C) is balanced if A is a asymptotically stable and A, B, C satisfy the following matrix Lyapunov equation:

$$A\Sigma + \Sigma A' + BB' = 0 \tag{7.1}$$

$$A'\Sigma + \Sigma A + C'C = 0 \tag{7.2}$$

where

$$\Sigma = \begin{bmatrix} \sigma_1 I_{r_1} & & & \\ & \sigma_2 I_{r_2} & & \\ & & \ddots & \\ & & & \sigma_m I_{r_m} \end{bmatrix} \tag{7.3}$$

with $\sigma_i > 0 \quad \forall i$ and $\sigma_i \neq \sigma_j, i \neq j$.

If in addition $\sigma_1 > \sigma_2 \ldots > \sigma_m > 0$, the realization is said to be an ordered balanced realization.

REMARKS
In the above definition r_i denotes the multiplicity of σ_i. The McMillan degree of the transfer matrix $C(sI - A)^{-1}B$ is given by

$$n = r_1 + r_2 + \cdots + r_m$$

In a balanced realization, we have chosen the coordinate reference frame in such a way that the basis vectors associated with the different modes are equally controllable and observable. The diagonal entries of Σ may be taken as a measure of the degree of controllability and observability of these modes. Hence the tail end of the sequence $\{\sigma_i\}$ represents modes which are relatively less controllable and observable.

Note that a balanced realization is asymptotically stable and minimal. Asymptotic stability is required to ensure that the gramians are finite. Minimality follows from the fact that $\sigma_i > 0$ for $i = 1, 2, \ldots m$. This implies that Σ is positive definite and coupled with the fact that A is stable, we have (A, B) and (C, A) are respectively controllable and observable and therefore (A, B, C) is a minimal realization.

■

Suppose (A, B, C) is an ordered balanced realization as described in definition 8.1. Partition Σ as

$$\Sigma = \begin{bmatrix} \Sigma_1 & \\ & \Sigma_2 \end{bmatrix} \tag{7.4}$$

with

$$\Sigma_1 = \begin{bmatrix} \sigma_1 I_{r_1} & & & \\ & \sigma_2 I_{r_2} & & \\ & & \ddots & \\ & & & \sigma_l I_{r_l} \end{bmatrix} \text{ and } \Sigma_2 = \begin{bmatrix} \sigma_{l+1} I_{r_{l+1}} & & \\ & \sigma_{l+2} I_{r_{l+2}} & \\ & & \sigma_m I_{r_m} \end{bmatrix}$$

The balanced system is conformably partitioned as

$$\begin{bmatrix} \dot{x}_1 \\ \dot{x}_2 \end{bmatrix} = \begin{bmatrix} A_{11} & A_{12} \\ A_{21} & A_{22} \end{bmatrix} \begin{bmatrix} x_1 \\ x_2 \end{bmatrix} + \begin{bmatrix} B_1 \\ B_2 \end{bmatrix} u \tag{7.5}$$

and

$$y = [\, C_1 \quad C_2 \,] \begin{bmatrix} x_1 \\ x_2 \end{bmatrix} \tag{7.6}$$

Let u_1 and u_2 be the $\mathcal{L}_2$ norms functions that drive the state from the origin to $\begin{bmatrix} x_1(\tau) \\ 0 \end{bmatrix}$ and $\begin{bmatrix} 0 \\ x_2(\tau) \end{bmatrix}$ respectively in the time interval $[0, \tau]$. Moore (1981) has shown that

$$\frac{\int_0^\tau ||u_2||^2 \mathrm{d}t}{\int_0^\tau ||u_1||^2 \mathrm{d}t} \geq \frac{\sigma_l}{\sigma_{l+1}} \cdot \frac{||x_2(\tau)||^2}{||x_1(\tau)||^2} \tag{7.7}$$

If we assume that $\sigma_l \gg \sigma_{l+1}$ and u_1 and u_2 have identical norms then from (7.7) it follows that

$$||x_2(\tau)|| \ll ||x_1(\tau)||$$

In other words; the x_2-part of the state is much less affected by the input than the x_1-part. Similarly if y_1 and y_2 are the zero input response from $\begin{bmatrix} x_1(0) \\ 0 \end{bmatrix}$ and $\begin{bmatrix} 0 \\ x_2(0) \end{bmatrix}$ respectively, then

$$\int_0^T ||y_2(t)||^2 \mathrm{d}t \ll \int_0^T ||y_1(t)||^2 \,\mathrm{d}t \tag{7.8}$$

provided $\sigma_l \gg \sigma_{l+1}$ and $||x_1(0)|| = ||x_2(0)||$.

This means that the x_2-part of the state affects the output much less than the x_1-part.

Hence, provided $\sigma_l \gg \sigma_{l+1}, x_2$-part has very little role to play in input-output relationship and therefore could be discarded. This forms the basis for truncation of a balanced system at the appropriate level to obtain a reduced order system. The truncated model of a balanced realization (A, B, C, D) is given by (A_{11}, B_1, C_1, D), which in turn is also a balanced realization. It may be noted that the D matrix has no role to play in arriving at balanced realization as the controllability and observability gramians do not depend on D.

The truncated reduced order model

$$\hat{\boldsymbol{G}}(s) : (A_{11}, B_1, C_1, D)$$

has a McMillan degree $r = (r_1 + r_2 + \cdots + r_l)$. A note worthy feature of system order reduction using truncated balanced realization is that an error norm could be specified for the approximation, namely

$$||\boldsymbol{G} - \hat{\boldsymbol{G}}||_\infty \leq 2(\sigma_{l+1} + \cdots + \sigma_m) \tag{7.9}$$

To prove that the truncated model realization (A_{11}, B_1, C_1, D) is balanced we need to first show that

$$A_{11}\Sigma_1 + \Sigma_1 A'_{11} + B_1 B'_1 = 0 \tag{7.10}$$

$$A'_{11}\Sigma_1 + \Sigma_1 A_{11} + C'_1 C_1 = 0 \tag{7.11}$$

This follows easily by appropriately partitioning the matrix in (7.1) and (7.2). We have also to show that A_{11} is asymptotically stable. This was proved by Pernebo and Silverman (1982).

By virtue of the fact that A_{11} is stable and Σ_1 is positive definite, (A_{11}, B_1) is controllable and (C_1, A_{11}) is observable. Hence (A_{11}, B_1, C_1, D) is an asymptotically stable minimal realization of $\hat{\boldsymbol{G}}$. The above results are stated in the following theorem

THEOREM 7.1
Assume a partitioning of Σ (which need not be ordered) into Σ_1 and Σ_2 with no diagonal entries in common between them. Then both the sub systems (A_{11}, B_1, C_1, D) and (A_{22}, B_2, C_2, D) are asymptotically stable.

REMARKS
1. The model order reduction procedure indicated above leads to balanced subsystems which are stable, observable and controllable. These subsystems may be considered as lower order approximations of the original system.
2. Both $\boldsymbol{G}$ and $\hat{\boldsymbol{G}}$ have perfect matching at $s = \infty$ because $\boldsymbol{G}(\infty) = \hat{\boldsymbol{G}}(\infty) = D$. However the steady state values of $\boldsymbol{G}$ and $\hat{\boldsymbol{G}}$ will differ from one another.
3. The inequality $||\boldsymbol{G} - \hat{\boldsymbol{G}}||_\infty \leq 2(\sigma_{l+1} \ldots \sigma_m)$ holds good irrespective of the ordering of σ_is. The right hand side of (7.9) takes into account the neglected σ_is.

7.2.1 Coordinate Transformation for Balanced Realizations

We now seek an appropriate coordinate transformation to obtain a balanced realization. Like any other state variable transformation, this does not affect the D-matrix. Let T be the required transformation connecting the old state variable x with the new state variable z such that $z = Tx$. In the new reference frame (A, B, C) matrices are transformed to (TAT^{-1}, TB, CT^{-1}). As already shown, balanced realization is possible if and only if A is stable and the realization (A, B, C) is minimal. It follows that the controllability and observability gramians of (A, B, C) as defined by

$$AP + PA' + BB' = 0 \tag{7.12}$$

$$A'Q + QA + C'C = 0 \tag{7.13}$$

and denoted by P and Q are positive definite.

Under the transformation T, P gets transformed to TPT' and Q gets transformed to $(T')^{-1}Q(T^{-1})$. This can be directly seen from (7.12) and (7.13) respectively. The product PQ gets transformed into

$$TPT'\,(T')^{-1}Q(T^{-1}) = TPQT^{-1}$$

It follows that the eigen values of PQ remain invariant under the state variable transformation. It will now be shown that by a suitable choice of T, both P and Q may be made diagonal and equal. The step by step procedure to obtain such a transformation is outlined below.

1. Given a stable minimal realization (A, B, C) compute P and Q using (7.12) and (7.13) respectively.
2. Carry out a Cholesky factorization $P = RR'$.
3. $R'QR$ is symmetric and positive definite and can be diagonalized by an orthogonal transformation. Hence choose an orthogonal matrix U such that

$$R'QR = U\Sigma^2 U'$$

where Σ is diagonal.

4. Let $T = \Sigma^{\frac{1}{2}} U' R^{-1}$. The required transformation is given by T. We now demonstrate that T indeed diagonalizes both P and Q. We have

$$\begin{aligned} TPT' &= \Sigma^{\frac{1}{2}} U' R^{-1} RR'(R')^{-1} U \Sigma^{\frac{1}{2}} \\ &= \Sigma \end{aligned}$$

Similarly we have:

$$\begin{aligned} (T')^{-1} Q T^{-1} &= (\Sigma^{-\frac{1}{2}} U' R') Q (R U \Sigma^{-\frac{1}{2}}) \\ &= \Sigma \end{aligned}$$

Balanced realization of (A, B, C) is therefore obtained as (TAT^{-1}, TB, CT^{-1}).

The ordering of σ_is can be changed by permuting the state variables and this gives rise to a new balanced realization. Hence the realization is unique only up to an ordering of σ_is.

■

Suppose (A, B, C) is a balanced realization with Σ having a particular ordering. The question is whether there exists a state transformation S which will preserve this ordering. In other words we require that

$$S\Sigma S' = \Sigma \text{ and } (S^{-1})'\Sigma(S^{-1}) = \Sigma$$

Hence $S\Sigma^2 = \Sigma^2 S$, which implies that

$$S\Sigma = \Sigma S (\text{since } \sigma_i > 0)$$

We also have $\Sigma = S\Sigma S' = \Sigma SS' \Rightarrow SS' = I$.

This shows that if we insist on preserving the same order of σ_i, then the balanced realization is unique up to an arbitrary orthogonal transformation S.

7.3 Best Approximation In Hilbert Space

In many engineering applications, there often arises a need for approximating a known function belonging to a certain class by another function which is a subset of this class. The criteria for approximation is usually the distance between the two functions measured by whatever norm that is considered appropriate. In this section this general problem is posed in a Hilbert space setting. Because of its rich geometric structure, Hilbert space approximation has an intuitive appeal for scientists and engineers. Briefly, the problem may be posed as follows.

Given a function $\phi \in \mathcal{L}_2$, find a function $f \in \mathcal{H}_2$ such that

$$||\phi - f||_2$$

is minimized.

Recall that $\mathcal{L}_2$ denotes the Hilbert space of vector valued square integrable function on the imaginary axis. The Hardy space $\mathcal{H}_2$ is a closed sub-space of $\mathcal{L}_2$ and $\mathcal{H}_2^{\perp}$ is its orthogonal complement in $\mathcal{L}_2$. We thus have

$$\mathcal{L}_2 = \mathcal{H}_2 \oplus \mathcal{H}_2^{\perp}$$

Let $x = y + z$ where $x \in \mathcal{L}_2, y \in \mathcal{H}_2$, and $z \in \mathcal{H}_2^{\perp}$. Further y and z are uniquely defined for a given x.

Let the orthogonal projections $\mathcal{P}_+$ and $\mathcal{P}_-$ map $\mathcal{L}_2$ to $\mathcal{H}_2$ and $\mathcal{H}_2^{\perp}$ respectively.

In Figure 7.1, OA represents ϕ. Now choose some $f \in \mathcal{H}_2$. Let it be represented by OD. The distance between ϕ and f is geometrically represented by AD. Clearly this distance is a minimum when f happens to lie directly below A along $\mathcal{H}_2$ i.e. f is a projection of ϕ on $\mathcal{H}_2$. Hence $f^* = \mathcal{P}_+\phi$.

This method may be applied to minimization problems involving rational functions.

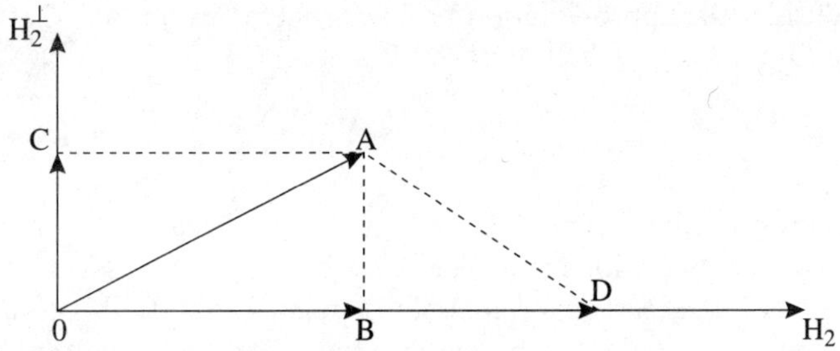

Figure 7.1 Minimization in Hilbert space.

EXAMPLE 7.1
Let $\phi(z) = \dfrac{z+3}{(z+4)(z+0.25)}$ be the pulse transfer function of a discrete time system. It is required to find an $f \in \mathcal{H}_2$ which is the closest approximation to ϕ.

Expanding $\phi(z)$ in terms of its partial fractions:

$$\begin{aligned}\phi(z) &= \frac{z+3}{(z+4)(z+0.25)} \\ &= \frac{A}{z+4} + \frac{B}{z+0.25} \\ &= \frac{4}{15}\frac{1}{z+4} + \frac{11}{15}\frac{1}{z+0.25}\end{aligned}$$

The first term on the right hand side is bounded and analytic in the closed unit disk. Therefore it belongs to $\mathcal{H}_2$. Hence we have:

$$(\mathcal{P}_+\phi)(z) = \frac{4}{15(z+4)}$$

Note. The reader is cautioned that in the conventional z-transform definition the interior of the unit disk corresponds to the open left half plane. In the above problem the closed right half complex plane corresponds to the closed unit disk and hence differs from the conventional z-transform definition.

7.4 The Hankel Operator

In the previous section, we saw the innate simplicity of $\mathcal{L}_2/\mathcal{H}_2$ approximation in Hilbert space. However best approximation problem with respect to norms which are not Hilbert space norms are much more difficult. For example consider the approximation problem where given a $\phi \in \mathcal{L}_\infty$, we are required to find a $f \in \mathcal{H}_\infty$ such that

$$||\phi - f||_\infty$$

is minimized.

The earlier procedure cannot be adopted here because the notion of orthogonality and projection cannot be applied to $\mathcal{L}_\infty$ space. However, even though $\mathcal{L}_\infty$ is not a Hilbert space, it could be related to it. In this context, an elegant theorem proposed by Nehari (1957) states that the minimum distance from a given function belonging to $\mathcal{L}_\infty$ to a function in $\mathcal{H}_\infty$ is given by the norm of an operator called *Hankel operator* which maps $\mathcal{H}_2$ on to $\mathcal{H}_2^{\perp}$. We shall presently see how the Hankel operator plays a key role in approximation and model reduction problems.

Some standard conventions
Before proceeding further, we recall some standard notation and conventions.

Corresponding to $\boldsymbol{G} \in \mathcal{L}_\infty$, we define a continuous linear operator M_G which is also called a multiplication operator or a *Laurent operator*. We have

$$M_G : \mathcal{L}_2 \to \mathcal{L}_2$$

We recall here the standard result that $||\boldsymbol{G}||_\infty$ is the induced norm of M_G as well as the *Toeplitz operator*

$$\mathcal{P}_+ M_G : \mathcal{H}_2 \to \mathcal{H}_2$$

Hence if $M_G(w) = \boldsymbol{G}w = z$ for $w \in \mathcal{L}_2$

$$\begin{aligned} ||\boldsymbol{G}||_\infty &= \sup_{w \in \mathcal{B}\mathcal{L}_2} ||z||_2 \\ &= \sup_{w \in \mathcal{B}\mathcal{L}_2^+} ||\mathcal{P}_+ z||_2 \\ &= \sup_{w \in \mathcal{B}\mathcal{H}_2} ||\mathcal{P}_+ M_G(w)||_2 \end{aligned}$$

where prefix $\mathcal{B}$ denotes the open unit ball. For any $x \in \mathcal{H}_\infty$, the operator M_x leaves the subspace $\mathcal{H}_2$ of $\mathcal{L}_2$ invariant, namely if $f \in \mathcal{H}_2$ then $M_X f = xf \in \mathcal{H}_2$. Denoting by $\mathcal{P}_-$ the orthogonal projection from $\mathcal{L}_2$ to $\mathcal{H}_2^{\perp}$ i.e. $\mathcal{P}_- : \mathcal{L}_2 \to \mathcal{H}_2^{\perp}$ the above result is equivalent to the statement

$$\mathcal{P}_- M_x | \mathcal{H}_2 = 0 \tag{7.14}$$

where '$|\mathcal{H}_2$' means that the domain of the operator $\mathcal{P}_- M_x$ is restricted to $\mathcal{H}_2$.

Given a normed space S and if A is a sub space of S then for any x belonging to S the distance of x from A in S is denoted by dist $_S(x, A)$.

We have

$$\text{dist}_{\,S}(x, A) = \operatorname*{Inf}_{y \in A} ||x - y||_S$$

Applying the above definition, for a given $R \in \mathcal{L}_\infty$

$$\begin{aligned} \text{dist}_{\,\mathcal{L}_\infty}(R, \mathcal{H}_\infty) &= \min\{||R - X||_\infty ; X \in \mathcal{H}_\infty\} \\ &= \min\{||M_R - M_X||; \quad X \in \mathcal{H}_\infty\} \\ &\geq \min\{||\mathcal{P}_-(M_R - M_X)|\mathcal{H}_2|| : X \in \mathcal{H}_\infty\} \\ &= ||\mathcal{P}_- M_R | \mathcal{H}_2|| \end{aligned} \tag{7.15}$$

The final step in (7.15) is obtained from the previous step by applying the result stated in (7.14). It can be shown that the inequality defined by (7.15) is indeed an equality leading to the result.

$$\text{dist}_{\mathcal{L}_\infty}(R, \mathcal{H}_\infty) = \mathcal{P}_- M_R|\mathcal{H}_2 \tag{7.16}$$

The above result is formally stated in the theorem that follows.

THEOREM 7.2 (Nehari's theorem)
The $\mathcal{L}_\infty$ distance from a matrix $R \in \mathcal{L}_\infty$ to the nearest matrix in $\mathcal{H}_\infty$ equals the norm of the Hankel operator $\mathcal{P}_- M_R|\mathcal{H}_2$.

Nehari's theorem is the principal tool used in solving the 1-block $\mathcal{H}_\infty$ optimization problem. The operator $\mathcal{P}_- M_R|\mathcal{H}_2$ *is known as the Hankel operator with symbol R.* It is compactly represented by Γ_R. Note that $\Gamma_R : \mathcal{H}_2 \to \mathcal{H}_2^\perp$.

EXAMPLE 7.2
Let $\boldsymbol{r}(s) = \dfrac{1}{s-3}$
Clearly $\boldsymbol{r} \in \mathcal{L}_\infty$. We now demonstrate the action of the Hankel operator Γ_r.
Let $g \in \mathcal{H}_2$. Then $M_r g = \boldsymbol{r}g$

$$\boldsymbol{r}(s)g(s) = \frac{1}{s-3}g(3) + \frac{1}{s-3}[g(s) - g(3)]$$

In the above expansion, the first term on the right hand side belongs to $\mathcal{H}_2^\perp$ and the second term belongs to $\mathcal{H}_2$. Hence the Hankel operator maps $g(s) \in \mathcal{H}_2$ to $\dfrac{g(3)}{(s-3)} \in \mathcal{H}_2^\perp$.

■

By choosing an appropriate basis in function space, we may represent a linear operator as a matrix operator. The matrix may be finite or infinite dimensional depending on whether the space on which it operates is finite or infinite dimensional.

The Hankel operator has a special matrix structure associated with it. Denoting the associated matrix by Γ, for a 3×3 Hankel matrix, we have the following structure;

$$\Gamma = \begin{bmatrix} a_1 & a_2 & a_3 \\ a_2 & a_3 & a_4 \\ a_3 & a_4 & a_5 \end{bmatrix}$$

We may generalize this and say that the *Hankel matrix* is a finite or infinite matrix which is constant along cross diagonals ($i + j =$ constant). For example in the above matrix, the elements in the (3,1), (2,2) and (1,3) positions are identical and equal to a_3. This corresponds to $i + j = 4$.

EXAMPLE 7.3
This example demonstrates a method to obtain the matrix representation of a Hankel operator.

Consider a function $\phi \in \mathcal{L}_\infty$ on the circular boundary of the unit disk. The function is periodic in 2π and can be identified with a Fourier expansion on the boundary circle. We have

$$\phi(z) = \sum_{n=-\infty}^{\infty} a_n z^n \text{ where } z = e^{j\theta} \tag{7.17}$$

With ϕ defined as in (7.17), $\{z^0, z^1, z^2, \ldots\}$ constitute a basis in $\mathcal{H}_2$ space and $\{z^{-1}, z^{-2}, z^{-3} \ldots\}$ constitute a basis in $\mathcal{H}_2^\perp$ space.

To obtain a matrix representation Γ for Γ_ϕ select each of the basis vectors z^0, z^1 etc. associated with $\mathcal{H}_2$ and perform the operation

$$\Gamma_\phi(z^j) \text{ for } j = 0, 1, 2, \ldots$$

We have

$$\begin{aligned}
\Gamma_\phi(z^j) &= \mathcal{P}_-(\phi z^j) \\
&= \mathcal{P}_-\left[\sum_{n=-\infty}^{\infty} a_n z^{n+j}\right] \\
&= \mathcal{P}_-\left[\sum_{m=-\infty}^{\infty} a_{m-j} z^m\right] \\
&= \sum_{m=-\infty}^{-1} a_{m-j} z^m
\end{aligned}$$

Thus the jth column of matrix Γ given by

$$[a_{-j-1} \quad a_{-j-2} \ldots]^T \qquad j = 0, 1, 2, \ldots$$

and the Γ matrix has the structure

$$\Gamma = \begin{bmatrix} a_{-1} & a_{-2} & a_{-3} & \cdots \\ a_{-2} & a_{-3} & a_{-4} & \cdots \\ a_{-3} & a_{-4} & a_{-5} & \cdots \\ \vdots & \vdots & \vdots & \vdots \end{bmatrix}$$

■

The time domain interpretation of Hankel operator may be stated as follows.

Assuming that $\boldsymbol{R}(s)$ is a rational function belonging to $\mathcal{L}_2$, it will be analytic in a narrow strip containing the imaginary axis. This strip constitutes the region of convergence for $r(t)$, which is the inverse bilateral Laplace transform of $\boldsymbol{R}(s)$. The linear system with impulse response $r(t)$ is therefore $\mathcal{L}_2(-\infty, \infty)$ stable. But it is not causal in general. The Hankel operator in the time domain maps any function u in $\mathcal{L}_2[0, \infty)$ into a function y in $\mathcal{L}_2(-\infty, 0]$.

The mapping is via the convolution integral

$$y(t) = \int_0^\infty r(t-\tau)u(\tau)d\tau \qquad t \leq 0$$

The analogy between the frequency domain and time domain versions of Hankel operator will now be clear. The mapping from $\mathcal{H}_2 \rightarrow \mathcal{H}_2^\perp$ in the frequency domain corresponds to the mapping $\mathcal{L}_2[0, \infty) \rightarrow \mathcal{L}_2(-\infty, 0]$ in the time domain. The bilateral Laplace transform is an isomorphism from $\mathcal{L}_2[0, \infty)$ on to $\mathcal{H}_2$ and from $\mathcal{L}_2(-\infty, 0]$ on to $\mathcal{H}_2^\perp$. Hence the Hankel operator in the two domains have identical norms.

If the system is causal, the input belonging to $\mathcal{L}_2[0, \infty)$ is mapped into an output belonging to $\mathcal{L}_2[0, \infty)$ and hence its Hankel operator is zero. In the light of the above explanation, we may give the following interpretation to Nehari's theorem in the time domain.

The distance from a non-causal system with impulse response $r(t)$ to the nearest causal system equals the norm of the Hankel operator. Since the distance is zero when $r(t)$ is causal (because the norm of the associated Hankel operator is zero) we may consider the norm of the Hankel operator as a measure of causality. The distance here is the norm of the error system considered as a mapping on $\mathcal{L}_2(-\infty, \infty)$.

7.4.1 The Hankel Norm

Having defined the Hankel operator, the next step is to define the norm of this operator. The Hankel norm of a system with transfer matrix $\boldsymbol{G}(s)$ is denoted by $||\boldsymbol{G}||_H$. It is the $\mathcal{L}_2[0, \infty)$ induced norm of the associated Hankel operator Γ_G. We thus have

$$||\boldsymbol{G}||_H = ||\Gamma_G||$$

Here we consider Γ_G as an operator which maps past inputs $u(\cdot)$ into future outputs $y(\cdot)$ (This is analogous to mapping future inputs into past outputs as was discussed earlier).

We have

$$||G||_H = \sup \frac{||y||_2}{||u||_2} \quad \begin{cases} y(t) = 0, & t < 0 \\ u(t) = 0, & t > 0 \end{cases}$$

The Hankel norm thus gives the $\mathcal{L}_2$ gain from past inputs to future outputs.

Consider an arbitrary unit energy input in $\mathcal{L}_2(-\infty, 0]$ such that $||u||_2 = 1$. Then

$$||G||_H = \sup ||y||_2$$

Hence $||G||_H^2$ represents the least upper bound of the future output energy.

Using similar notation but now considering unit energy input with $u \in \mathcal{L}_2(-\infty, \infty)$ and $y \in \mathcal{L}_2(-\infty, \infty)$, we have by definition

$$||\boldsymbol{G}||_\infty = \sup ||y||_2$$

Here $||\boldsymbol{G}||_\infty^2$ is seen as the least upper bound of the energy contained in the output y in the time interval $(-\infty, \infty)$. Clearly this energy must be equal to or greater than the energy represented by $||\boldsymbol{G}||_H^2$. Hence

$$||\boldsymbol{G}||_H \leq ||\boldsymbol{G}||_\infty \tag{7.18}$$

Consider an anticausal system $\boldsymbol{F}$ (a system causal in reverse time) which has the property that $u \in \mathcal{L}_2(-\infty, 0]$ implies that $y \in \mathcal{L}_2(-\infty, 0]$. If $\boldsymbol{F}$ is now added on to $\boldsymbol{G}$ which is a causal system, then the future output of the combined system with $u \in \mathcal{L}_2(-\infty, 0]$ is the same as the future output of G. It therefore follows that for any anticausal F

$$||G||_H \leq ||\boldsymbol{G} - \boldsymbol{F}||_\infty \tag{7.19}$$

Note that if $\boldsymbol{F} \in \mathcal{RH}_\infty^-$ then $\boldsymbol{F}$ is anticausal and all of its poles are in the open right half plane. Nehari's result (Theorem 7.2) could now be rephrased as

$$||\boldsymbol{G}||_H = \min_{\boldsymbol{F} \in \mathcal{H}_\infty^-} ||\boldsymbol{G} - \boldsymbol{F}||_\infty \tag{7.20}$$

Roughly speaking, (7.20) means that if one wishes to approximate a causal transfer function $\boldsymbol{G}$ by an anticausal transfer function $\boldsymbol{F}$, then the smallest $\mathcal{L}_\infty(-\infty, \infty)$ error that can be achieved is precisely the Hankel norm of $\boldsymbol{G}$. This result forms the basis for Glover's (1984) method which is discussed later on in this chapter.

7.4.2 Hankel Singular Values

Hankel operators have wide applications in system theory and approximation theory. A detailed discussion is beyond the scope of this book. Interested readers may refer to Partington (1988). Hankel operator Γ_G is a compact operator on Hilbert space and can be expanded in the form

$$\Gamma_G x = \sum_{i=1}^{n} \sigma_i \langle x, \nu_i \rangle w_i \tag{7.21}$$

where $\sigma_1 \geq \sigma_2 \geq \cdots \geq \sigma_n > 0$ and ν_i and w_i are orthogonal sequences with $\langle \nu_i, \nu_j \rangle = \langle w_i, w_j \rangle = 0$ for $i \neq j$ and $||\nu_i||_2^2 = ||w_i||_2^2 = 1 \; \forall \; i$. σ_is are called the *singular values* and the pair (ν_i, w_i) corresponding to σ_i is known as *Schmidt pair.* (7.21) represents the Schmidt decomposition of the Hankel operator Γ_G. This is nothing but the singular value decomposition of Γ_G. For a transfer matrix $\boldsymbol{G}$ characterized by its minimal realization (A, B, C, D) the Hankel singular values $\sigma_i s$ are given by

$$\sigma_i = \lambda_i^{\frac{1}{2}}(PQ) \qquad i = 1, 2, \ldots n \tag{7.22}$$

where P and Q are the controllability and observability gramians associated with the minimal realization. The proof is omitted (Refer to Green and Limebeer 1995). It may be noted that Hankel singular values are independent of the feed-through matrix D. The norm of the Hankel operator may now be defined as follows.

Definition 7.2 (Hankel Norm)
Let $\boldsymbol{G} = C(sI - A)^{-1}B$ and let A be stable. Then Hankel norm of G is defined as

$$||G||_H = \overline{\sigma}(\Gamma_G) = \lambda_{\max}^{\frac{1}{2}}(PQ) \tag{7.23}$$

The rank of the Hankel operator Γ_G is equal to the McMillan degree of its symbol $\boldsymbol{G}$.

7.5 Hankel Norm Approximation Problem

The Hankel norm approximation problem may be posed as follows.

Given a stable transfer matrix $\boldsymbol{G}$, with McMillan degree n, and given an integer $m < n$ which denotes the desired order of approximation for $\boldsymbol{G}$, find another stable transfer matrix $\hat{\boldsymbol{G}}$ of McMillan degree m such that

$$J(\hat{\boldsymbol{G}}) = ||\boldsymbol{G} - \hat{\boldsymbol{G}}||_H$$

is minimized over all stable transfer matrices of McMillan degree m. $\hat{\boldsymbol{G}}_{\text{opt}}$ thus obtained is the best mth order approximation of the given nth order system $\boldsymbol{G}$. The distance between $\boldsymbol{G}$ and $\hat{\boldsymbol{G}}$ is taken as the Hankel norm of their difference. We have the following fundamental result in this connection.

THEOREM 7.3 (Hankel norm approximation)
The Hankel norm distance between $\boldsymbol{G}$ and the set of all stable transfer matrices of McMillan degree m (where $m < n$, the McMillan degree of $\boldsymbol{G}$) is given by σ_{m+1}, the $(m+1)$th Hankel singular value of Γ_G, i.e. $\sigma_{m+1}[\boldsymbol{G}]$

Proof: Omitted (see Glover 1984).

REMARKS
1. The Hankel norm of the error incurred in approximating $\boldsymbol{G}$ by $\hat{\boldsymbol{G}}$ of McMillan degree m is at least as large as σ_{m+1}. This lower bound could actually be attained by selecting a suitable $\hat{\boldsymbol{G}}$.
2. If $\hat{\boldsymbol{G}}$ is so chosen that $||\boldsymbol{G} - \hat{\boldsymbol{G}}||_H = \sigma_{m+1}[\boldsymbol{G}]$ then it can be shown that for this particular $\hat{G}$

$$||G - \hat{G}||_\infty \leq 2(\sigma_{m+1} + \sigma_{m+2} \cdots + \sigma_n)$$

The above inequality states that the $\mathcal{L}_\infty$ distance between $\boldsymbol{G}$ and $\hat{\boldsymbol{G}}$ cannot exceed twice the sum the 'neglected' Hankel singular values of Γ_G.

7.6 Glover's Method

In a definitive paper, Glover (1984) discussed the Hankel norm approximation problem from several angles. Specifically, he delineated a method for solving the following Hankel norm approximation problem:

$$\underset{F \in \mathcal{RH}_\infty^-}{\text{Minimize}} ||\boldsymbol{G} - \boldsymbol{F}||_\infty, \boldsymbol{G} \in \mathcal{RH}_\infty \text{ and } \boldsymbol{G} \text{ is square}$$

The scalar version of the above problem, in the discrete time case was solved by Nehari (1957) and its multivariable version completely solved by Adamjan *et al.* (1978). A related problem was also solved by Sarason (1967). The multivariable continuous time case was solved by Glover (1984) who provided an explicit straightforward solution to the problem. In this section, we consider this solution in some detail. However the proof is omitted as it is fairly lengthy.

The salient steps in arriving at Glover's solution are as follows.

First it is shown that $||\boldsymbol{G}||_H \leq ||\boldsymbol{G} - \boldsymbol{F}||_\infty$ for any $F \in \mathcal{RH}_\infty^-$. It is then shown that the lower bound is actually achieved by a suitable choice of $\boldsymbol{F}$ as generated by Glover's algorithm. This algorithm proposes a constructive procedure which is central to the computation of the optimal $\boldsymbol{F}$. It is further shown that for the optimally constructed $\hat{\boldsymbol{F}}$ with realization $(\hat{A}, \hat{B}, \hat{C}, \hat{D})$. $\frac{1}{\sigma_1}[\boldsymbol{G}(s) - \hat{\boldsymbol{F}}(s)]$ is an all pass function. If we denote the error $[\boldsymbol{G}(s) - \hat{\boldsymbol{F}}(s)]$ by $\boldsymbol{E}$(s) than we have $\boldsymbol{E}(s)\boldsymbol{E}^{\sim}(s) = \sigma_1^2 I$ where σ_1 is the maximum Hankel singular value of $\boldsymbol{G}$. The procedure requires that a balanced realization of $\boldsymbol{G}$ be first obtained. This makes the construction of $\hat{\boldsymbol{F}}$ quite explicit. The algorithm is developed for a square transfer matrix $\boldsymbol{G}$.

ALGORITHM

1. Obtain a balanced realization of $\boldsymbol{G} : (A, B, C, D)$.
2. Obtain the Hankel singular values of $\boldsymbol{G}$. They are the diagonal entries of Σ namely $\sigma_1, \sigma_2, \ldots \sigma_n$ and all of them are positive. Let $r_1, r_2, \ldots, r_n$ be the multiplicities of the respective singular values.
3. Permute the state variables if necessary and let Σ have the form

$$\Sigma = \begin{bmatrix} \sigma_m I_{r_m} & 0 \\ 0 & \Sigma_2 \end{bmatrix}$$

where σ_m is the mth singular value with multiplicity r_m.

4. Partition A, B, C in conformity with the partitioning of Σ. We have

$$A = \begin{bmatrix} A_{11} & A_{12} \\ A_{21} & A_{22} \end{bmatrix}; \quad B = \begin{bmatrix} B_1 \\ B_2 \end{bmatrix}; \quad C = [\,C_1 \quad C_2\,]$$

With the indicated partitioning the following Lyapunov equations are satisfied.

$$A_{11}\Sigma_1 + \Sigma_1 A'_{11} + B_1 B'_1 = 0$$
$$A'_{11}\Sigma_1 + \Sigma_1 A_{11} + C'_1 C_1 = 0$$

where $\Sigma_1 = \sigma_m I_{r_m}$

5. Construct an orthogonal matrix U such that

$$B_1 = -C'_1 U$$

This is possible because using the two Lyapunov equations and substituting $\sigma_m I_{r_m}$ for Σ_1 we obtain $B_1 B'_1 = C'_1 C_1$ from which the above relationship could be deduced.

6. Define $\Gamma = \Sigma_2^2 - \sigma_m^2 I$ (where $\boldsymbol{I}$ has the same dimension as Σ_2) and form the system

$$\hat{\boldsymbol{F}}_m = (\hat{A}_m, \hat{B}_m, \hat{C}_m, \hat{D}_m)$$

which may or may not be stable.

$$\hat{A}_m = \Gamma^{-1}(\sigma_m^2 A'_{22} + \Sigma_2 A_{22}\Sigma_2 - \sigma_m C'_2 U B'_2)$$
$$\hat{B}_m = \Gamma^{-1}(\Sigma_2 B_2 + \sigma_m C'_2 U)$$
$$\hat{C}_m = C_2\Sigma_2 + \sigma_m U B'_2$$
$$\hat{D}_m = D - \sigma_m U$$

Note that for $m = 1, \hat{F}_m \in \mathcal{RH}_\infty^-$ i.e. $\hat{F}_m$ contains only unstable part. For $m > 1$, $\hat{F}_m$ may have both stable and unstable parts.

7. Obtain $\hat{G}_m$ as the stable part of $\hat{F}_m$ (applicable for $m > 1$).
The following results can be deduced from the algorithm.
(i) $||G - \hat{G}_m||_H = \sigma_m[G]$ for $m = 2, 3, \dots n$ where $\hat{G}_m$ is the $(m - 1)$ order optimal Hankel norm approximation of G.
(ii) In the special case when $\Sigma_1 = \sigma_1 I_{r_1}$ corresponding to $m = 1$

$$\hat{F}_1 = \hat{C}_1(sI - \hat{A}_1)^{-1}\hat{B}_1 + \hat{D}_1$$

gives the optimal solution to the problem

$$\underset{F \in \mathcal{RH}_\infty^-}{\text{minimize}} ||G - F||_\infty, G \in \mathcal{RH}_\infty$$

Further, $\hat{F}_1$ is anticausal. $\hat{F}_1$ thus obtained is known as the *zeroth order Hankel norm approximation of G.*
We have

$$||G - \hat{F}_1||_\infty = ||G||_H = \sigma_1[G]$$

(iii) Let $(G - \hat{F}_m) = E_m$. We have

$$E_m E_m^\sim = \sigma_m^2 I \qquad \text{m} = 1, 2, \dots \text{ n}$$

■

If the transfer matrix G is not square, Glover's method may still be applied by adding appropriate rows/columns of zeros to make it square. Let the augmented matrix be denoted by G_a. We may now apply the algorithm to obtain a square matrix $F^* \in \mathcal{H}_\infty^-$ such that

$$||G_a - F^*||_\infty = ||G_a||_H = ||G||_H$$

To illustrate, let G be augmented by a column of zeros. By partitioning F^* suitably, we have

$$\begin{aligned} ||G_a - F^*||_\infty &= ||\, G - F_1^* \quad 0 - F_2^* \,||_\infty \\ &= ||G||_H \end{aligned}$$

Using the property that the infinity norm of a submatrix will always be equal to or less than the infinity norm of the entire matrix, we have

$$||\boldsymbol{G} - \boldsymbol{F}_1^*||_\infty \leq ||\boldsymbol{G}||_H$$

But by Nehari's theorem $||\boldsymbol{G} - \boldsymbol{F}_1||_\infty \geq ||G||_H \quad \forall\, \boldsymbol{F}_1 \in \mathcal{H}_\infty^-$. Hence we conclude that

$$||\boldsymbol{G} - \boldsymbol{F}_1^*||_\infty = ||\boldsymbol{G}||_H$$

Therefore $\boldsymbol{F}_1^*$ is a solution to the zeroth order Hankel norm approximation problem for the non square matrix G.

7.7 Summary

Discussions in this chapter reveal the close affinity between balanced realization, $\mathcal{H}_\infty$ optimization and Hankel-norm approximation. Starting with balanced realization we note that this type of realization is always possible provided the system under consideration is stable and minimal. Balanced realization exhibits some useful properties which suggest that it could be used for model reduction purposes. For example, one could truncate the system to yield a lower dimensional model of whatever size one wishes to have. Furthermore, the truncated reduced order model is minimal and most importantly stable. The infinity norm error bound associated with the truncated model gives a clear indication of the approximation involved.

Design of systems via approximation in Banach and Hilbert spaces might have sounded esoteric a couple of decades back, but now it is considered an important tool in the control engineer's tool box. The Hilbert space approximation is particularly attractive because of the notions of orthogonality and projection embedded in this approach. The $\mathcal{H}_2$ optimization problem which makes use of this approach is simple when compared to the $\mathcal{H}_\infty$ optimization problem. The central issue in the latter case is how to choose an appropriate stabilizing controller, which minimizes the $\mathcal{L}_\infty$ norm distance between a known causal function and a function in $\mathcal{H}_\infty^-$ subspace. A solution to this problem is achieved via Nehari's theorem. This brings us to the study of compact operators in Hilbert space known as Hankel operators. It has wide applications in functional analysis, approximation theory, system analysis and control theory. The topics of interest here are the norm of the operator and its singular values. Briefly stated, in the language of system theory the Hankel operator maps past inputs into future outputs, with time, $t = 0$ being reckoned as the present moment. Its norm could be thought of as the square root of the maximum possible amplification of energy in the system between inputs which are zero in positive time and outputs which are zero in negative time. The Hankel norm sets the lower bound for the $\mathcal{H}_\infty$ norm of a stable transfer matrix. There is a further tie-up between the Hankel norm and $\mathcal{H}_\infty$ norm. Thus if $\boldsymbol{F}$ is anticausal, then we can arrive at the following inequality:

$$||\boldsymbol{G}||_H \leq ||\boldsymbol{G} - \boldsymbol{F}||_\infty \quad \forall F \in \mathcal{H}_\infty^-$$

However, Nehari's theorem asserts that there exists an $\hat{\boldsymbol{F}} \in \mathcal{H}_\infty^-$ such that

$$||\boldsymbol{G}||_H = ||\boldsymbol{G} - \hat{\boldsymbol{F}}||_\infty$$

The lower bound in the above inequality is thus achievable. Such an $\hat{\boldsymbol{F}}$ can be obtained by the application of Glover's algorithm.

Another interesting relationship concerns the Hankel norm distance between the original system $\boldsymbol{G}$ and the reduced order model $\boldsymbol{G}_k$ of order $\boldsymbol{k}$. Here we have the well known result

$$\text{Inf } ||\boldsymbol{G} - \boldsymbol{G}_k||_H \geq \sigma_{k+1}[\boldsymbol{G}]$$

where $\boldsymbol{G}_k$ ranges over all stable kth order models of $\boldsymbol{G}$. This implies that the closest that $\boldsymbol{G}_k$ can approach $\boldsymbol{G}$ (in the Hankel norm sense) is governed by the σ_{k+1}th Hankel singular value of $\boldsymbol{G}$. We can in addition show that there exists an optimal $\hat{\boldsymbol{G}}_k$ for which the lower bound is actually achieved. Thus

$$||\boldsymbol{G} - \hat{\boldsymbol{G}}_k||_H = \sigma_{k+1}[\boldsymbol{G}]$$

Developing further on the above result, let $\boldsymbol{F} \in \mathcal{H}_\infty^-$. Then:

$$||\boldsymbol{G} - \hat{\boldsymbol{G}}_k + \boldsymbol{F} - \boldsymbol{F}||_H = \sigma_{k+1}[G]$$

we have

$$||(\boldsymbol{G} - \hat{\boldsymbol{G}}_k - \boldsymbol{F}) + \boldsymbol{F}||_\infty \leq ||\boldsymbol{G} - \hat{\boldsymbol{G}}_k - \boldsymbol{F}||_\infty + ||\boldsymbol{F}||_\infty$$

Note that $(\boldsymbol{G} - \hat{\boldsymbol{G}}_k)$ is stable. Let $\hat{\boldsymbol{F}}$ be the optimal value of $\boldsymbol{F}$ which minimizes $||\boldsymbol{G} - \hat{\boldsymbol{G}}_k - \boldsymbol{F}||_\infty$. Hence by Nehari's result, $||(\boldsymbol{G} - \hat{\boldsymbol{G}}_k) - \hat{\boldsymbol{F}}||_\infty = ||\boldsymbol{G} - \hat{\boldsymbol{G}}_k||_H$.

But $||\boldsymbol{G} - \hat{\boldsymbol{G}}_k||_H = \sigma_{k+1}[\boldsymbol{G}]$.

Hence:

$$||\boldsymbol{G} - \hat{\boldsymbol{G}}_k||_\infty \leq \sigma_{k+1}[\boldsymbol{G}] + ||\hat{\boldsymbol{F}}||_\infty$$

Another upper bound for $||\boldsymbol{G} - \hat{\boldsymbol{G}}_k||_\infty$ is given by:

$$||\boldsymbol{G} - \hat{\boldsymbol{G}}_k||_\infty \leq 2(\sigma_{k+1} + \sigma_{k+2} + \cdots \sigma_n)$$

This interplay of $\mathcal{H}_\infty$ norms and Hankel norms is a common feature of all Hankel norm approximation and $\mathcal{H}_\infty$ optimization problems.

Notes and Additional References

Model reduction based on measures of controllability and observability was first suggested by Moore (1981). Subsequently Pernebo and Silverman (1982) discussed model reduction procedures based on balanced state space representation. They considered both continuous time and discrete time systems. The computational aspects of balanced realization and singular value decomposition are covered in Klema and Laub (1980) and Laub *et al.* (1987). Hankel operators have a rich and varied literature. Partington (1988) gives a rigorous account of Hankel operators and their use in approximation theory and system analysis. The first attempt at deriving a closed form Hankel norm optimal solution for multivariable system reduction problems is reported by Kung and Lin (1981). Glover (1984) made a detailed study of Hankel norm approximation and their $\mathcal{L}_\infty$ error bounds. Specifically, he proposed a state-space method based on balanced realization for solving the optimal Hankel norm approximation problem. Suboptimal Hankel norm approximation is considered in Green and Limebeer (1995). Adamjan et al. (1978), in a celebrated paper, obtained a generalization of the Hankel norm approximation problem. They deal with the multivariable case and do not restrict their discussion to only rational functions. Other noteworthy contributions in this area are those made by Nehari (1957) and Sarason (1967).

Exercises

1. A minimal realization of a stable transfer function is given by $(A, B, C, 0)$ where

$$A = \begin{bmatrix} 0 & 1 & 0 \\ 0 & 0 & 1 \\ -50 & -45 & -12 \end{bmatrix} \quad B = \begin{bmatrix} 0 \\ 0 \\ 1 \end{bmatrix} \quad C = [250 \quad 105 \quad 39]$$

   (i) Calculate the controllability and observability gramians associated with the system.
   (ii) Obtain the transformation T which achieves balanced realization.
   (iii) Compute the balanced realization pertaining to the system. Is this realization unique?

2. Suppose (A, B) satisfies the Lyapunov equation $PA' + AP = -BB'$ for some positive definite P. Then it is well known that A is asymptotically stable if and only if (A, B) is controllable.

However, if the system is not symptotically stable prove that the eigenvalues of A that are not in the open left half plane are on the imaginary axis.

3. Let $\boldsymbol{T}_1, \boldsymbol{T}_2$, and $\boldsymbol{T}_3$ belong to $\mathcal{RH}_2$ and satisfy the following conditions.
   (i) $\boldsymbol{T}_1$ is strictly proper.
   (ii) $\boldsymbol{T}_2$ has full row rank at all points on the imaginary axis and at infinity.
   (iii) $\boldsymbol{T}_3$ is square and non singular at all points on the imaginary axis and at infinity.

   Define $J(\boldsymbol{Q}) = ||\boldsymbol{T}_1 - \boldsymbol{T}_2\boldsymbol{Q}\boldsymbol{T}_3||_2$ and let the inner-outer factorization of $\boldsymbol{T}_2$ and $\boldsymbol{T}_3$ be.

$$\boldsymbol{T}_2 = (\boldsymbol{T}_2)_i(\boldsymbol{T}_2)_0;\ \boldsymbol{T}_3 = (\boldsymbol{T}_3)_0(\boldsymbol{T}_3)_i$$

   Show that:

$$\min_{\boldsymbol{Q}\in\mathcal{RH}_2} J(\boldsymbol{Q}) = ||[(\boldsymbol{T}_2)_i^{\sim}\boldsymbol{T}_1(\boldsymbol{T}_3)_i]_-||_2$$

   Further show that the minimum is obtained by a $\boldsymbol{Q}$ satisfying $(\boldsymbol{T}_2)_0\boldsymbol{Q}(\boldsymbol{T}_3)_0 = [(\boldsymbol{T}_2)_i^{\sim}\boldsymbol{T}_1(\boldsymbol{T}_3)_i^{\sim}]_+$ where $(\cdot)_+$ and $(\cdot)_-$ denote the stable and unstable parts of the argument.
   [Hint: Apply projection theorem in Hilbert space.]

4. Obtain the best approximation of ϕ in the $\mathcal{L}_2$ norm by a $\mathcal{H}_2$ function where

$$\phi(e^{i^{\theta}}) = -j(\pi - \theta) \qquad 0 \leq \theta < 2\pi$$

   [Hint: obtain the Fourier expansion of ϕ namely $\phi(z) = \Sigma_{-\infty}^{\infty} a_n z^n$ and evaluate the Fourier coefficients $\{a_n\}$. Let $\mathcal{P}_+$ be the projection operator and note that $\mathcal{P}_+\left[\sum_{-\infty}^{\infty} a_n z^n\right] = \sum_{0}^{\infty} a_n z^n$ where $a_n = 0$ if $n = 0$ and $a_n = \dfrac{-1}{n}$ if $n \neq 0$.

5. Let $\phi(z) = \dfrac{1}{z - \alpha}$ where z lies on the unit circle and α belongs to the interior of the unit circle.
   i) What is the best approximation of ϕ in $\mathcal{H}_2$ with respect to the $\mathcal{L}_2$ norm? [Ans : zero]
   ii) What is the best approximation of ϕ in $\mathcal{H}_\infty$ with respect to the $\mathcal{L}_\infty$ norm? $\left[\text{Ans : } \dfrac{\overline{\alpha}}{1 - |\alpha|^2}\right]$

6. Let $\phi(e^{i^{\theta}}) = -j(\pi - \theta) \quad 0 \geq \theta < 2\pi$ show that the matrix of the Hankel operator Γ_ϕ with respect to the standard basis of $\mathcal{H}_2$ and $\mathcal{H}_2^{\perp}$

is given by

$$\Gamma = \begin{bmatrix} 1 & \frac{1}{2} & \frac{1}{3} & \cdots \\ \frac{1}{2} & \frac{1}{3} & \frac{1}{4} & \cdots \\ \frac{1}{3} & \frac{1}{4} & \frac{1}{5} & \cdots \\ \vdots & \vdots & \vdots & \cdots \end{bmatrix}$$

Γ is known as Hilbert's Hankel matrix.

7. The norm of the Hankel operator Γ_ϕ for ϕ as given in problem 6 is $||\Gamma_\phi|| = ||\Gamma||$ where Γ is the matrix of Γ_ϕ with respect to orthogonal bases. We have $||\Gamma_\phi|| = ||\phi||_H \geq ||\phi||_\infty$ (because the Hankel norm is less than or equal to the infinity norm). Hence show that $||\Gamma|| \geq \pi$.
8. Prove that the singular values of the Hankel operator Γ_G where $\boldsymbol{G}$ is a stable matrix is given by $\sigma_i[\boldsymbol{G}] = \lambda_i^{\frac{1}{2}}\ (PQ)$ where P and Q are the controllability and observability gramians of the miñimal realization of G.
9. Use Glover's algorithm to obtain the optimal anti-causal approximation of $\boldsymbol{g}\ (s) = \dfrac{2s+3}{s^3+5s^2+11s+6}$. Evaluate the $\mathcal{L}_\infty$ norm of the error.
10. For $\boldsymbol{g}(s)$ as given in problem 9, obtain its first and second order approximations using Glover's algorithm. Evaluate the corresponding $\mathcal{L}_\infty$ norm of the errors.

8

$\mathcal{H}_2$ and $\mathcal{H}_\infty$ Optimization

8.1 Introduction

This chapter is written with two objectives in mind. The first is to present a rapid-fire survey of the LQG problem with a view towards highlighting its points of contact with its counterpart, namely the $\mathcal{H}_\infty$ optimization problem. The second is to discuss two classical approaches adopted in solving the optimization problem using the $\mathcal{H}_\infty$ norm.

The LQG theory, built over a period of three decades, is rich in literature. What is attempted here is to merely review the basics of LQG theory sufficiently to draw an analogy with $\mathcal{H}_\infty$ optimization studies. The Riccati equation is the kingpin on which both LQG theory and $\mathcal{H}_\infty$ theory revolves. Hence some space is devoted to the discussion of the solution of the Riccati equation and the various properties associated with it. As part of the second objective, we take up for discussion the NP interpolation problem. Both the scalar and matrix versions are discussed. It is shown how the NP approach can be used to solve the model matching problem. Finally, we take a look at the operator-theoretic approach in a Hilbert space setting. It is an approach characterized by deep theoretical underpinnings and therefore it holds much promise for the future. Here, by necessity, our treatment is at an elementary level, limiting the discussion only to SISO systems.

8.2 LQG Methodology

Ever since its introduction in the early sixties, primarily through the seminal contributions of Kalman, LQG approach has remained one of the cornerstones of modern control theory. The LQG theory formalizes in a very elegant manner, a specific design situation in a stochastic environment. It involves the synthesis of a feedback controller for linear finite dimensional plant models with stability and least square performance as design objectives. The breadth and sweep of LQG methodology makes it a particularly useful design tool in many control applications.

Broadly speaking, the method requires that we first select stochastic models (of the additive type) for disturbance and sensor noise and then define weighted mean square error criteria as a standard of goodness for the design. As an outcome of the design we obtain a controller $\boldsymbol{K}$ which not only stabilizes the system but also optimizes the criteria of goodness laid down. In the sequel we first state the LQG problem and then indicate its solution without actually proving any of the results. The proof is available in most of the present-day textbooks on control theory.

8.2.1 The Separation Principle

Consider the following plant model in state space, governed by the equations

$$\dot{x} = Ax + Bu + Mw \tag{8.1}$$

$$y = Cx + v \tag{8.2}$$

where x is the state, u the input, y the output and w and v are white noise Gaussian stochastic processes with zero mean and covariances as defined below:

$$\mathcal{E}[w(t)w'(t+\tau)] = W\delta(\tau), \quad W \geq 0$$
$$\mathcal{E}[v(t)v'(t+\tau)] = V\delta(\tau), \quad V > 0$$

and

$$\mathcal{E}[w(t)v'(t+\tau)] = 0$$

where $\mathcal{E}$ is the expectation operator and $\delta(\cdot)$ is the Dirac delta function.

The problem is to derive a controller $u = \boldsymbol{K}y$, which minimizes the performance index

$$J_1 = \lim_{T\to\infty} \mathcal{E}\left\{\frac{1}{T}\int_0^T (z'Qz + u'Ru)\mathrm{d}t\right\} \tag{8.3}$$

Where $z = Lx$, Q and R are weighting matrices which are constant with $Q = Q' \geq 0$ and $R = R' > 0$. The solution of the LQG problem is based on a principle, known in literature as the *separation principle.* According to this principle, the LQG problem separates itself into two parts – a solution of the linear quadratic deterministic problem and a solution of the linear Gaussian estimation problem.

The solution of the deterministic problem is first obtained by minimizing J_1, assuming that the disturbance noises w and v are absent. The solution to this deterministic problem leads to a linear control law $u = Fx$, where F is a constant matrix known as control gain matrix. Since we do not know x, the next step is to obtain an estimate $\hat{x}$, of state x, which is optimal in the least square sense. We therefore seek to minimize

$$J_2 = \lim_{T\to\infty} \mathcal{E}\left\{\frac{1}{T}\int_0^T (\hat{x} - x)'(\hat{x} - x)\mathrm{d}t\right\} \tag{8.4}$$

by an optimal choice of $\hat{x}$.

Note that x is not known and only $\hat{x}$ is available for control purpose. It turns out that the optimal control gain matrix, obtained if we were to use $\hat{x}$ instead of x in (8.3) is identical to the control gain matrix obtained, as a solution to the purely deterministic problem. The separation principle thus reduces the LQG problem into two sub-problems for each of which solutions are separately known. It should be noted that the separation principle is valid only if the model is linear, the performance index J_1 is quadratic and w and v are white noise gaussian processes. The solution of the LQG problem crucially depends upon the solution of associated Riccati equations. In the next section, we take a deeper look at these equations.

8.2.2 The Algebraic Riccati Equation

The Riccati equation makes its appearance in the solution of several problems in optimal control and estimation. Basically it is a matrix quadratic equation, with a quadratic term, two linear terms and a constant term. The linear part comprises of two terms because of the non commutative nature of matrix multiplication. The algebraic matrix Riccati equation has the form

$$A'X + XA + X'RX + Q = 0 \tag{8.5}$$

Where X is the unknown matrix and A, R and Q are $n \times n$ real matrices, Further R and Q are assumed to be symmetric.

We may associate a $2n \times 2n$ real matrix H with the Riccati equation. We define H as

$$H = \begin{bmatrix} A & R \\ -Q & -A' \end{bmatrix} \tag{8.6}$$

The Hamiltonian matrix H has the property that if λ is an eigenvalue of H then so is $-\lambda$, occurring with the same degree of multiplicity as λ. Specifically, if H has no eigenvalues on the imaginary axis, then n of the eigenvalues of H are in the open left half plane and the remaining n are in the open right half plane.

Let U be a matrix of eigen vectors of H ordered in such a way that the first n columns starting from the left correspond to eigenvalues in Re $s < 0$ and the remaining n columns correspond to eigenvalues in Re $s > 0$. Partitioning U into four $n \times n$ block matrices we have

$$U = \begin{bmatrix} U_{11} & U_{12} \\ U_{21} & U_{22} \end{bmatrix} \tag{8.7}$$

If U_{11} is non-singular, define $X = U_{21}U_{11}^{-1}$. It can be shown that X is the solution of the Riccati equation (8.5) provided U_{11} is non-singular. X is uniquely defined by H. We are interested in a special class of Hamiltonian matrices exhibiting the following twin properties:

(i) H has no eigen values on the imaginary axis.
(ii) U_{11} is non-singular.

Now define an operator called the Riccati operator ((Ric) operator) whose domain is a Hamiltonian matrix possessing the twin properties referred to above. These ideas are expressed in a cryptic form as follows:

$H \in$ *Dom* (Ric) $\Rightarrow$ H satisfies the two properties listed above.

$X =$ *Ric* $(H) \Rightarrow X$ is a solution of the Riccati equation (8.5) with $H \in$ *Dom* (Ric).

The following lemmas (see Doyle *et al.* 1989) are relevant in our study of Riccati equation. Proofs are omitted.

LEMMA 8.1
Consider a Hamiltonian matrix H defined in (8.6).
If $H \in$ *Dom* (Ric) and $X = Ric(H)$, then the following are true.

i) X is symmetric.
ii) X is the solution of $A'X + XA + X'RX + Q = 0$.
iii) $A + RX$ is stable.

The above lemma essentially spells out the properties of the solution matrix X

LEMMA 8.2
If i) H has no eigenvalues on the imaginary axis.
ii) R is either positive or negative semidefinite.
iii) (A, R) is stabilizable.

Then $H \in$ *Dom* (Ric).
This lemma suggests an alternative way of checking whether $H \in$ *Dom* (Ric) or not.

Let B and C be full rank matrices with

$$R = BB' \text{ with rank of } R = \text{ rank of } B$$
$$Q = C'C \text{ with rank of } Q = \text{ rank of } C$$

Now define H as

$$H = \begin{bmatrix} A & BB' \\ -C'C & -A' \end{bmatrix} \tag{8.8}$$

We have then the following result

LEMMA 8.3
Suppose H is as given in (8.8) with (A, B) stabilizable and (C, A) detectable. Then

$$H \in Dom \text{ (Ric) and } X = Ric(H) \geq 0 \quad \text{i.e. } X \text{ is positive semidefinite.}$$

further, if (C, A) is observable, then

$$X = Ric(H) > 0 \quad \text{i.e. } X \text{ is positive definite.}$$

The above lemma shows that if certain assumptions are satisfied we have an easy way of checking whether $H \in Dom$ (Ric). Moreover the solution X exhibits sign definite properties.

For proofs of the above lemmas, the reader may consult Kucera (1972) and Potter (1966). These references also contain some earlier results which throw light on the nature of solutions of Riccati equation. They are reproduced below for easy reference. The results are stated with reference to the Riccati equation

$$A'X + XA + XBB'X + C'C = 0 \tag{8.9}$$

In this connection Kalman (1960, 1963, 1964, 1965) has proved the following *sufficiency conditions.*

1. If (A, B) is controllable, then there exists *a solution* $X \geq 0$ for (8.9).
2. If (A, B) is controllable and (C, A) is observable then there exists *a unique solution* $X > 0$ for (8.9) and further it is a *stabilizing solution* in the sense that $A + BB'X$ is asymptotically stable. Stronger results were obtained by Wonham (1968) using weaker hypothesis. These results are as follows.
3. If (A, B) is stabilizable, then there exists *a solution* $X \geq 0$ to (8.9).
4. If (A, B) is stabilizable and (C, A) detectable, then there exists *a solution* $X \geq 0$ to (8.9). This solution is also a *stabilizing solution.*
5. If (A, B) is stabilizable and (C, A) observable, then there exists a *unique solution* $X > 0$ to (8.9) which is also a *stabilizing solution.*

Necessary and sufficient conditions for the solution of Riccati equation were first reported by Kucera (1972). They are as follows. The Hamiltonian matrix referred to in the lemmas have the following structure viz.

$$H = \begin{bmatrix} A & BB' \\ -C'C & -A' \end{bmatrix}$$

LEMMA 8.4
The following conditions are equivalent.

(i) H has no eigen values on the imaginary axis and (A, B) is stabilizable.
(ii) There exists *a solution* $X \geq 0$ and $A + BB'X$ is *asymptomatically stable.*

LEMMA 8.5
Given that H has no eigen values on the imaginary axis and (A, B) stabilizable, the following conditions are equivalent.

(i) There exists a *unique solution* $X \geq 0$.
(ii) (C, A) detectable.

LEMMA 8.6
The following conditions are equivalent.
(i) (A, B) stabilizable and (C, A) detectable.
(ii) There exists a *unique solution* $X \geq 0$ and $(A + BB'X)$ is *asymptotically stable.*

8.2.3 The Solution of the LQG Problem

In Section 8.2.1 we posed the LQG problem and indicated how the separation principle can be applied to solve it. In this section we go into more detail regarding the solution of the problem. We first assume that the disturbance and noise are not present. We then have the deterministic problem

$$\dot{x} = Ax + Bu$$
$$y = Cx$$
$$z = Lx$$

and the performance index to be minimized is

$$J_1 = \int_0^\infty (x'L'QLx + u'Ru)\mathrm{d}t$$

We are seeking an optimal u^* which minimizes J_1.

It can be shown that

$$u^* = Fx \text{ where } F = -R^{-1}B'X$$

and X satisfied the Riccati equation

$$A'X + XA - XBR^{-1}B'X + L'QL = 0 \tag{8.10}$$

Now reverting to the optimal estimation problem let $\hat{x}$ be the optimal estimate of x in the sense that

$$J_2 = \lim_{T\to\infty} \mathcal{E}\left\{\frac{1}{T}\int_0^T (\hat{x}-x)'(\hat{x}-x)dt\right\}$$

is minimized subject to the Kalman filter equation

$$\dot{\hat{x}} = A\hat{x} + Bu + H(C\hat{x} - y)$$

being satisfied.

The solution of this filter problem enables us to obtain the Kalman filter gain matrix H. This is expressed as

$$H = -YC'V^{-1}$$

where Y is obtained as the solution of the Riccati equation

$$AY + YA' - YC'V^{-1}CY + MWM' = 0 \tag{8.11}$$

The solution of the LQG problem thus involves the solution of two matrix Riccati equations. In turn they determine the control gain matrix F and the filter gain matrix H.

The filter equations may now be rewritten as

$$\dot{\hat{x}} = A\hat{x} + Bu + H(C\hat{x} - y)$$

$$\text{But } u = F\hat{x}$$

$$\text{Hence } \dot{\hat{x}} = (A + BF + HC)\hat{x} - Hy$$

The stabilizing controller $\boldsymbol{K}$ with input y and output u has a realization

$$\boldsymbol{K} : (A + BF + HC, \quad -H, \quad F)$$

A schematic representation of the dynamic compensator is given in Figure 8.1.

To recapitulate, the two relevant Riccati equations are

$$A'X + XA - XBR^{-1}B'X + L'QL = 0$$

$$AY + YA' - YC'V^{-1}CY + MWM' = 0$$

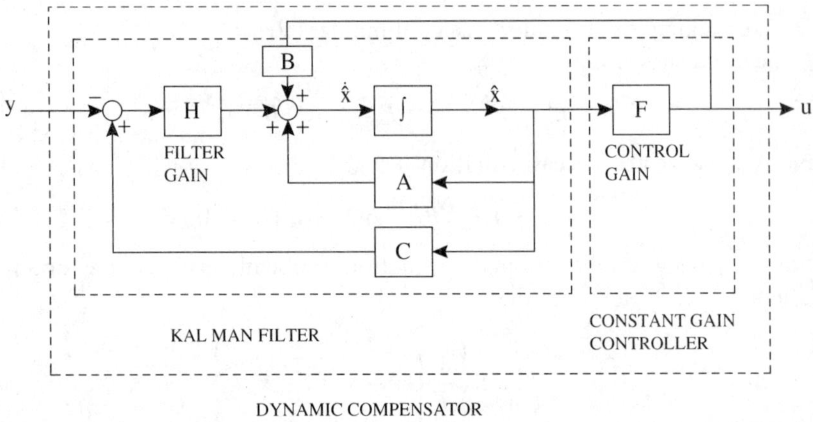

Figure 8.1 Dynamic compensator for the LQG problem.

We also note that $R > 0$; $Q \geq 0$; $V > 0$; $W \geq 0$. The respective Hamiltonian matrices associated with the two Riccati equations are

$$H_1 = \begin{bmatrix} A & -BR^{-1}B' \\ -L'QL & -A' \end{bmatrix}; \quad H_2 = \begin{bmatrix} A' & -C'V^{-1}C \\ -MWM' & -A \end{bmatrix}$$

Now apply Lemma 8.6 to the first Riccati equation. In order that a unique positive semi definite solution may exist, which also provides internal stability $(A, BR^{-\frac{1}{2}})$ has to be stabilizable and $(Q^{\frac{1}{2}}L, A)$ has to be detectable.

Since R is positive definite so is $R^{-\frac{1}{2}}$ and hence $(A, BR^{-\frac{1}{2}})$ stabilizable is implied by (A, B) being stabilizable. The same simplification cannot be effected for $(Q^{\frac{1}{2}}L, A)$ because $Q^{\frac{1}{2}}$ is only positive semidefinite.

Hence we surmise that for a unique internally stabilizing solution to exist for the first Riccati equation, $(A, B, Q^{\frac{1}{2}}L)$ be stabilizable and detectable. Using the duality principle we can similarly deduce that for a unique internally stabilizing solution to exist for the second Riccati equation $(A, MW^{\frac{1}{2}}, C)$ be stabilizable and detectable

8.2.4 Robustness Properties of the LQG Solution

Safonov and Athans (1977) showed that the Liner Quadratic Regulator (LQR) exhibits very good robustness properties with a gain margin in the range $[\frac{1}{2}, \infty)$ and a phase margin of at least 60° at the plant input. By robustness we mean that any unstructured plant perturbation within an bound does not destabilize the plant. Thus, provided the state is available

LQ regulator is preferable to state feedback pole assignment technique from the standpoint of robustness as in the latter case no robustness guarantees are available. However Doyle (1978) demonstrated that the LQG regulators unlike LQ regulators have no intrinsic robustness properties and can at times exhibit poor stability margins. This is one of the reasons why $\mathcal{H}_\infty$ regulator with its known property of robustness is preferred to LQG regulator in certain control applications.

8.3 $\mathcal{H}_\infty$ Optimization Techniques

The central problem in $\mathcal{H}_\infty$ optimization is the design of a stabilizing controller that minimizes or at least imposes an upper bound on the $\mathcal{H}_\infty$ norm of the closed loop transfer function. As shown in Chapter 6 and further elaborated in Chapter 9, most of the $\mathcal{H}_\infty$ optimization problems could be reduced to the so called model-matching problem. The solution techniques available for solving the model matching problem can be listed as follows:

1. The classical function-theoretic approach of Nevanlinna and Pick (Delsarte *et al.* 1979 and Delsarte *et al.* 1981)
2. The operator-theoretic methods of Sarason (1967), Adamjan *et al.* (1978) and Ball and Helton (1983)
3. State space method in which the model matching problem is first reduced to Nehari's problem and then solved by Glover's (1984) method.
4. Statespace solution by Doyle *et al.* (1989) using the separation principle approach.

The method cited under item 3 above does indeed provide a complete solution to the 4-block problem. However the associated computational effort appears daunting. The method listed under item 4 seems to be as of now, the most attractive method to solve $\mathcal{H}_\infty$ optimization problems. This method is discussed in Chapter 10. Glover's method listed under item 3, has already been explained in Chapter 7. Therefore in this chapter, we shall confine our attention to the two methods listed under items 1 and 2.

8.4 The Nevanlinna–Pick Interpolation Problem

The NP problem is solved by following a frequency domain approach. The method is conceptually clear and computationally simple. Further, it can be tailored to solve problems involving irrational functions which are not covered by state space methods.

One of the classic problems in interpolation theory is the interpolation of a finite set of points in the interior of a unit disk U by a function analytic and bounded in U. A criterion for the existence of such a function was proposed by Pick as early as 1916. Later Nevanlinna in 1929 proposed an algorithm for the construction of such a function, if it exists. The significance of this result in relation to engineering problems was realized much later. In control theory, applications of NP approach started appearing in the eighties (Chang and Pearson 1984, Kimura 1984, Vidyasagar and Kimura 1986). Since then several modifications to the method have been suggested.

In this chapter, we first discuss the scalar version of the NP problem. The matrix version is taken up subsequently.

8.4.1 NP Problem – Scalar Case

Problem Statement

Given n pairs of complex numbers $(\alpha_i, \beta_i) i = 1, 2, \ldots, n$ with $|\alpha_i| < 1$ and $|\beta_i| < 1$, find an analytic function ϕ, analytic in the closed unit disk U and of modulus not greater than unity in U, satisfying the following interpolation conditions:

$$\phi(\alpha_i) = \beta_i \quad i = 1, 2, \ldots, n$$

Remarks

1. The points $(\alpha_1, \alpha_2 \ldots \alpha_n)$ need not be distinct. If the points are not distinct, them the problem could still be solved at the expense of extra algebraic manipulation.
2. A rational function $\phi(z)$ analytic in U and having a modulus nc greater than unity in U is said to belong to *Schur* class of functions
3. There is no guarantee that a Schur function satisfying the giver interpolation constraints exists. Such a function exists if and only i the Pick matrix:

$$P = [P_{ij}] \text{ where } P_{ij} = \frac{1 - \overline{\beta}_i \beta_j}{1 - \overline{\alpha}_i \alpha_j} \quad i, j = 1, 2, \ldots, n$$

is positive definite if $\|\phi\|_\infty < 1$ and positive semi definite i $\|\phi\|_\infty \leq 1$. This is known as Pick's theorem.

4. If the region of interest is the closed right half plane, then the NP problem may be posed as follows.

   Obtain an analytic function $\phi(s)$, analytic in the closed right half plane such that

$$\phi(\alpha_i) = \beta_i \qquad i = 1, 2, \ldots, n$$
$$\|\phi\|_\infty \leq 1$$

where $(\alpha_1, \alpha_2, \ldots, \alpha_n)$ are a set of points in the open right half plane. The $|\beta_i|$s have necessarily to be equal to or less than unity. Otherwise the maximum modulus theorem would be violated. Even in the special case of $|\beta_i| = 1$ for some i, by applying the above mentioned theorem, $|\phi(s)| \equiv 1$ which implies that ϕ is a constant throughout the right half plane and no interpolation is possible. Hence $|\beta_i|$ is constrained to be less than unity for $i = 1, 2, \ldots, n$. The Pick matrix for the problem is

$$P = [P_{ij}] \text{ where } P_{ij} = \frac{1 - \beta_i \overline{\beta}_j}{\alpha_i + \overline{\alpha}_j} \qquad i, j = 1, 2, \ldots, n$$

for $\|\phi\|_\infty \leq 1$, NP problem is solvable if and only if $P \geq 0$.
for $\|\phi\|_\infty < 1$, NP problem is solvable if and only if $P > 0$.

Solution of the NP Problem

Let $\phi(z)$ be a Schur function with

$$\phi(\alpha_i) = \beta_i \quad i = 1, 2, \ldots, n \text{ and } |\alpha_i| < 1 \text{ and } |\beta_i| < 1 \quad i = 1, 2, \ldots, n$$

for notational convenience denote $\phi(z)$ by $\phi_1(z)$ and β_i by β_i^1, $i = 1, 2, \ldots, n$. In the new notation we have

$$\phi_1(\alpha_i) = \beta_i^1 \quad i = 1, 2, \ldots, n$$

By assumption, $\phi_1(z)$ belongs to the Schur class of functions. Let T be a transformation such that

$$\phi_2(z) = T[\phi_1(z)] \tag{8.12}$$

The transformation T is assumed to possess an inverse such that

$$\phi_1(z) = T^{-1}\phi_2(z) \tag{8.13}$$

Further the transformation is such that if ϕ_1 belongs to Schur class of functions, so is ϕ_2 and vice versa. Details of T will be spelt out at a later stage. We shall presently see that $\phi_1(z)$ satisfies the interpolation pair (α_1, β_1^1) independent of the value of the function $\phi_2(z)$ at α_1. We have $\phi_1(\alpha_1) = \beta_1^1$. Note that $\phi_1(z)$ satisfies the rest of the interpolation pairs

$(\alpha_2, \beta_2^1), (\alpha_3, \beta_3^1) \ldots (\alpha_n, \beta_n^1)$ in an indirect manner through the assignment of suitable interpolation pairs for $\phi_2(z)$. Thus

$$\phi_2(\alpha_i) = \beta_i^2 \quad i = 2, 3, \ldots n$$

The $(n-1)$ interpolation pairs associated with $\phi_2(z)$ are $(\alpha_2, \beta_2^2), (\alpha_3, \beta_3^2) \ldots (\alpha_n, \beta_n^2)$. Proceeding in a similar fashion

$$\phi_3(z) = T[\phi_2(z)] \text{ and } \phi_2(z) = T^{-1}[\phi_3(z)]$$

Every successive transformation, results in the number of interpolation pairs getting reduced one at a time. Finally we have

$$\phi_n(z) = T[\phi_{n-1}(z)]$$

Satisfying the conditions

$$\phi_n(\alpha_n) = \beta_n^n \tag{8.14}$$

Assuming that we know how to calculate β_n^n the problem is how to find a function belonging to Schur class such that (8.14) is satisfied. There are infinitely many functions from which we may make a choice. The simplest choice will be to assume $\phi_n(z)$ a constant function equal to β_n^n. Knowing $\phi_n(z)$ we may now proceed backwards and by making use of the inverse transformation T^{-1} successively generate $\phi_{n-1}(z), \phi_{n-2}(z) \ldots \phi_1(z)$.

The required function $\phi_1(z)$ solves the given NP problem. Note that $\phi_1(z)$ is not unique as it ultimately depends upon the choice of $\phi_n(z)$ which is not unique.

In the above presentation, we have assumed that $|\beta_i^i| < 1$. For $i = 1, 2, \ldots n$. This is the so called *non degenerate* condition for which NP problem has infinitely many solutions. If for some k we have $|\beta_k^k| = 1$ then we have the *degenerate case* and the method described above needs modification. The reader may refer to Delsarte *et al.* (1981) for details.

We now proceed to examine in more detail the structure of the transformation T involved. We have

$$\phi_1(z) = \frac{\beta_1^1(1 - \overline{\alpha}_1 z) + (z - \alpha_1)\phi_2(z)}{(1 - \overline{\alpha}_1 z) + \overline{\beta}_1^1(z - \alpha_1)\phi_2(z)} \tag{8.15}$$

It is readily verified that

$$\phi_1(\alpha_1) = \beta_1^1$$

Inverting (8.15) we have

$$\phi_2(z) = \frac{1 - \overline{\alpha}_1 z}{z - \alpha_1} \; \frac{\phi_1(z) - \beta_1^1}{1 - \overline{\beta}_1^1 \phi_1(z)}$$

The associated interpolation pairs$(\alpha_2, \beta_2^2) \dots (\alpha_n, \beta_n^2)$ could be obtained from

$$\phi_2(\alpha_i) = \beta_i^2 = \frac{1-\bar{\alpha}_1\alpha_i}{\alpha_i - \alpha_1} \quad \frac{\beta_i^1 - \beta_1^1}{1-\overline{\beta}_1^1\beta_i^1} \qquad i = 2, 3, \dots, n$$

From the known values of β_i^1 on the right hand side of the above equation, $\beta_2^2, \beta_3^2, \dots \beta_n^2$ could be readily calculated.

Generalizing the above relationship we have

$$\beta_i^k = \frac{1-\bar{\alpha}_{k-1}\alpha_i}{\alpha_i - \alpha_{k-1}} \cdot \frac{\beta_i^{k-1} - \beta_{k-1}^{k-1}}{1-\overline{\beta}_{k-1}^{k-1}\beta_i^{k-1}} \qquad i = k, k+1, \dots, n \tag{8.16}$$

For notational convenience β_k^k may be represented by γ_k. We now have a general expression for $\phi_k(z)$ in terms of $\phi_{k+1}(z)$

$$\phi_k(z) = \frac{\gamma_k(1-\bar{\alpha}_k z) + (z-\alpha_k)\phi_{k+1}(z)}{(1-\overline{\alpha_k} z) + \bar{\gamma}_k(z-\alpha_k)\phi_{k+1}(z)} \cdots \tag{8.17}$$

Knowing $\phi_n(z)$, using (8.17) we may work backwards and generate $\phi_{n-1}(z), \dots, \phi_1(z)$.

8.4.2 NP Problem – Matrix Version

Problem Statement

Let $(\alpha_1, \alpha_2 \dots \alpha_n)$ be distinct complex numbers and $(F_1, F_2, \dots F_n)$ be complex matrices satisfying the conditions

$$|\alpha_i| < 1 \text{ and } \|F_i\| < 1 \quad i = 1, 2, \dots, n$$

We require an analytic function Φ analytic in the closed unit disk, with $\|\Phi\|_\infty < 1$ and satisfying the conditions

$$\Phi(\alpha_i) = F_i \quad i = 1, 2, \dots, n$$

Such a function Φ exists if and only if the Pick matrix defined below is positive definite

$$P = [P_{ij}] \text{ where } P_{ij} = \frac{1}{1-\bar{\alpha}_i\alpha_j}(I - F_i^* F_j) \quad i = 1, 2, \dots, n$$

Solution

The principal steps in the matrix case are similar to the steps taken for the scalar case. Following Chang and Pearson (1984) we define a transformation matrix L which is a function of a constant matrix E

with $\overline{\sigma}[E] < 1$. By square root of a matrix we mean the Hermitian square root. For example $M^{\frac{1}{2}}$ is the Hermitian square root of M, i.e. $M = RR^*$ where R denotes $M^{\frac{1}{2}}$. We now define $L(E)$ as follows

$$L(E) = \begin{bmatrix} (I - EE^*)^{\frac{1}{2}} & -(I - EE^*)^{-\frac{1}{2}}E \\ -(I - E^*E)^{-\frac{1}{2}}E^* & (I - E^*E)^{-\frac{1}{2}} \end{bmatrix} = \begin{bmatrix} A & B \\ C & D \end{bmatrix} \tag{8.18}$$

Now define a Linear Fractional Transformation (LFT) as follows

$$\Gamma_{L(E)} = X \to (AX + B)(CX + D)^{-1}$$

Also define

$$y(\zeta, z) = \begin{cases} \dfrac{\zeta - z}{1 - \overline{\zeta} z} \cdot \dfrac{|\zeta|}{\zeta} & \text{if } \zeta \neq 0 \\ -z & \text{if } \zeta = 0 \end{cases} \tag{8.19}$$

We have

$$\Gamma_{L(E),y(\zeta,z)} : X \to Y = y(\zeta, z)^{-1}(AX + B)(CX + D)^{-1} \tag{8.20}$$

and the inverse transformation

$$\Gamma^{-1}_{L(E),y(\zeta,z)} : Y \to X = [y(\zeta, z)YC - A]^{-1}[B - y(\zeta, z)YD] \tag{8.21}$$

The NP algorithm is implemented via the following steps.

Step 1: Represent $\Psi(z)$, the solution of the NP problem by $\Psi_1(z)$ and denote $F_1, F_2 \ldots F_n$ by $F_1^1, F_2^1, \ldots F_n^1$.

$$\text{Set } \Psi_2(z) = \Gamma_{L(F_1^1),y(\alpha_1,z)}\,[\Psi_1(z)]$$

Note that if $\overline{\sigma}(F_1^1) < 1$ then $\Psi_2(z)$ is a function belonging to the Schur class if and only if $\Psi_1(z)$ belongs to the same class and $\Psi_1(\alpha_1) = F_1^1$

$$\text{Define } \Psi_2(\alpha_i) = F_i^2 \quad i = 2, 3, \ldots, n$$

Step 2: Set $\Psi_3(z) = \Gamma_{L(F_2^2),y(\alpha_2,z)}\,[\Psi_2(z)]$

$$\text{and define } \Psi_3(\alpha_i) = F_i^3 \quad i = 3, 4, \ldots n$$

Step 3: Continue with the iteration and obtain

$$\Psi_k(z) = \mathbf{\Gamma}_{L(F_{k-1}^{k-1}),y(\alpha_{k-1},z)}\,[\Psi_{k-1}(z)]$$

and

$$F_i^k = \Psi_k(\alpha_i) \quad i = k, k+1, \ldots n$$

Step 4: Finally we have

$$\Psi_n(z) = \mathbf{\Gamma}_{L(F_{n-1}^{n-1}),y(\alpha_{n-1},z)}\,[\Psi_{n-1}(z)]$$

and

$$F_n^n = \Psi_n(\alpha_n)$$

The problem is thus ultimately reduced to that of finding a function $\Psi_n(z)$ belonging to the Schur Class with $\Psi_n(\alpha_n) = F_n^n$. The most direct choice, in this case will be to assume $\Psi_n(z)$ to be a constant matrix equal to F_n^n.

Step 5: Proceeding backwards, we obtain

$$\Psi_{n-1}(z) = \mathbf{\Gamma}^{-1}_{L(F_{n-1}^{n-1}),y(\alpha_{n-1},z)}\,[\Psi_n(z)]$$

$$\Psi_{k-1}(z) = \mathbf{\Gamma}^{-1}_{L(F_{k-1}^{k-1}),y(\alpha_{k-1},z)}\,[\Psi_k(z)]$$

$$\Psi_1(z) = \mathbf{\Gamma}^{-1}_{L(F_1^1),y(\alpha_1,z)}\,[\Psi_2(z)]$$

where the inverse transformations are implemented using (8.21). As in the scalar case, degeneracy may set in if for some k, $\overline{\sigma}[F_k^k] = 1$. The method has then to be modified as explained in Chang and Pearson (1984). Even though the NP algorithm in the matrix case looks computationally formidable in reality it is not so. Steps 1 to 4 involve nothing more than matrix multiplication and step 5 involves multiplication and addition of polynomial matrices.

8.4.3 Solution of Model Matching Problem

In this section we explain how a typical model matching problem can be cast in the NP format. This has already been elaborated in Chapter 6, in the context of sensitivity minimization and robust specialization. Here we explore how sub-optimal controllers could be generated for the model matching problem using the NP method.

Consider the model matching problem, where we are seeking sub-optimal stabilizing controllers which will result in

$$\|\boldsymbol{T}_1 - \boldsymbol{T}_2\boldsymbol{Q}\boldsymbol{T}_3\|_\infty < \gamma \tag{8.22}$$

where $\boldsymbol{T}_1, \boldsymbol{T}_2, \boldsymbol{T}_3$ and $\boldsymbol{Q}$ belong to $\mathcal{RH}_\infty$ and γ is chosen such that $\gamma > \mu$, where

$$\mu = \operatorname*{Inf}_{\boldsymbol{Q}\in\mathcal{H}_\infty} \|\boldsymbol{T}_1 - \boldsymbol{T}_2\boldsymbol{Q}\boldsymbol{T}_3\|_\infty$$

From (8.22) we have

$$\left\|\frac{1}{\gamma}(\boldsymbol{T}_1 - \boldsymbol{T}_2\boldsymbol{Q}\boldsymbol{T}_3)\right\|_\infty < 1$$

Let $\Psi = \frac{1}{\gamma}\boldsymbol{T}_1 - \frac{1}{\gamma}\boldsymbol{T}_2\boldsymbol{Q}\boldsymbol{T}_3$. We assume for the present that $\boldsymbol{T}_2$ and $\boldsymbol{T}_3$ are inner square matrices. Let $\delta_1 = \det \boldsymbol{T}_2$ and $\delta_2 = \det \boldsymbol{T}_3$. We note that δ_1 and δ_2 are inners and hence $\delta = \delta_1\delta_2$ is also an inner. Further adj $\boldsymbol{T}_i$ for $i = 2$ and 3 are inners. Pre multiplying $\boldsymbol{\Phi}$ with adj $\boldsymbol{T}_2$ and postmultiplying it by adj $\boldsymbol{T}_3$ and noting that adj $\boldsymbol{T}_i = \frac{\boldsymbol{T}_i^{-1}}{det\boldsymbol{T}_i} (i = 2, 3)$ we have

$$\begin{aligned}\|\Psi\|_\infty &= \|\text{adj}\,\boldsymbol{T}_2\Psi\text{adj}\,\boldsymbol{T}_3\|_\infty \\ &= \left\|\frac{1}{\gamma}(\text{adj}\,\boldsymbol{T}_2)\boldsymbol{T}_1(\text{adj}\,\boldsymbol{T}_3) - \frac{\delta}{\gamma}\boldsymbol{T}_2^{-1}\boldsymbol{T}_2\boldsymbol{Q}\boldsymbol{T}_3\boldsymbol{T}_3^{-1}\right\|_\infty \\ &= \left\|\frac{1}{\gamma}\boldsymbol{A} - \frac{\delta}{\gamma}\boldsymbol{Q}\right\|_\infty\end{aligned}$$

where $\boldsymbol{A} = (\text{adj}\,\boldsymbol{T}_2)\boldsymbol{T}_1(\text{adj}\,\boldsymbol{T}_3)$.

Let $(\alpha_1, \alpha_2 \ldots \alpha_m)$ be the distinct zeros of δ in the open right half plane by virtue of δ being an inner. As already explained in Chapter 6, the stability of $\boldsymbol{Q}$ is assured if

$$\Psi(\alpha_i) = \frac{1}{\gamma}\boldsymbol{A}(\alpha_i) \quad i = 1, 2, \ldots m$$

for in that case there could be no cancellation of the zeros of δ with the unstable poles of $\boldsymbol{Q}$ and thus making $\delta\boldsymbol{Q}$ stable without $\boldsymbol{Q}$ being stable. We are therefore seeking a function $\Psi \in \mathcal{RH}_\infty$ such that $\|\boldsymbol{\Psi}\|_\infty < 1$ and $\boldsymbol{\Psi}(\alpha_i) = \frac{1}{\gamma}A(\alpha_i) = i = 1, 2, \ldots m.$

Form the Pick matrix for the above interpolation problem. If the Pick matrix is not positive definite, the γ chosen is less than μ and hence the value of γ has to be increased. Through iteration we may obtain a value of γ sufficiently close to μ. The function $\boldsymbol{\Psi}$ may now be determined using the NP algorithm and we may proceed on to compute $\boldsymbol{Q}$ and the sub-optimal stabilizing controller $\boldsymbol{K}$.

8.5 Operator-Theoretic Methods

Operator-theoretic methods are important on two counts. Firstly, they offer an intutive geometric insight into the problem, which is extremely useful while searching for solutions. Secondly, they are much more general than e.g. state space method, currently the most widely used method for solving $\mathcal{H}_\infty$ optimization problems. Because of this generality, their future application potential is great. Unfortunately, operator-theoretic methods are not very popular among practising engineers because of their unfamiliarity with the mathematical tools used. In this section we restrict ourselves to an elementary exposition of the topic enough to convey some of the rich flavour associated with operator-theoretic methods.

8.5.1 $\mathcal{H}_\infty$ Optimization Problem – Scalar Case

The model matching problem was discussed at some length in Chapter 6. For the SISO case the problem may be stated as follows:

$$\min_{q \in \mathcal{H}_\infty} \|\boldsymbol{t}_1 - \boldsymbol{t}_2\boldsymbol{q}\|_\infty$$

where $\boldsymbol{t}_1$ and $\boldsymbol{t}_2$ belong to $\mathcal{RH}_\infty$. To solve the problem factorize $\boldsymbol{t}_2 = (\boldsymbol{t}_2)_i(\boldsymbol{t}_2)_0$

Set $(\boldsymbol{t}_2)_i = \boldsymbol{\delta}(s)$ where $\boldsymbol{\delta}(s)$ has simple zeros $(\alpha_1, \alpha_2, \ldots, \alpha_m)$ in Re $s > 0$. Assuming that $(t_2)_0$ has no zeros on the imaginary axis we may replace $(\boldsymbol{t}_2)_0\boldsymbol{q}$ by a free parameter $\boldsymbol{q}$ (because the inverse of $(t_2)_0$ exists in $\mathcal{H}_\infty$). The model matching problem hence reduces to $\min_{q \in \mathcal{H}_\infty} \|\boldsymbol{f} - \boldsymbol{\delta}\boldsymbol{q}\|_\infty$ where $\boldsymbol{f} \in \mathcal{RH}_\infty$.

Let $\delta\mathcal{H}_2$ be a subspace of $\mathcal{H}_2$ defined as

$$\delta\mathcal{H}_2 = \{\boldsymbol{\delta h} : \boldsymbol{h} \in \mathcal{H}_2\}$$

Let $\mathcal{K}$ be the orthogonal complement of $\delta\mathcal{H}_2$ in $\mathcal{H}_2$. We define $\mathcal{K}$ as

$$\mathcal{K} = \{\boldsymbol{h} \in \mathcal{H}_2 : \langle \boldsymbol{h}, \boldsymbol{\delta g}\rangle = 0, \forall \boldsymbol{g} \in \mathcal{H}_2\} \tag{8.23}$$

As a consequence

$$\mathcal{H}_2 = \mathcal{K} \oplus \delta\mathcal{H}_2 \tag{8.24}$$

Now define a bounded linear operator $\mathbf{\Lambda}_f$ on $\mathcal{K}, \mathbf{\Lambda}_f : \mathcal{K} \to \mathcal{K}$ and for any $k \in \mathcal{K}$ we have

$$\mathbf{\Lambda}_f(k) = \mathbf{\Pi}(fk)$$

where $\mathbf{\Pi}$ is an orthogonal projection operator mapping $\mathcal{H}_2$ on to $\mathcal{K}$. In fact we may write

$$\mathbf{\Lambda}_f = \mathbf{\Pi} M_f | \mathcal{K}$$

where M_f is the multiplication operator. $\mathbf{\Lambda}_{\hat{f}}$ therefore denotes an operator which is a restriction of $\mathbf{\Pi} M_f$ to $\mathcal{K}$. A result proved by Sarason (1967) regarding the operator $\mathbf{\Lambda}_f$ states that

$$\min_{q \in \mathcal{H}_\infty} \|\boldsymbol{f} - \boldsymbol{\delta q}\|_\infty = \|\mathbf{\Lambda}_{\hat{f}}\| \tag{8.25}$$

Recall that in the case of a multiplication operator $M_{\boldsymbol{h}} : \mathcal{H}_2 \to \mathcal{H}_2$ we have

$$\|M_{\boldsymbol{h}}\| = \|\boldsymbol{h}\|_\infty$$

Bearing this in mind we have

$$\|M_{\mathbf{f}-\delta q}\| = \|\mathbf{f} - \boldsymbol{\delta q}\|_\infty$$

$$\min_{\mathbf{q} \in \mathcal{H}_\infty} \|M_{\mathbf{f}-\delta q}\| = \min_{\mathbf{q} \in \mathcal{H}_\infty} \|\mathbf{f} - \boldsymbol{\delta q}\|_\infty$$

It therefore follows from (8.25) that

$$\begin{aligned} \min_{q \in \mathcal{H}_\infty} \|\boldsymbol{f} - \boldsymbol{\delta q}\|_\infty &= \min_{q \in \mathcal{H}_\infty} \|M_{f-\boldsymbol{\delta q}}\| \\ &= \|\mathbf{\Lambda}_f\| \end{aligned} \tag{8.26}$$

As $\boldsymbol{q}$ varies over $\mathcal{H}_\infty$, the operator $M_{f-\boldsymbol{\delta q}}$ ranges over different operators. However the operator $\mathbf{\Pi} M_{f-\boldsymbol{\delta q}}$ is the same for all q and in fact is equal to $\mathbf{\Pi} M_f$. This is shown as follows. Let $\boldsymbol{g} \in \mathcal{H}_2$. We have

$$\begin{aligned} \mathbf{\Pi} M_{f-\boldsymbol{\delta q}}(\boldsymbol{g}) &= \mathbf{\Pi}(f\boldsymbol{g} - \boldsymbol{\delta q g}) \\ &= \mathbf{\Pi}(f\boldsymbol{g}) - \mathbf{\Pi}(\boldsymbol{\delta q g}) \\ &= \mathbf{\Pi}(f\boldsymbol{g}) \\ &= \mathbf{\Pi} M_f(\boldsymbol{g}) \end{aligned} \tag{8.27}$$

In the above derivation, we make use of the fact that $\boldsymbol{\delta q g} \in \delta\mathcal{H}_\infty$ which is orthoganal to $\mathcal{K}$. Hence

$$\mathbf{\Pi}(\boldsymbol{\delta q g}) = 0$$

Sarason's result may be interpreted as follows:

As $\boldsymbol{q}$ is varied over $\mathcal{H}_\infty$, the operator $M_{f-\delta q}$ ranges over various operators. The minimum of $\|M_{f-\delta q}\|$ as $\boldsymbol{q}$ is varied, over $\mathcal{H}_\infty$ is given by $\|\Lambda_f\|$. By the definition of operator norm

$$\|\Lambda_f\| = \max\{\|\Lambda_f(\boldsymbol{k})\|_2 : \boldsymbol{k} \in \mathcal{K}, \|\boldsymbol{k}\|_2 = 1\} \tag{8.28}$$

The operator Λ_f has for its domain and range the sub space $\mathcal{K}$, which we will presently show in finite dimensional, its dimension being equal to the number of zeros of δ. Hence the matrix representation of the operator with reference to a chosen basis in $\mathcal{K}$ is also finite dimensional. This simplifies considerably the calculation of the norm of the operator Λ_f.

MATRIX REPRESENTATION OF THE OPERATOR Λ_f

To begin with we have to identify a set of basis functions in $\mathcal{K}$ which span the space. Such a set is given by the m functions $\dfrac{1}{s+\overline{\alpha}_i}$ $(i = 1, 2, \ldots m)$. To prove that they are indeed basis functions which span the space $\mathcal{K}$, we invoke *Cauchy's integral formula* which may be stated as follows:
Let $\boldsymbol{f}$ be analytic on a non intersecting closed contour $\mathcal{D}$ and also in its interior. Let α be a point in the interior region of $\mathcal{D}$. Then

$$\frac{1}{2\pi j}\oint_{\mathcal{D}} \frac{\boldsymbol{f}(s)}{(s-\alpha)}\, ds = \boldsymbol{f}(\alpha) \tag{8.29}$$

In Hilbert space $\mathcal{H}_2$, the inner product is defined as

$$\langle \boldsymbol{\Phi}, \boldsymbol{g}\rangle = \frac{1}{2\pi}\int_{-\infty}^{\infty} \boldsymbol{\Phi}^*(j\omega)\boldsymbol{g}(j\omega)d\omega$$

for $\boldsymbol{\Phi}, \boldsymbol{g} \in \mathcal{H}_2$

If $\boldsymbol{\Phi} = \dfrac{1}{s+\overline{\alpha}}$ than it can be shown using Cauchy's integral formula that for $\boldsymbol{g} \in \mathcal{H}_2$ and Re$\alpha > 0$

$$\langle \frac{1}{s+\overline{\alpha}}, \boldsymbol{g}\rangle = \boldsymbol{g}(\alpha)$$

Now δ has m right half plane zeros $\alpha_1, \alpha_2 \ldots \alpha_m$.
Hence for $\boldsymbol{h} \in \mathcal{H}_2$

$$\langle \frac{1}{s+\overline{\alpha}_i}, \delta \boldsymbol{h}\rangle \text{ is given by } \delta\ (\alpha_i)\boldsymbol{h}(\alpha_i) = 0 \text{ for } \quad i = 1.2.\ldots m$$

Therefore the m functions $\dfrac{1}{s+\overline{\alpha}_i}$, $i = 1, 2, \ldots m$ belong the $\mathcal{K}$ space. Further, it is easy to show that they are linearly independent. It remains to be shown that they span the $\mathcal{K}$ space.

Suppose $\boldsymbol{g} \in \mathcal{H}_2$ is orthogonal to every basis function $\dfrac{1}{s+\overline{\alpha}_i}$ $i = 1, 2, \ldots, m$. Then

$$\langle \frac{1}{s+\overline{\alpha}_i}, \boldsymbol{g} \rangle = 0 \quad i = 1, 2, \ldots, m$$

which implies that g has m zeros, $\alpha_1, \alpha_2, \ldots \alpha_m$ and hence $\boldsymbol{g} \in \delta\mathcal{H}_2$. Hence the orthogonal complement of the span of the basis functions $\dfrac{1}{s+\overline{\alpha}_i}$ belong to $\delta\mathcal{H}_2$ space which is what is required to be proved.

We also note that the chosen basis functions are not orthogonal because

$$\langle \frac{1}{s+\overline{\alpha}_i}, \frac{1}{s+\overline{\alpha}_j} \rangle = \frac{1}{\alpha_i + \overline{\alpha}_j} \neq 0 \text{ for } i \neq j$$

Having chosen a basis for $\mathcal{K}$, the next step is to obtain a matrix representation of the operator $\boldsymbol{\Lambda}_f$ with references to the chosen basis.

Let $\boldsymbol{k}_1, \boldsymbol{k}_2, \ldots \boldsymbol{k}_m$ be the chosen basis functions. We have $\boldsymbol{\Lambda}_f(\boldsymbol{k}_1) = (\boldsymbol{fk}_1)_{\mathcal{K}}$ where $(\boldsymbol{fk}_1)_{\mathcal{K}}$ means the projection of $(\boldsymbol{fk}_1)$ on to $\mathcal{K}$ space, i.e. $(\boldsymbol{fk}_1)_{\mathcal{K}} = \Pi(\boldsymbol{fk}_1)$.

Since $(\boldsymbol{fk}_1)_{\mathcal{K}}$ belongs to $\mathcal{K}$ space we have

$$(\boldsymbol{fk}_1)_{\mathcal{K}} = l_{11}\boldsymbol{k}_1 + l_{21}\boldsymbol{k}_2 \ldots + l_{m1}\boldsymbol{k}_m \tag{8.30}$$

Figure 8.2 gives a geometric interpretation of the action of $\boldsymbol{\Lambda}_f$ operator for a 2-dimensional $\mathcal{K}$ space.

In the figure, vector $\boldsymbol{fk}_1$ i.e. OA is projected on to the $\boldsymbol{k}_1 - \boldsymbol{k}_2$ space. This is denoted by OB. The vector OB has two components, OC and OD along $\boldsymbol{k}_1$ and $\boldsymbol{k}_2$ respectively. If $\mathcal{K}$ space is m dimensional, then

$$OB = l_{11}\boldsymbol{k}_1 + l_{21}\boldsymbol{k}_2 + \ldots + l_{m1}\boldsymbol{k}_m$$
$$OC = < \boldsymbol{k}_1, l_{11}\boldsymbol{k}_1 + \ldots + l_{m1}\boldsymbol{k}_m >$$
$$OD = < \boldsymbol{k}_2, l_{11}\boldsymbol{k}_1 + \ldots + l_{m1}\boldsymbol{k}_m >$$

An alternative way of obtaining OC and OD is by projecting OA directly on to $\boldsymbol{k}_1$ and $\boldsymbol{k}_2$ respectively. Thus

$$OC = < \boldsymbol{k}_1, l_{11}\boldsymbol{k}_1 + \ldots + l_{m1}\boldsymbol{k}_m > = < \boldsymbol{k}_1, \boldsymbol{fk}_1 >$$
$$OD = < \boldsymbol{k}_2, l_{11}\boldsymbol{k}_1 + \ldots + l_{m1}\boldsymbol{k}_m > = < \boldsymbol{k}_2, \boldsymbol{fk}_1 >$$

Form the square matrix

$$P = \begin{bmatrix} < \boldsymbol{k}_1, \boldsymbol{\Lambda}_f(\boldsymbol{k}_1) > & \ldots & < \boldsymbol{k}_1, \boldsymbol{\Lambda}_f(\boldsymbol{k}_m) > \\ \vdots & \vdots & \vdots \\ < \boldsymbol{k}_m, \boldsymbol{\Lambda}_f(\boldsymbol{k}_1) > & \ldots & < \boldsymbol{k}_m, \boldsymbol{\Lambda}_f(\boldsymbol{k}_m) > \end{bmatrix} \tag{8.31}$$

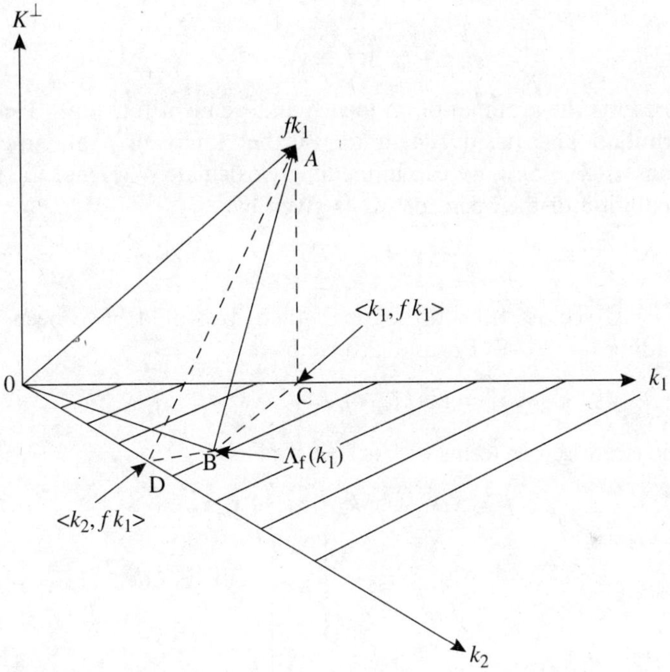

Figure 8.2 Geometric interpretation of Λ_f.

The first column of (8.31) represents the projections of $\boldsymbol{\Lambda}_f(\boldsymbol{k}_1)$ on $\boldsymbol{k}_1, \boldsymbol{k}_2, \ldots \boldsymbol{k}_m$ and the last column of (8.31) represents the projection of $\boldsymbol{\Lambda}_f(\boldsymbol{k}_m)$ on $\boldsymbol{k}_1, \boldsymbol{k}_2, \ldots \boldsymbol{k}_m$. As already explained an alternative way of obtaining P is by writing

$$N = \begin{bmatrix} <\boldsymbol{k}_1, \boldsymbol{fk}_1> & \ldots & <\boldsymbol{k}_1, \boldsymbol{fk}_m> \\ \vdots & \vdots & \vdots \\ <\boldsymbol{k}_m, \boldsymbol{fk}_1> & \ldots & <\boldsymbol{k}_m, \boldsymbol{fk}_m> \end{bmatrix} \tag{8.32}$$

We have $P = N$.
further P could be expressed as

$$P = \begin{bmatrix} <\boldsymbol{k}_1, \boldsymbol{k}_1> & \ldots & <\boldsymbol{k}_1, \boldsymbol{k}_m> \\ \vdots & \vdots & \vdots \\ <\boldsymbol{k}_m, \boldsymbol{k}_1> & \ldots & <\boldsymbol{k}_m, \boldsymbol{k}_m> \end{bmatrix} \begin{bmatrix} l_{11} & l_{1m} \\ \vdots & \vdots \\ l_{m1} & l_{mm} \end{bmatrix} \tag{8.33}$$

Denoting the matrices on the right hand side of (8.33) by M and L, we have $P = ML$.

Hence

$$ML = N \tag{8.34}$$

M represents the gramian of m linearly independent functions. Hence M is Hermitian and positive definite. Further knowing $\boldsymbol{f}$ and the basis functions $\boldsymbol{k}_1, \boldsymbol{k}_2 \ldots \boldsymbol{k}_m$, we can immediately calculate N. Hence the matrix representation of the operator $\boldsymbol{\Lambda}_f$ is given by

$$L = M^{-1}N$$

■

Had the basis vectors been orthogonal, then M would have been a unit matrix leading to $L = N$. From (8.28) we have

$$\|\boldsymbol{\Lambda}_f\| = \max\{\|\boldsymbol{\Lambda}_f(\boldsymbol{k})\|_2 : \|\boldsymbol{k}\|_2 = 1\}$$

k could be expressed in terms of basis vectors as

$$\begin{aligned} \boldsymbol{k} &= x_1\boldsymbol{k}_1 + x_2\boldsymbol{k}_2 + \ldots + \boldsymbol{x}_m\boldsymbol{k}_m \\ &= [\boldsymbol{k}_1\boldsymbol{k}_2\ldots\boldsymbol{k}_m]\begin{bmatrix} x_1 \\ x_2 \\ \vdots \\ x_m \end{bmatrix} \end{aligned}$$

Hence

$$\|\boldsymbol{k}\|_2^2 = x^*Mx$$

Similarly

$$\begin{aligned} \boldsymbol{\Lambda}_f(\boldsymbol{k}) &= \boldsymbol{\Lambda}_f[x_1\boldsymbol{k}_1 + \ldots x_m\boldsymbol{k}_m] \\ & [\boldsymbol{\Lambda}_f(\boldsymbol{k}_1)\ldots\boldsymbol{\Lambda}_f(\boldsymbol{k}_m)]\begin{bmatrix} x_1 \\ \vdots \\ x_m \end{bmatrix} \\ &= [\boldsymbol{k}_1\boldsymbol{k}_2\ldots\boldsymbol{k}_m]Lx \end{aligned}$$

Hence

$$\begin{aligned} \|\boldsymbol{\Lambda}_f(\boldsymbol{k})\|_2^2 &= x^*L^*MLx \\ &= x^*N^*M^{-1}Nx \end{aligned}$$

Substituting for $\|\boldsymbol{\Lambda}_f(\boldsymbol{k})\|_2$ and $\|\boldsymbol{k}\|_2$ in (8.28), we have

$$\|\boldsymbol{\Lambda}_f\| = \max\left\{(x^*N^*M^{-1}Nx)^{\frac{1}{2}} : x^*Mx = 1\right\} \tag{8.35}$$

(8.35) is a standard quadratic maximization problem with quadratic constraints. It can be shown that the maximum of the quadratic function on the right hand side of (8.35) is given by the square root of largest value of λ such that

$$\det(N^* M^{-1} N - \lambda M) = 0 \tag{8.36}$$

The problem of finding λ can be reduced to either a generalized eigenvalue problem or a singular value problem for which computer subroutines are available. Hence.

$$\|\boldsymbol{\Lambda}_f\| = \sqrt{\lambda}$$

The corresponding $\tilde{x}$ that achieves the maximum value in (8.35) could also be determined as

$$\tilde{x} = [\tilde{x}_1 \tilde{x}_2 \ldots \tilde{x}_m]^T$$

We have

$$\widetilde{\boldsymbol{k}} = \tilde{x}_1 \boldsymbol{k}_1 + \tilde{x}_2 \boldsymbol{k}_2 + \ldots + \tilde{x}_m \boldsymbol{k}_m$$

Let $\widetilde{\boldsymbol{q}}$ be the optimum value of $\boldsymbol{q}$ for which

$$\min \|\boldsymbol{f} - \boldsymbol{\delta q}\|_\infty = \|\Lambda_f\|$$

let $(\boldsymbol{f} - \boldsymbol{\delta}\tilde{\boldsymbol{q}}) = \widetilde{\boldsymbol{y}}$.

Hence

$$\begin{aligned}\boldsymbol{\Lambda}_{\tilde{y}}(\widetilde{\boldsymbol{k}}) &= (\widetilde{\boldsymbol{y}}\widetilde{\boldsymbol{k}})_{\mathcal{K}} = [(\boldsymbol{f} - \boldsymbol{\delta}\tilde{\boldsymbol{q}})\widetilde{\boldsymbol{k}}]_{\mathcal{K}} \\ &= (\boldsymbol{f}\widetilde{\boldsymbol{k}})_{\mathcal{K}} - [(\boldsymbol{\delta}\tilde{\boldsymbol{q}})\widetilde{\boldsymbol{k}}]_{\mathcal{K}} \\ &= (\boldsymbol{f}\widetilde{\boldsymbol{k}})_{\mathcal{K}} \\ &= \boldsymbol{\Lambda}_f(\widetilde{\boldsymbol{k}}) \end{aligned} \tag{8.37}$$

Hence

$$\boldsymbol{\Lambda}_f = \boldsymbol{\Lambda}_{\widetilde{y}} \tag{8.38}$$

We next show that $\widetilde{\boldsymbol{y}}\widetilde{\boldsymbol{k}} \in \mathcal{K}$ as follows

$$\begin{aligned}\|\boldsymbol{\Lambda}_f\| &= \|\boldsymbol{\Lambda}_f(\widetilde{\boldsymbol{k}})\|_2 \\ &= \|\boldsymbol{\Lambda}_{\widetilde{y}}(\widetilde{\boldsymbol{k}})\|_2 \text{ from (8.38)} \\ &= \|(\widetilde{\boldsymbol{y}}\widetilde{\boldsymbol{k}})_{\mathcal{K}}\|_2 \\ &\le \|\widetilde{\boldsymbol{y}}\widetilde{\boldsymbol{k}}\|_2 && (8.39) \\ &\le \|\widetilde{\boldsymbol{y}}\|_\infty \\ &= \|\boldsymbol{\Lambda}_f\| \text{ (by Sarason's theorem)} && (8.40)\end{aligned}$$

from (8.40) we infer that the intermediate inequalities must be equalities. Hence we have

$$\Lambda_f(\widetilde{\boldsymbol{k}}) = (\widetilde{\boldsymbol{y}}\widetilde{\boldsymbol{k}})_{\mathcal{K}} = (\widetilde{\boldsymbol{y}}\widetilde{\boldsymbol{k}})$$

which shows that $\tilde{\boldsymbol{y}}\tilde{\boldsymbol{k}} \in \mathcal{K}$. Further

$$\widetilde{\boldsymbol{y}} = \frac{\Lambda_f(\widetilde{\boldsymbol{k}})}{\widetilde{\boldsymbol{k}}} \tag{8.41}$$

Recall that $\tilde{x}$ represents the coordinates of $\widetilde{\boldsymbol{k}}$ with respect the basis vectors $(\boldsymbol{k}_1, \boldsymbol{k}_2, \ldots \boldsymbol{k}_m)$. Since the matrix representation of the operator Λ_f is L, it is easy to see that

$$\Lambda_f(\widetilde{\boldsymbol{k}}) = \sum_{i=1}^{m} \tilde{z}_i \boldsymbol{k}_i \text{ where } \tilde{z} = L\tilde{x}.$$

From (8.41) we have

$$\widetilde{\boldsymbol{y}} = \frac{\sum_{i=1}^{m} \tilde{z}_i \boldsymbol{k}_i}{\sum_{i=1}^{m} \tilde{x}_i \boldsymbol{k}_i} \tag{8.42}$$

From (8.42) $\tilde{y}$ is determined and knowing $\tilde{y}$ and using the relation $\widetilde{\boldsymbol{y}} = \boldsymbol{f} - \delta\tilde{\boldsymbol{q}}$ we can obtain $\widetilde{\boldsymbol{q}}$.

The procedure for minimizing $\|\boldsymbol{f} - \delta\, \boldsymbol{q}\|_\infty$ from among the set of all $\boldsymbol{q} \in \mathcal{H}_\infty$ is summarized below.

Step 1: Determine the zeros on δ. All of them belong to the open right half plane. Identify the basis vectors for the space $\mathcal{K}$ namely $\boldsymbol{k}_i = \dfrac{1}{s + \overline{\alpha}_i}$ $i = 1, 2, \ldots, m$ where α_is are the zeros on δ. It is assumed that the zeros are simple and that there are no zeros on the imaginary axis

Step 2: Compute the following square matrices

$$M = [M_{ij}] = [< \boldsymbol{k}_i, \boldsymbol{k}_j >]$$
$$N = [N_{ij}] = [< \boldsymbol{k}_i, \boldsymbol{f}\boldsymbol{k}_j >]$$

and obtain $L = M^{-1}N$.

Step 3: Compute $\tilde{x}$ which minimizes $x^*N^*M^{-1}Nx$ subject to $x^*Mx = 1$.

Step 4: Obtain $\tilde{z} = L\tilde{x}$.
and hence compute

$$\widetilde{\boldsymbol{y}} = \frac{\sum_{i=1}^{m} \tilde{z}_i \boldsymbol{k}_i}{\sum_{i=1}^{m} \tilde{x}_i \boldsymbol{k}_i}$$

Step 5: From the relationship $\widetilde{\boldsymbol{y}} = \boldsymbol{f} - \delta\widetilde{\boldsymbol{q}}$ obtain $\widetilde{\boldsymbol{q}}$.
This completes the discussion on operator-theoretic method, to solve the SISO $\mathcal{H}_\infty$ optimization problem.

8.5.2 MIMO Systems

In any $\mathcal{H}_\infty$ optimization problem two quantities have to be determined. They are the optimal value of the $\mathcal{H}_\infty$ norm of the function and the determination of the minimizing function. In the SISO case, these two issues are resolved side by side using a single method. However in the MIMO Case they have to be evaluated separately using possibly different approaches. For example Chang and Pearson (1984) evaluated the optimal norm in the case of multivariable $\mathcal{H}_\infty$ optimization problem by a method similar to what we have explained in the SISO case. Sarason's theorem holds in this context also and we have $\inf_{\boldsymbol{Q}\in\mathcal{H}_\infty} \|\boldsymbol{F} - \delta\boldsymbol{Q}\|_\infty = \boldsymbol{\Lambda}_F$ where $\boldsymbol{F}$ and $\boldsymbol{Q}$ are square matrices belonging to $\mathcal{H}_\infty$ and δ is a scalar function with zeros in the open right half plane. The $\mathcal{K}$ space is again finite, with its dimension being equal to the number zeros of δ located in the open right half plane. Once again, following a procedure similar to the one adopted in the SISO case, a matrix representation of the operator $\boldsymbol{\Lambda}_F$ may be derived. This is used to establish the minimum norm. However Chang and Pearson (1984) uses NP interpolation theory to obtain the minimizing function. In contrast to this Francis, Helton and Zames (1984) use the theory of Ball and Helton (1983) to solve the multivariable $\mathcal{H}_\infty$ optimization problem of a restricted type. The more general case exemplified in the 4-block problem, involves the minimization of the norm of a mixed Hankel-Toeplitz operator. Unfortunately there is no direct procedure at present to evaluate this norm.

8.6 Summary

The solution of the LQG problem is based on the 'separation principle' according to which the optimal control and optimal estimation problems are treated separately before they are combined to obtain the total solution. In chapter 10 we shall extend this principle to solve the $\mathcal{H}_\infty$ optimization problem.

The Riccati equation plays a pivotal role in both $\mathcal{H}_2$ and $\mathcal{H}_\infty$ optimization problems. In this context, the properties of the solutions of this equation are important. Specifically, we are interested in the sign definiteness, uniqueness and stabilizing nature of the solutions. These aspects are explained through a series of theorems and lemmas. As a performance measure the $\mathcal{H}_2$ norm has excellent characteristics. However, as exemplified in the LQG approach, it suffers from lack of robustness. In this contest, design based on $\mathcal{H}_\infty$ norm is basically robust and even though it has other shortcomings, offers a viable alternative approach. These issues will be discussed in more detail in Chapter 12.

The main topics discussed in this chapter pertain to two different approaches to solve the $\mathcal{H}_\infty$ optimization problem. They are the NP interpolation method and the operator theoretic method. From the computational viewpoint both these methods have been somewhat eclipsed by state space methods. However these methods cannot be ignored as they are classical in content and are based on deep theoretical foundations. This point of view has prevailed on us while presenting these methods. It is shown how the NP problem arises naturally in $\mathcal{H}_\infty$ optimization studies. Both the scalar and matrix versions of the NP problem are discussed. As far as operator-theoretic methods are concerned, the presentation is confined to an elementary level. Only SISO systems are discussed with an indication of how MIMO systems could be tackled. A deeper presentation is not attempted as it is beyond the scope of this book.

Notes and Additional References

LQG Theory is fully explained in many recently published books on feedback control e.g. Kwakernaak and Sivan (1991). A comprehensive survey of LQG theory explaining its role and usefulness in feedback design is given by Athans (1971a). The IEEE Automatic control special issue (Athans 1971b) is entirely devoted to LQG theory and contains useful information. Robustness properties of LQG design are discussed by Safonov and Athans (1977), Lehtomaki *et al.* (1981), and Stein and Athans (1987).

A modern treatment of NP interpolation problems with particular reference to circuit and system theory may be found in Delsarte *et al.* (1981). Control engineering applications of NP method are given in Chang and Pearson (1984), Young and Francis (1985), Kimura (1984) and Vidyasagar and Kimura (1986). Operator-theoretic methods are reported in Francis *et al.* (1984), Francis and Zames (1984) and Ball and Helton (1983).

Exercises

1. In an optimal control problem the state and output equations are

$$\dot{x} = Ax + Bu$$
$$z = C_1 x + D_{12} u$$

The performance index to be minimized subject to the above equations is

$$J = \int_0^\infty z' z dt$$

Note that $z'z$ involves cross terms in x and u. Show that the optimal control law is $u = -Fx$ where $F = D_{12}' C_1 + B_2' X$, where X is the solution of an appropriately defined Riccati equation. Note that when $D_{12}' C = 0$ then $u^* = -B_2' X$, which is the usual solution when cross coupling terms are not present.

2. Solve the LQG problem for which the relevant equations are

$$\begin{bmatrix} \dot{x}_1 \\ \dot{x}_2 \end{bmatrix} = \begin{bmatrix} 1 & 1 \\ 0 & 1 \end{bmatrix} \begin{bmatrix} x_1 \\ x_2 \end{bmatrix} + \sqrt{5} \begin{bmatrix} 1 & 0 \\ 1 & 0 \end{bmatrix} w + \begin{bmatrix} 0 \\ 1 \end{bmatrix} u$$

$$z = \begin{bmatrix} \sqrt{12}(x_1 + x_2) \\ u \end{bmatrix}$$

$$y = [1 \quad 0] \begin{bmatrix} x_1 \\ x_2 \end{bmatrix} + [0 \quad 1] w$$

obtain the optimal controller.

$$\left[\text{Ans. } \boldsymbol{k}(s) = \frac{30(1-2s)}{(s^2+9s+3)}\right]$$

3. In an NP problem, the interpolation points are $\alpha_1 = 1, \alpha_2 = 2$ and $\alpha_3 = 3$ and the corresponding values are $\beta_1 = \frac{1}{2}, \beta_2 = \frac{1}{3}$ and $\beta_3 = \frac{1}{4}$.

Check for the solvability of the problem by forming the Pick matrix and examining whether it is positive definite or not.

4. In a model matching problem, it is required to find a $\boldsymbol{q}$ such that

$$\|\boldsymbol{t}_1 - \boldsymbol{t}_2\boldsymbol{q}\|_\infty \leq \gamma$$

Let $\frac{1}{\gamma}\boldsymbol{t}_1 - \frac{1}{\gamma}\boldsymbol{t}_2 q = \boldsymbol{g}$

The problem now reduces to that of obtaining an analytic function $\boldsymbol{g}$ such that $\|\boldsymbol{g}\|_\infty \leq 1$ satisfying certain interpolation conditions. Let $\alpha_1, \alpha_2, \ldots \alpha_n$ be the zeros of $\boldsymbol{t}_2$ in the open right-half plane. Then

$$\boldsymbol{g}(\alpha_i) = \frac{1}{\gamma}\boldsymbol{t}_1(\alpha_i) = \frac{1}{\gamma}\beta_i \quad i = 1, 2, \ldots n$$

The Pick matrix for the above problem is $P = [P_{ij}]$

$$\text{where } P_{ij} = \frac{1}{\alpha_i + \overline{\alpha}_j} - \frac{1}{\gamma^2}\frac{\beta_i\overline{\beta}_j}{\alpha_i + \overline{\alpha}_j}$$

$$\text{Hence } P = A - \gamma^{-2}B$$

where both A and B are hermitian and further A is positive definite (because the a_is are distinct)

Let the smallest value of γ for which $A - \gamma^{-2}B$ ceases to be positive semi-definite be γ_{opt}. Prove that γ_{opt} is the square root of the largest eigenvalue of the matrix $A^{-1/2}BA^{-1/2}$. What does γ_{opt} signify?

5. Let λ be a complex number with $|\lambda| < 1$ and let k be an integer greater than or equal to 1. Define

$$\boldsymbol{\Phi}(z) = \frac{z^{k-1}}{(1 - \overline{\lambda}z)^k}$$

Show that $\boldsymbol{\Phi} \in \mathcal{H}_2$, and then prove that for any $g \in \mathcal{H}_2$, the inner product:

$$< \boldsymbol{\Phi}, \boldsymbol{g} >= \frac{g^{(k-1)}(\lambda)}{(k-1)!}$$

where $\boldsymbol{g}^{(k-1)}$ denotes the $(k-1)$ derivative of $\boldsymbol{g}(z)$.

6. It is required to minimize the $\mathcal{H}_\infty$ norm of a weighted sensitivity operator $\boldsymbol{w}(1 - \boldsymbol{gk})^{-1}$ where

$$\text{the weighting function } \boldsymbol{w}(s) = \frac{1}{s+1} \text{ and}$$

$$\text{plant transfer function } \boldsymbol{g}(s) = \frac{(s-1)}{(s-3)(s+2)(s+3)}$$

Note that the plant is strictly proper and non-minimum phase. Transform the above problem into a model matching problem via Youla parametrization and obtain

$$\min_{q \in \mathcal{H}_\infty} \|\boldsymbol{f} - \delta \boldsymbol{q}\|_\infty \text{ where } \boldsymbol{f} \in \mathcal{RH}_\infty$$

Note that in the above formulation $\delta(s)$ has simple zeros, located in the open right half plane. In this setting obtain a matrix representation of the Sarason's operator Λ_f in terms of the basis vectors of the space $\mathcal{K}$. Hence derive the optimal controller which minimizes the $\mathcal{H}_\infty$ norm of the weighted sensitivity function.

9

Generalized Plant–Controller Configuration

9.1 Introduction

Ever since the definitive contributions of Zames (1981), in connection with the application of $\mathcal{H}_\infty$ optimization to feedback control problems, a number of related problems have been posed and solved by many researchers working in this area. As a result, a powerful $\mathcal{H}_\infty$ framework has been developed. The generalized plant–controller configuration discussed in this chapter provides an ideal setting for applying existing theory and to prove new results. We show in this chapter how most of the control optimization problems could be reduced to the so-called standard form. In this representation, the plant to be controlled has two sets of inputs and two sets of outputs. The first input set comprises of exogenous signals and the second input set comprises of control signals. The first output set consists of signals to be regulated and the second output set consists of measured signals used for control purpose. The controller accepts the measured signals and then generates the control signals which are fed into the plant. For such a general configuration (which subsumes most of the particular cases) a criteria for stabilizability is derived in Section 9.3. In the next section the parameterization of all stabilizing controllers for the generalized plant is considered. To tackle problems of this generality, the Linear Fractional Transformation (LFT) comes in handy and this transformation is used throughout the rest of this chapter. In Section 9.5, it is shown how to reduce some common $\mathcal{H}_\infty$ optimization problems to the standard form. Section 9.6 discusses how to recast problems posed in the standard form into the model matching format. It is shown how the complexity of the model matching problem depends upon certain rank conditions of the associated transfer matrics. However, solution of the model matching problem on lines reminiscent of the LQG problem is postponed to the next chapter.

9.2 The Standard Configuration

In multivariable feedback, it is possible to envisage a variety of plant controller configurations. In this context it would be advantageous to work with a standard configuration so that a common theoretical framework could be envolved to cover different situations. Such a standard configuration is shown in Figure 9.1.

In the figure, $\boldsymbol{G}$ represents the plant transfer function and $\boldsymbol{K}$ the controller transfer function. The vector valued signals, w, u, z and y represent the following:

- w – Exogenous signals representing disturbances, sensor noise, reference input etc.
- u – The control signal.
- z – The signal which is to be regulated (may include tracking errors, plant outputs, weighted actuator inputs etc).
- y – The measured signal.

The classification of signals as described above is quite broad and does not preclude the same signal appearing in more than one classification. For example, measured signal which normally belongs to the signal set y may also be a member of the signal set z.

The plant transfer function $\boldsymbol{G}$ is assumed to be real, rational and proper. Plant includes not only the actual system to be controlled but also the associated actuators and sensors, including the weighting functions used for design purpose. We are seeking a real rational proper matrix transfer function K which minimizes the $\mathcal{H}_\infty$ norm of the transfer matrix connecting w and z, under the constraint that internal stability be guaranteed.

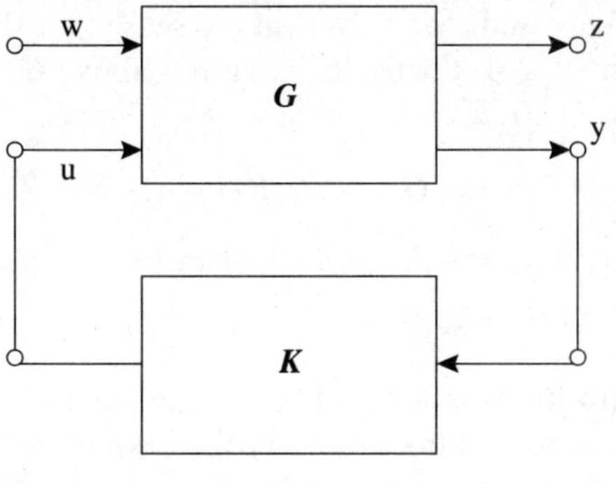

Figure 9.1 The standard configuration.

In conformity with the dimensions of input and output vector signals, the plant transfer matrix G may be partitioned as

$$\boldsymbol{G} = \begin{bmatrix} \boldsymbol{G}_{11} & \boldsymbol{G}_{12} \\ \boldsymbol{G}_{21} & \boldsymbol{G}_{22} \end{bmatrix}$$

This leads to the following relationships for the feedback configuration in Figure 9.1.

$$\begin{aligned} z &= \boldsymbol{G}_{11} w + \boldsymbol{G}_{12} u \\ y &= \boldsymbol{G}_{21} w + \boldsymbol{G}_{22} u \\ u &= \boldsymbol{K} y \end{aligned}$$

Eliminating u and y between the above equations

$$z = \left[\boldsymbol{G}_{11} + \boldsymbol{G}_{12} K (I - \boldsymbol{G}_{22} \boldsymbol{K})^{-1} \boldsymbol{G}_{21}\right] w \tag{9.1}$$

To ensure that $(I - \boldsymbol{G}_{22}\boldsymbol{K})^{-1}$ is a real rational proper matrix, $\lim_{s \to \infty} (I - \boldsymbol{G}_{22}\boldsymbol{K})^{-1}$ should be finite. A sufficient condition for this to happen is that $\boldsymbol{G}_{22}(\infty) = 0$, i.e. that $\boldsymbol{G}_{22}$ be strictly proper. Hence we shall make this assumption throughout unless a statement is made to the contrary. The transformation associated with

$$\boldsymbol{G}_{11} + \boldsymbol{G}_{12} \boldsymbol{K} (I - \boldsymbol{G}_{22} \boldsymbol{K})^{-1} \boldsymbol{G}_{21}$$

may be identified as a LFT which may be expressed as

$$\boldsymbol{T}_{zw} = \mathcal{F}_l \left(\begin{bmatrix} \boldsymbol{G}_{11} & \boldsymbol{G}_{12} \\ \boldsymbol{G}_{21} & \boldsymbol{G}_{22} \end{bmatrix}, \boldsymbol{K} \right) \tag{9.2}$$

or in a more cryptic form as

$$\boldsymbol{T}_{zw} = \mathcal{F}_l (\boldsymbol{G}, \boldsymbol{K})$$

The subscript l of $\mathcal{F}$ stands for 'lower' in the sense that the feedback path is below $\boldsymbol{G}$ (see Figure 9.1). If the feedback path is above $\boldsymbol{G}$, then we denote the transformation by $\mathcal{F}_u(\boldsymbol{G}, \boldsymbol{K})$ where u stands for 'upper'. We then have

$$\mathcal{F}_u(\boldsymbol{G}, \boldsymbol{K}) = \boldsymbol{G}_{22} + \boldsymbol{G}_{21} \boldsymbol{K} (I - \boldsymbol{G}_{11} \boldsymbol{K})^{-1} \boldsymbol{G}_{12} \tag{9.3}$$

In LFT format the $\mathcal{H}_\infty$ optimization problem may be posed as

$$\text{minimize } ||\mathcal{F}_l(\boldsymbol{G}, \boldsymbol{K})||_\infty$$

where the minimization is carried over all realizable controllers (i.e. controllers which an real, rational and proper) which make the closed loop system internally stable. Internal stability may be checked by injecting two fictitious inputs v_1 and v_2 as shown in Figure 9.2.

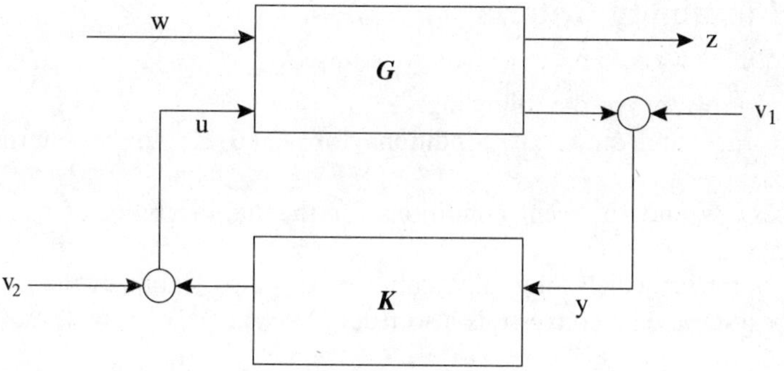

Figure 9.2 Feedback system with fictitious inputs.

For internal stability, the nine transfer functions connecting inputs w, v_1 and v_2 and outputs z, y and u should exist and further they should be proper and stable. By assumption $(I - \boldsymbol{G}_{22}\boldsymbol{K})$ is invertible as a real rational proper matrix. Hence the nine transfer matrices mentioned above indeed exist and they are real and proper. If these matrices are in addition stable, then internal stability is assured. Later on in this chapter, we show that stability of the input–output transfer function is synonymous with exponential stability of the closed loop system under certain mild restrictions imposed on $\boldsymbol{G}$ and $\boldsymbol{K}$.

Let (A, B, C, D) be a realization of $\boldsymbol{G}$. This may be represented by using the following compact notation. Thus we have

$$\boldsymbol{G} = \begin{bmatrix} \boldsymbol{G}_{11} & \boldsymbol{G}_{12} \\ \boldsymbol{G}_{21} & \boldsymbol{G}_{22} \end{bmatrix} = \left[\begin{array}{c|cc} A & B_1 & B_2 \\ \hline C_1 & D_{11} & D_{12} \\ C_2 & D_{21} & D_{22} \end{array}\right] \tag{9.4}$$

where

$$\boldsymbol{G}_{ij} = C_i(sI - A)^{-1}B_j \quad i,j = 1,2 \tag{9.5}$$

Note that the plant $\boldsymbol{G}$ need not always be stabilizable in the sense that there always exists a $\boldsymbol{K}$ such that

$$\boldsymbol{G}_{11} + \boldsymbol{G}_{12}\boldsymbol{K}(I - \boldsymbol{G}_{22}\boldsymbol{K})^{-1}\boldsymbol{G}_{21}$$

is internally stable. For example take a simple case where $\boldsymbol{G}_{12} = \boldsymbol{G}_{21} = \boldsymbol{G}_{22} = 0$ and $\boldsymbol{G}_{11}$ is unstable. Then $\boldsymbol{T}_{zw} = \boldsymbol{G}_{11}$, and no choice of $\boldsymbol{K}$ can make the feedback system stable. Hence it is relevant to enquire into the conditions under which the feedback system represented in Figure 9.1 is stabilizable.

9.3 Stabilizability Criteria

In this section we prove the following:

i) Necessary and sufficient conditions for $\mathcal{F}_l(\boldsymbol{G}, \boldsymbol{K})$ to be internally stable.
ii) Necessary and sufficient conditions for the stabilizability of transfer matrix $\boldsymbol{G}$.
iii) A controller which internally stabilizes $\boldsymbol{G}_{22}$ also stabilizes the closed loop system. The converse is also true.

THEOREM 9.1 (Internal stability)
The feedback system in Figure 9.2 characterized by the LFT $\mathcal{F}_l(\boldsymbol{G}, \boldsymbol{K})$ is internally stable if and only if the 'A matrix' associated with the closed loop system is asymptotically stable.

Before proceeding with the proof of the theorem we have the following definition for internal stability.

DEFINITION 9.1
The closed loop system represented in Figure 9.2 is *internally stable* if all the nine transfer matrices relating inputs w, v_1, v_2 to the outputs z, y and u are stable.

We assume in the sequel that the realizations of $\boldsymbol{G}$ and $\boldsymbol{K}$ are stabilizable and detectable. This is a standard assumption and implies that there are no hidden unstable modes associated with the realizations of either $\boldsymbol{G}$ or $\boldsymbol{K}$. Without this assumption it is not possible to ascertain whether the closed loop system is stable or not.

Let the stabilizable and detectable realizations of $\boldsymbol{G}$ and $\boldsymbol{K}$ be represented as follows:

$$\boldsymbol{G} = \left[\begin{array}{c|cc} A & B_1 & B_2 \\ \hline C_1 & 0 & D_{12} \\ C_2 & D_{21} & 0 \end{array}\right] \tag{9.6}$$

$$\boldsymbol{K} = \left[\begin{array}{c|c} \bar{A} & \bar{B} \\ \hline \bar{C} & \bar{D} \end{array}\right] \tag{9.7}$$

Note that in (9.6) we have assumed $D_{11} = 0$ and $D_{22} = 0$. These assumptions have been made for the sake of simplicity. Relaxing them would make the controller synthesis complicated. It has been shown that in the output feedback case, an equivalent problem can always be constructed with $D_{11} = 0$ and $D_{22} = 0$. Hence there is no loss of generality in making these assumptions.

Let x and $\hat{x}$ be the states of $\boldsymbol{G}$ and $\boldsymbol{K}$ respectively. Then the state and output equations of the closed loop system shown in Figure 9.2 are

$$\begin{bmatrix} \dot{x} \\ \dot{\hat{x}} \end{bmatrix} = \begin{bmatrix} A + B_2\tilde{D}C_2 & B_2\tilde{C} \\ \tilde{B}C_2 & \tilde{A} \end{bmatrix} \begin{bmatrix} x \\ \hat{x} \end{bmatrix} + \begin{bmatrix} B_1 + B_2\tilde{D}D_{21} & B_2\tilde{D} & B_2 \\ \tilde{B}D_{21} & \tilde{B} & 0 \end{bmatrix} \begin{bmatrix} w \\ v_1 \\ v_2 \end{bmatrix} \tag{9.8}$$

$$\begin{bmatrix} z \\ y \\ u \end{bmatrix} = \begin{bmatrix} C_1 + D_{12}\tilde{D}C_2 & D_{12}\tilde{C} \\ C_2 & 0 \\ \tilde{D}C_2 & \tilde{C} \end{bmatrix} \begin{bmatrix} x \\ \hat{x} \end{bmatrix} + \begin{bmatrix} D_{12}\tilde{D}D_{21} & D_{12}\tilde{D} & D_{12} \\ D_{21} & I & 0 \\ \tilde{D}D_{21} & \tilde{D} & I \end{bmatrix} \begin{bmatrix} w \\ v_1 \\ v_2 \end{bmatrix} \tag{9.9}$$

To prove Theorem 9.1, all that we need prove is that the feedback system shown in Figure 9.2 is both stabilizable and detectable. If this is proved then all the unstable modes are present in the transfer matrix. If the closed loop system is internally stable, then by definition 9.1 the system transfer matrix is stable, which considered along with stabilizability and detectability implies that the system has no unstable modes. It therefore follows that the system 'A matrix' is stable. The converse is also true, as a stable 'A matrix' implies that all the nine transfer matrices are stable.

Proof of Theorem 9.1: The closed loop system 'A matrix' is given by

$$\hat{A} = \begin{bmatrix} A + B_2\tilde{D}C_2 & B_2\tilde{C} \\ \tilde{B}C_2 & \tilde{A} \end{bmatrix} \tag{9.10}$$

To show that the closed loop system is detectable we apply the *PBH* test on observability (see appendix B). Suppose the closed loop system is not detectable then for some λ such that Re $\lambda \geq 0$ there exists a non-zero vector $[\,w_1' \quad w_2'\,]'$ such that

$$\begin{bmatrix} A + B_2\tilde{D}C_2 - \lambda I & B_2\tilde{C} \\ \tilde{B}_2C_2 & \tilde{A} - \lambda I \\ C_1 + D_{12}\tilde{D}C_2 & D_{12}\tilde{C} \\ C_2 & 0 \\ \tilde{D}C_2 & \tilde{C} \end{bmatrix} \begin{bmatrix} w_1 \\ w_2 \end{bmatrix} = \begin{bmatrix} 0 \\ 0 \\ 0 \\ 0 \\ 0 \end{bmatrix} \tag{9.11}$$

In (9.11), the fourth equation yields $C_2 w_1 = 0$ and the fifth equation yields $\tilde{D} C_2 w_1 + \tilde{C} w_2 = 0$. Together they imply that $\tilde{C} w_2 = 0$.

From the first equation we have

$$(A - \lambda I) w_1 + B_2 \tilde{D} C_2 w_1 + B_2 \tilde{C} w_2 = 0$$

Since $C_2 w_1$ and $\tilde{C} w_2$ are both zero we have

$$(A - \lambda I) w_1 = 0$$

Similarly from the third equation we can show that $C_1 w_1 = 0$.

The above results may be combined to yield

$$\begin{bmatrix} A - \lambda I \\ C_1 \\ C_2 \end{bmatrix} w_1 = 0 \tag{9.12}$$

By hypothesis $\left(A, \begin{bmatrix} C_1 \\ C_2 \end{bmatrix} \right)$ is detectable. Hence for Re $\lambda \geq 0$ and for non-zero w_1 (9.12) cannot be true which implies that $w_1 = 0$. Similarly making use of the fact that $(\tilde{A}, \tilde{C})$ is detectable we conclude that $w_2 = 0$. Hence for Re $\lambda \geq 0$, (9.11) can be true if and only if w_1 and w_2 are both zero, which proves that the closed loop system is indeed detectable. The stabilizability of the closed loop system may be proved in a similar manner. The proof of Theorem 9.1 is now complete.

THEOREM 9.2 (Stabilizability)
The closed loop system shown in Figure 9.2 is stabilizable if and only if the realization (A, B_2, C_2) of the sub-matrix $\boldsymbol{G}_{22}$ is stabilizable and detectable.

Proof: Suppose the closed loop system is stabilizable. Then there exists a feedback controller

$$\boldsymbol{K} = \left[\begin{array}{c|c} \tilde{A} & \tilde{B} \\ \hline \tilde{C} & \tilde{D} \end{array} \right]$$

such that $\hat{A}$ in (9.10) is stable.

Hence the matrix

$$\begin{bmatrix} A + B_2 \tilde{D} C_2 & B_2 \tilde{C} \\ \tilde{B} C_2 & \tilde{A} \end{bmatrix} = \begin{bmatrix} A & B_2 \tilde{C} \\ 0 & \tilde{A} \end{bmatrix} + \begin{bmatrix} B_2 \tilde{D} \\ \tilde{B} \end{bmatrix} [C_2 \quad 0] \tag{9.13}$$

is stable. From (9.13) it follows that the matrix pair

$$\left(\begin{bmatrix} A & B_2\tilde{C} \\ 0 & \tilde{A} \end{bmatrix}, [C_2 \quad 0]\right) \tag{9.14}$$

is detectable. We have now to show that (A, C_2) is detectable. The proof is by contradiction. If (A, C_2) is not detectable, then for Re $\lambda \geq 0$ there exists a non-zero vector w_1 such that

$$\begin{bmatrix} A - \lambda I \\ C_2 \end{bmatrix} w_1 = 0 \tag{9.15}$$

Now apply the *PBH* test on observability to (9.14) and make use of (9.15). We then have

$$\begin{bmatrix} A - \lambda I & B_2\tilde{C}. \\ 0 & \tilde{A} - \lambda I \\ C_2 & 0 \end{bmatrix} \begin{bmatrix} w_1 \\ 0 \end{bmatrix} = \begin{bmatrix} 0 \\ 0 \end{bmatrix} \tag{9.16}$$

for some λ with Re $\lambda \geq 0$ and a non zero vector $[w_1' \quad 0]'$. This is impossible since we know that the matrix pair in (9.14) is detectable. Hence (A, C_2) is detectable. Similarly factoring (9.10) as

$$\begin{bmatrix} A + B_2\tilde{D}C_2 & B_2C \\ \tilde{B}C_2 & \tilde{A} \end{bmatrix} = \begin{bmatrix} A & 0 \\ \tilde{B}C_2 & \tilde{A} \end{bmatrix} + \begin{bmatrix} B_2 \\ 0 \end{bmatrix} [\tilde{D}C_2 \quad \tilde{C}] \tag{9.17}$$

it can be proved that (A, B_2) is stabilizable. Thus internal stability implies that (A, B_2, C_2) is stabilizable and detectable.

To prove the converse, we assume that (A, B_2, C_2) is stabilizable and detectable. Hence there exist constant matrices F and H such that $(A + B_2F)$ and $(A + HC_2)$ are stable matrices. Now we may construct an observer based controller (see Chapter 5) whose equations are given by

$$\begin{aligned} \dot{\zeta} &= A\zeta + B_2u + H(C_2\zeta - y) \\ u &= F\zeta \end{aligned}$$

which on further simplification leads to

$$\begin{aligned} \dot{\zeta} &= (A + B_2F + HC_2)\zeta - Hy \\ u &= F\zeta \end{aligned}$$

The state space realization of an observer based controller with input y and output u is given by

$$\left[\begin{array}{c|c} \tilde{A} & \tilde{B} \\ \hline \tilde{C} & \tilde{D} \end{array}\right] = \left[\begin{array}{c|c} A + B_2F + HC_2 & -H \\ \hline F & 0 \end{array}\right] \tag{9.18}$$

Substituting the values of $(\tilde{A}, \tilde{B}, \tilde{C}, \tilde{D})$ as obtained in (9.18) in (9.10) we get

$$\hat{A} = \begin{bmatrix} A & B_2F \\ -HC_2 & A + B_2F + HC_2 \end{bmatrix}$$

$$= \begin{bmatrix} I & O \\ -I & I \end{bmatrix} \begin{bmatrix} A + B_2F & B_2F \\ 0 & A + HC_2 \end{bmatrix} \begin{bmatrix} I & O \\ I & I \end{bmatrix} \quad (9.19)$$

Since $(A + B_2$ F) and $(A + HC_2$) are stable it follows that $\hat{A}$ is stable and the proof is complete. Thus stabilizability and detectability of $(A,\ B_2,\ C_2)$ ensures that the closed loop system matrix $\hat{A}$ is stable. ■

Our next aim is to prove that any $\boldsymbol{K}$ which internally stabilizes the closed loop system also internally stabilizes the subsystem (A, B_2, C_2) and vice versa. This is stated in Theorem 9.3.

THEOREM 9.3
The closed loop system shown in Figure 9.2 is internally stabilized by a controller $\boldsymbol{K}$ if and only if it internally stabilizes the subsystem $\boldsymbol{G}_{22}$ with realization (A, B_2, C_2).

Proof: Figure 9.3 shows that subsystem $\boldsymbol{G}_{22}$ with feedback controller $\boldsymbol{K}$ and fictitious inputs v_1 and v_2.

The governing equations for the system are

$$G_{22} : \dot{x} = Ax + B_2u$$
$$y' = C_2x$$
$$K : \dot{\tilde{x}} = \tilde{A}\tilde{x} + \tilde{B}y$$
$$u' = \tilde{C}\tilde{x} + \tilde{D}y$$

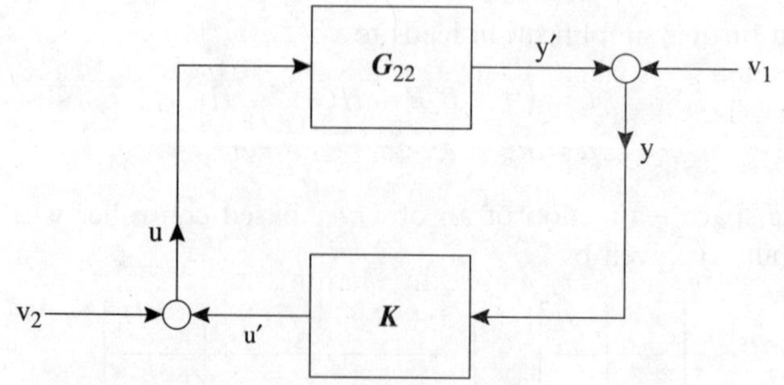

Figure 9.3 Subsystem feedback with fictitious inputs.

and

$$y' = y - v_1$$
$$u' = u - v_2$$

Eliminating y' and u' from the above equations, we can obtain the state space description of the composite system. It turns out as expected that the 'A matrix' of the composite system is identical to the 'A matrix' of the closed loop system shown in Figure 9.2. Hence if $\boldsymbol{K}$ is an internally stabilizing controller for $\boldsymbol{G}_{22}$ it is also an internally stabilizing controller for $\boldsymbol{G}$. The converse is trivially true because if $\boldsymbol{K}$ internally stabilizes $\boldsymbol{G}$, then all the nine transfer matrices are stable. We note that the four transfer matrices connecting v_1 and v_2 with y and u are a subset of these nine transfer matrices referred to earlier and the result follows.

■

To recapitulate, if $\boldsymbol{G}$ and $\boldsymbol{K}$ are stabilizable and detectable, than input-output stability is equivalent to exponential stability for the closed loop system. This is stated in Theorem 9.1. As a direct consequence of Theorem 9.2 if the realization of $\boldsymbol{G}_{22}$, namely (A, B_2, C_2), is stabilizable and detectable, then $\boldsymbol{G}$ can always be stabilized by an appropriate choice of the controller $\boldsymbol{K}$. Further, according to Theorem 9.3, it is enough if we design a controller to stabilize the subsystem $\boldsymbol{G}_{22}$. The same controller will also stabilize $\boldsymbol{G}$. This results in simplification of controller design as $\boldsymbol{G}_{22}$ has a smaller dimension than $\boldsymbol{G}$. The next section explains how to design a controller $\boldsymbol{K}$ which internally stabilizes $\boldsymbol{G}_{22}$.

9.4 Parametrization of Stabilizing Controllers for G_{22}

We consider here the feedback configuration given in Figure 9.1. The plant transfer matrix $\boldsymbol{G}$ has a state space realization as given in (9.6). Further the subsystem (A, B_2, C_2) is assumed to be stabilizable and detectable. Our aim here is to design a controller $\boldsymbol{K}$ which internally stabilizes $\boldsymbol{G}_{22}$. Under the assumptions made, such a controller always exists and further it also internally stablizes $\boldsymbol{G}$.

Let the right and left coprime fractional factorizations of $\boldsymbol{G}_{22}$ be

$$\boldsymbol{G}_{22} = \boldsymbol{N}_r\boldsymbol{D}_r^{-1} = \boldsymbol{D}_l^{-1}\boldsymbol{N}_l \tag{9.20}$$

The generalized Bezout identity for the system is given by

$$\begin{bmatrix} \boldsymbol{V}_r & \boldsymbol{U}_r \\ -\boldsymbol{N}_l & \boldsymbol{D}_l \end{bmatrix} \begin{bmatrix} \boldsymbol{D}_r & -\boldsymbol{U}_l \\ \boldsymbol{N}_r & \boldsymbol{V}_l \end{bmatrix} = \begin{bmatrix} I & 0 \\ 0 & I \end{bmatrix} \tag{9.21}$$

The eight submatrics on the left hand side of (9.21) all belong to $\mathcal{RH}_\infty$. The parametrized version of controller K which internally stabilizes $\boldsymbol{G}_{22}$ is given by (refer to Chapter 5) either the right coprime factorization

$$\boldsymbol{K} = (\boldsymbol{D}_r\boldsymbol{Q} - \boldsymbol{U}_l)(\boldsymbol{N}_r\boldsymbol{Q} + \boldsymbol{V}_l)^{-1} \tag{9.22}$$

or the left coprime factorization

$$\boldsymbol{K} = (\boldsymbol{V}_r + \boldsymbol{Q}\boldsymbol{N}_l)^{-1}(\boldsymbol{Q}\boldsymbol{D}_l - \boldsymbol{U}_r) \tag{9.23}$$

From (9.21) we get

$$\boldsymbol{D}_r = \boldsymbol{V}_r^{-1} - \boldsymbol{V}_r^{-1}\boldsymbol{U}_r\boldsymbol{N}_r \tag{9.24}$$

and

$$\boldsymbol{V}_r^{-1}\boldsymbol{U}_r = \boldsymbol{U}_l\boldsymbol{V}_l^{-1} \tag{9.25}$$

Substituting (9.24) and (9.25) in (9.22) we get

$$\begin{aligned}\boldsymbol{K} &= (\boldsymbol{V}_r^{-1}\boldsymbol{Q} - \boldsymbol{V}_r^{-1}\boldsymbol{U}_r\boldsymbol{N}_r\boldsymbol{Q} - \boldsymbol{U}_l)(\boldsymbol{I} + \boldsymbol{V}_l^{-1}\boldsymbol{N}_r\boldsymbol{Q})^{-1}\boldsymbol{V}_l^{-1}\\ &= (\boldsymbol{V}_r^{-1}\boldsymbol{Q} - \boldsymbol{U}_l\boldsymbol{V}_l^{-1}\boldsymbol{N}_r\boldsymbol{Q} - \boldsymbol{U}_l)(\boldsymbol{I} + \boldsymbol{V}_l^{-1}\boldsymbol{N}_r\boldsymbol{Q})^{-1}\boldsymbol{V}_l^{-1}\\ &= [\boldsymbol{V}_r^{-1}\boldsymbol{Q} - \boldsymbol{U}_l(\boldsymbol{I} + \boldsymbol{V}_l^{-1}\boldsymbol{N}_r\boldsymbol{Q})](\boldsymbol{I} + \boldsymbol{V}_l^{-1}\boldsymbol{N}_r\boldsymbol{Q})^{-1}\boldsymbol{V}_l^{-1}\\ &= -\boldsymbol{U}_l\boldsymbol{V}_l^{-1} + \boldsymbol{V}_r^{-1}\boldsymbol{Q}(\boldsymbol{I} + \boldsymbol{V}_l^{-1}\boldsymbol{N}_r\boldsymbol{Q})^{-1}\boldsymbol{V}_l^{-1}\end{aligned} \tag{9.26}$$

From (9.26) the underlying LFT is immediately identified as

$$\boldsymbol{K} = \left(\begin{bmatrix} -\boldsymbol{V}_r^{-1}\boldsymbol{U}_r & \boldsymbol{V}_r^{-1} \\ \boldsymbol{V}_l^{-1} & -\boldsymbol{V}_l^{-1}\boldsymbol{N}_r \end{bmatrix}, \boldsymbol{Q}\right) \tag{9.27}$$

$$= \mathcal{F}_l(\boldsymbol{K}_0, \boldsymbol{Q}) \tag{9.28}$$

Figure 9.1 may now be redrawn as shown in Figure 9.4(a). The controller $\boldsymbol{K}$ is represented as an LFT with $\boldsymbol{Q} \in \mathcal{RH}_\infty$ assigned the role of a free parameter. Thus

$$\mathcal{F}_l(\boldsymbol{G}, \boldsymbol{K}) = \mathcal{F}_l(\boldsymbol{G}, \mathcal{F}_l(\boldsymbol{K}_0, \boldsymbol{Q})) \tag{9.29}$$

$$= \mathcal{F}_l(\boldsymbol{T}, \boldsymbol{Q}) \tag{9.30}$$

In (9.30), $\boldsymbol{T}$ represents the composition of two LFTs, which in turn gives rise to another LFT. Figure 9.4(b) shows the composite LFT, along with the controller $\boldsymbol{Q}$.

At this stage we seek answers to the following questions. Given the state space realization of $\boldsymbol{G}$ obtain:

1. State space realization of $\boldsymbol{K}_0$
2. State space realization of $\boldsymbol{T}$.
3. State space realization of $\mathcal{F}_l(\boldsymbol{T}, \boldsymbol{Q})$.

They are relevant in the context of $\mathcal{H}_\infty$ optimization by state space methods

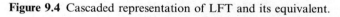

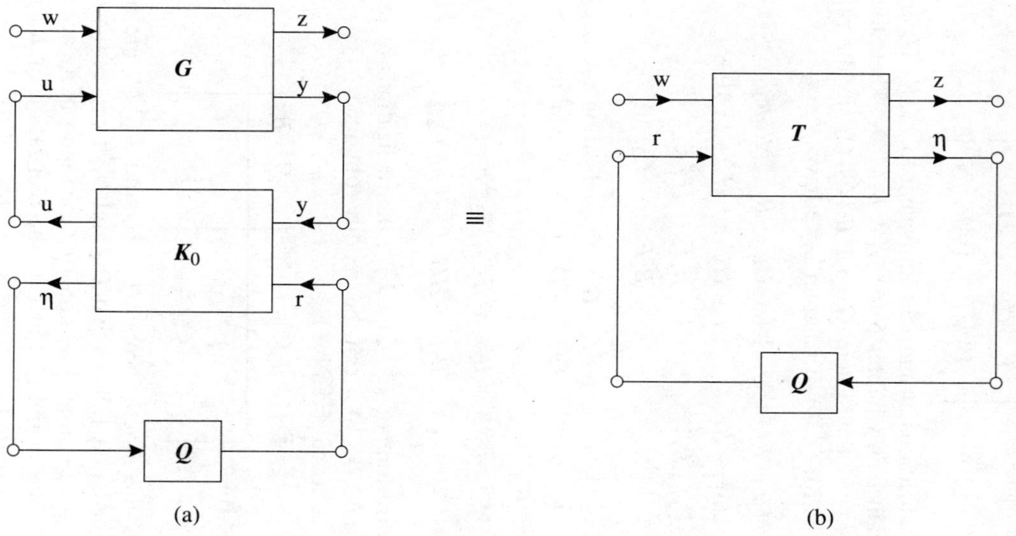

Figure 9.4 Cascaded representation of LFT and its equivalent.

9.4.1 State Space Realizations

We first take up the question of state space realization of $\boldsymbol{K}_0$. In the previous section it was shown that (refer to (9.27))

$$\boldsymbol{K}_0 = \begin{bmatrix} -\boldsymbol{V}_r^{-1}\boldsymbol{U}_r & \boldsymbol{V}_r^{-1} \\ \boldsymbol{V}_l^{-1} & -\boldsymbol{V}_l^{-1}\boldsymbol{N}_r \end{bmatrix} \tag{9.31}$$

The state space realizations of matrices equivalent to $\boldsymbol{V}_r, \boldsymbol{V}_l, \boldsymbol{U}_r$ and $\boldsymbol{N}_r$ have already been obtained in Chapter 5 for $\boldsymbol{G}(A, B, C, D)$. Their formulae can readily be adapted for the system $\boldsymbol{G}_{22}$ (A, B_2, C_2, O) in the present case. Comparing (9.21) with (5.30) of Chapter 5 we have

$$\begin{aligned} \boldsymbol{V}_r = \boldsymbol{R}_l &\quad : \quad (A + HC_2,\ B_2,\ -F,\ I) \\ \boldsymbol{U}_r = -\boldsymbol{S}_l &\quad : \quad (A + HC_2,\ -H,\ -F,\ 0) \\ \boldsymbol{V}_l = \boldsymbol{R}_r &\quad : \quad (A + B_2F,\ -H,\ C_2,\ I) \\ \boldsymbol{N}_r = \boldsymbol{N}_r &\quad : \quad (A + B_2F,\ B_2,\ C_2,\ 0) \end{aligned}$$

Recall that the inverse of a system $\boldsymbol{G}$ with realization (A, B, C, D) has a realization $\boldsymbol{G}^{-1} : (A - BD^{-1}C, BD^{-1}, -D^{-1}C, D^{-1})$. Making use of this well known formula we have

$$\begin{aligned} \boldsymbol{V}_r^{-1} &: \quad (A + B_2F + HC_2,\ B_2,\ F,\ I) \\ \boldsymbol{V}_l^{-1} &: \quad (A + B_2F + HC_2,\ -H,\ -C_2,\ I) \end{aligned}$$

Knowing the realizations of $\boldsymbol{V}_r^{-1}$ and $\boldsymbol{V}_l^{-1}$, the realizations of the cascaded systems $\boldsymbol{V}_r^{-1}\boldsymbol{U}_r$ and $\boldsymbol{V}_l^{-1}\boldsymbol{N}_r$ can be readily obtained. The state space realization of K_0 is hence given by

$$\boldsymbol{K}_0 = \left[\begin{array}{c|cc} A + B_2F + HC_2 & -H & B_2 \\ \hline F & 0 & I \\ -C_2 & I & 0 \end{array}\right] \tag{9.32}$$

Since $\boldsymbol{U}_r$ and $\boldsymbol{N}_r$ are strictly proper $\boldsymbol{V}_r^{-1}\boldsymbol{U}_r$ and $\boldsymbol{V}_l^{-1}\boldsymbol{N}_r$ are also strictly proper and it immediately follows that D_{11} and D_{22} terms of the D matrix in (9.32) are zero. From Figure 9.4 we note that the inputs of $\boldsymbol{K}_0$ are y and r and the outputs of $\boldsymbol{K}_0$ are u and η. Further, the realization of $\boldsymbol{K}_0$ is given by (9.32). The state and output equations associated with $\boldsymbol{K}_0$ are

$$\begin{aligned} \dot{\xi} &= (A + B_2F + HC_2)\xi - Hy + B_2r \\ u &= F\xi + r \\ \eta &= -C_2\xi + y \\ r &= \boldsymbol{Q}\eta \end{aligned}$$

Rearranging the above equations we obtain

$$\dot{\xi} = A\xi + B_2 u - H(y - C_2\xi) \tag{9.33}$$
$$u = F\xi + \boldsymbol{Q}(y - C_2\xi) \tag{9.34}$$

The term η which is the difference between the actual output y and its estimated value $C_2\xi$ is sometimes referred to as the 'innovations' signal. Strictly speaking this term is used only when H happens to be the Kalman gain for the stochastic system.

The usual observer based controller identifies only a specific controller which internally stabilizes the plant. The present approach using Yaoula Parameterization delineates the set of all stabilizing controllers. Note further that we have $u = F\xi + \boldsymbol{Q}\eta$ from (9.34) where the additional term $\boldsymbol{Q}\eta$ is responsible for the generalization leading to all stabilizing controllers. Note further that $\boldsymbol{Q}$ is a free parameter belonging to $\mathcal{RH}_\infty$. A schematic diagram of the controller is given in Figure 9.5.

To obtain the state space realization of $\boldsymbol{T}$ we proceed as follows:

Assume that

$$\boldsymbol{G} = \left[\begin{array}{c|cc} A & B_1 & B_2 \\ \hline C_1 & 0 & D_{12} \\ C_2 & D_{21} & 0 \end{array}\right] \quad \text{and } \boldsymbol{K}_o = \left[\begin{array}{c|cc} \tilde{A} & \tilde{B}_1 & \tilde{B}_2 \\ \hline \tilde{C}_1 & 0 & I \\ \tilde{C}_2 & I & 0 \end{array}\right]$$

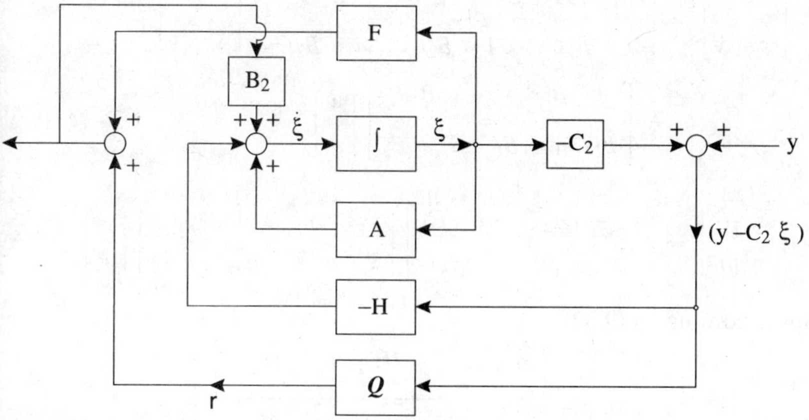

Figure 9.5 Generalization of stabilizing controller.

The corresponding state space and output equations are

$$
G \quad : \quad \dot{x} = Ax + B_1 w + B_2 u
$$
$$
z = C_1 x + D_{12} u
$$
$$
y = C_2 x + D_{21} w
$$
$$
K_0 \quad : \quad \dot{\tilde{x}} = \tilde{A}\tilde{x} + \tilde{B}_1 y + \tilde{B}_2 r
$$
$$
u = C_1 \tilde{x} + r
$$
$$
\eta = C_2 \tilde{x} + y
$$

As can be seen from Figure 9.4, the composition of G and K_0 gives rise to a 4-terminal transfer matrix T with w and r as inputs and z and η as outputs. To obtain the state space description of T, eliminate u and y from the above equations. On simplification we obtain

$$
\begin{bmatrix} \dot{x} \\ \dot{\tilde{x}} \end{bmatrix} = \begin{bmatrix} A & B_2 \tilde{C}_1 \\ \tilde{B}_1 C_2 & \tilde{A} \end{bmatrix} \begin{bmatrix} x \\ \tilde{x} \end{bmatrix} + \begin{bmatrix} B_1 & B_2 \\ \tilde{B}_1 D_{21} & \tilde{B}_2 \end{bmatrix} \begin{bmatrix} w \\ r \end{bmatrix} \tag{9.35}
$$

$$
\begin{bmatrix} z \\ \eta \end{bmatrix} = \begin{bmatrix} C_1 & D_{12} \tilde{C}_1 \\ C_2 & \tilde{C}_2 \end{bmatrix} \begin{bmatrix} x \\ \tilde{x} \end{bmatrix} + \begin{bmatrix} 0 & D_{12} \\ D_{21} & 0 \end{bmatrix} \begin{bmatrix} w \\ r \end{bmatrix} \tag{9.36}
$$

We now transform the old set of state variables $[x' \ \tilde{x}']'$ to obtain a new set of state variables $[x' \ (\tilde{x} - x)']'$. We have

$$
\begin{bmatrix} x \\ \tilde{x} - x \end{bmatrix} = \begin{bmatrix} I & 0 \\ -I & I \end{bmatrix} \begin{bmatrix} x \\ \tilde{x} \end{bmatrix} \tag{9.37}
$$

The transformed state and output equations are

$$
\begin{bmatrix} \dot{\tilde{x}} \\ \dot{\tilde{x}} - \dot{x} \end{bmatrix} = \begin{bmatrix} A + B_2 \tilde{C}_1 & B_2 \tilde{C}_1 \\ \tilde{A} + \tilde{B}_1 C_2 - A - B_2 \tilde{C}_1 & \tilde{A} - B_2 \tilde{C}_1 \end{bmatrix} \begin{bmatrix} x \\ \tilde{x} - x \end{bmatrix}
$$
$$
+ \begin{bmatrix} B_1 & B_2 \\ \tilde{B}_1 D_{21} - B_1 & \tilde{B}_2 - B_2 \end{bmatrix} \begin{bmatrix} w \\ r \end{bmatrix} \tag{9.38}
$$

$$
\begin{bmatrix} z \\ \eta \end{bmatrix} = \begin{bmatrix} C_1 + D_{12} \tilde{C}_1 & D_{12} \tilde{C}_1 \\ C_2 + \tilde{C}_2 & \tilde{C}_2 \end{bmatrix} \begin{bmatrix} x \\ \tilde{x} - x \end{bmatrix} + \begin{bmatrix} 0 & D_{12} \\ D_{21} & 0 \end{bmatrix} \begin{bmatrix} w \\ r \end{bmatrix} \tag{9.39}
$$

But according to (9.32)

$$
K_0 = \left[\begin{array}{c|cc} A + B_2 F + HC_2 & -H & B_2 \\ \hline F & 0 & I \\ -C_2 & I & 0 \end{array} \right]
$$

We may identify $\tilde{A}, \tilde{B}_1, \tilde{B}_2, \tilde{C}_1$ and $\tilde{C}_2$ with the corresponding entries in K_o above and then substitute in (9.38) to yield

$$T = \left[\begin{array}{cc|cc} A + B_2F & B_2F & B_1 & B_2 \\ 0 & A + HC_2 & -(B_1 + HD_{21}) & 0 \\ \hline C_1 + D_{12}F & D_{12}F & 0 & D_{12} \\ 0 & -C_2 & D_{21} & 0 \end{array}\right] \tag{9.40}$$

From (9.40), the state space realizations of the sub matrices T_{11}, T_{12}, T_{21} and T_{22} could be readily obtained. Thus we have

$$T_{11} : \left(\begin{bmatrix} A + B_2F & B_2F \\ 0 & A + HC_2 \end{bmatrix}, \begin{bmatrix} B_1 \\ -(B_1 + HD_{21}) \end{bmatrix}, [C_1 + D_{12}F \quad D_{12}F] \right) \tag{9.41}$$

$$T_{12} : \; ((A + B_2F), \; B_2, \; (C_1 + D_{12}F), D_{12}) \tag{9.42}$$

$$T_{21} : \; ((A + HC_2), -(B_1 + HD_{21}), -C_2, D_{21}) \tag{9.43}$$

T_{22} : It is easily shown on substitution that T_{22} vanishes and hence

$$T_{22} = 0 \tag{9.44}$$

Referring to Figure 9.4(b). We have

$$\begin{aligned} T_{zw} &= \mathcal{F}_l(T, Q) \\ &= T_{11} + T_{12}Q(I - T_{22}Q)^{-1}T_{21} \\ &= T_{11} + T_{12}QT_{21} \quad (\text{since} \, T_{22} = 0) \end{aligned} \tag{9.45}$$

We have thus demonstrated through (9.45) that even in the case of generalized plant–controller configuration, the closed loop transfer matrix T_{zw} is affine in the Youla parameter Q.

■

Before concluding this section we take a deeper look at some of the properties of the composite transfer function $\mathcal{F}_l(G, K)$ with particular reference to its uncontrollable and unobservable modes. Let

$$G = \left[\begin{array}{c|cc} A & B_1 & B_2 \\ \hline C_1 & D_{11} & D_{12} \\ C_2 & D_{21} & D_{22} \end{array}\right] \quad \text{and} \quad K = \left[\begin{array}{c|c} \tilde{A} & \tilde{B} \\ \hline \tilde{C} & \tilde{D} \end{array}\right]$$

We further assume that the realization of K is minimal.

As was done earlier in a slightly different context, we first write down the state equations of G and K and then eliminate u and y between these equations to obtain the state space description of the composite closed loop system.

Thus we have

$$\mathcal{F}_l(\boldsymbol{G}, \boldsymbol{K}) = \left[\begin{array}{cc|c} A + B_2\tilde{D}M C_2 & B_2 I + \tilde{D}MD_{22}\tilde{C} & B_1 + B_2\tilde{D}MD_{21} \\ \tilde{B}MC_2 & \tilde{A} + \tilde{B}MD_{22}\tilde{C} & \tilde{B}MD_{21} \\ \hline C_1 + D_{12}\tilde{D}MC_2 & D_{12}(I + \tilde{D}MD_{22})\tilde{C} & D_{11} + D_{12}\tilde{D}MD_{21} \end{array}\right] \tag{9.46}$$

where $M = (I - D_{22}\tilde{D})^{-1}$ is non-singular. We are concerned here about the uncontrollable and unobservable modes of the realization given in (9.46). It turns out that these modes are precisely those values of λ for which the matrices

$$\begin{bmatrix} A - \lambda I & B_2 \\ C_1 & D_{12} \end{bmatrix} \text{ loses full column rank}$$

and

$$\begin{bmatrix} A - \lambda I & B_1 \\ C_2 & D_{21} \end{bmatrix} \text{ loses full row rank}$$

We shall briefly outline the proof. Apply the *PBH* observability test to the system defined in (9.46). If λ is an unobservable mode, then there exists a non-zero vector $[w_1^* \quad w_2^*]^*$ such that

$$\begin{bmatrix} A + B_2\tilde{D}MC_2 - \lambda I & B_2(I + \tilde{D}MD_{22})\tilde{C} \\ \tilde{B}MC_2 & \tilde{A} + \tilde{B}MD_{22}\tilde{C} - \lambda I \\ C_1 + D_{12}\tilde{D}MC_2 & D_{12}(I + \tilde{D}MD_{22})\tilde{C} \end{bmatrix} \begin{bmatrix} w_1 \\ w_2 \end{bmatrix} = 0 \tag{9.47}$$

The first equation in (9.47) yields

$$(A - \lambda I)w_1 + B_2[\tilde{D}MC_2 w_1 + (I + \tilde{D}MD_{22})\tilde{C}w_2] = 0$$

and the last equation in (9.47) yields

$$C_1 w_1 + D_{12}[\tilde{D}MC_2 w_1 + (I + \tilde{D}MD_{22})\tilde{C}w_2] = 0$$

Replacing the quantity inside the square brackets by z we have

$$\begin{bmatrix} (A - \lambda I) & B_2 \\ C_1 & D_{12} \end{bmatrix} \begin{bmatrix} w_1 \\ z \end{bmatrix} = 0 \tag{9.48}$$

We now show that

$$[w_1^* \quad w_2^*]^* \neq 0 \Rightarrow [w_1^* \quad z^*]^* \neq 0$$

Suppose the contrary were to be true, then for $[w_1^* \quad w_2^*]^* \neq 0$ we have both $w_1 = 0$ and $z = 0$. Hence

$$z = \tilde{D}MC_2w_1 + (I + \tilde{D}MD_{22})\tilde{C}w_2 = 0$$

which implies that $(I + \tilde{D}MD_{22})\tilde{C}w_2 = 0$. But $(I + \tilde{D}MD_{22}) = (I - \tilde{D}D_{22})^{-1}$ which is non-singular by hypothesis. Hence $\tilde{C}w_2 = 0$.
The second equation in (9.47) reads

$$\tilde{B}MC_2w_1 + (\tilde{A} - \lambda I)w_2 + \tilde{B}MD_{22}\tilde{C}w_2 = 0$$

In view of the fact that $\tilde{C}w_2$ and w_1 are zero we have

$$(\tilde{A} - \lambda I)w_2 = 0$$

We may now write

$$\begin{bmatrix} \tilde{A} - \lambda I \\ \tilde{C} \end{bmatrix} w_2 = 0$$

which goes against the hypothesis that $(\tilde{A}, \tilde{B}, \tilde{C}, \tilde{D})$ is a minimal realization and therefore controllable and observable. Hence there exists a non-zero vector $[w_1^* \quad z^*]^*$ for which (9.48) is satisfied. Therefore precisely for those value of λ for which the matrix in (9.47) loses rank, the matrix in (9.48) also loses rank. This shows that the unobservable modes of the composite feedback system are the same as the Smith zeros of the realization of $\boldsymbol{G}_{12}$ characterized (A, B_2, C_1, D_{12}). The condition regarding uncontrollable modes could be similarly proved.

These results are summarized in the following theorem.

THEOREM 9.4
Let $\boldsymbol{G}$ be the plant transfer matrix with the realization

$$\boldsymbol{G} = \left[\begin{array}{c|cc} A & B_1 & B_2 \\ \hline C_1 & D_{11} & D_{12} \\ C_2 & D_{21} & D_{22} \end{array}\right]$$

and with inputs w and u and the outputs z and y. Further let $\boldsymbol{K}$ be the feedback controller with *minimal* realization:

$$\boldsymbol{K} = \left[\begin{array}{c|c} \tilde{A} & \tilde{B} \\ \hline \tilde{C} & \tilde{D} \end{array}\right]$$

connected to $\boldsymbol{G}$ as in Figure 9.6.

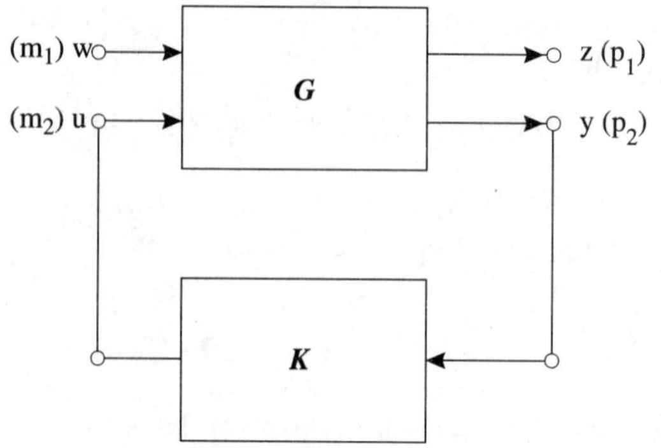

Figure 9.6 MIMO feedback configuration.

Let m_1 and m_2 be the sizes of the input vectors w and u and p_1 and p_2 the sizes of the output vectors z and y with $m_1 \geq p_2$ and $m_2 \leq p_1$ and satisfying the well posedness condition det $(I - D_{22}\tilde{D}) \neq 0$. The following statements are true.

(a) $\mathcal{F}_l(\boldsymbol{G}, \boldsymbol{K})$ has a realization given by (9.46).

(b) Every unobservable mode pertaining to the realization defined in (a) is a value of λ for which

$$\begin{bmatrix} A - \lambda I & B_2 \\ C_1 & D_{12} \end{bmatrix} \text{ has less than full column rank}$$

(c) Every uncontrollable mode pertaining to the realization defined in (a) is a value of λ for which

$$\begin{bmatrix} A - \lambda I & B_1 \\ C_2 & D_{21} \end{bmatrix} \text{ has less than full row rank}$$

Remarks

1. Under the assumptions stated in Theorem 9.4 no unobservable modes of the composite system are located on the imaginary axes, provided

$$\begin{bmatrix} A - j\omega I & B_2 \\ C_1 & D_{12} \end{bmatrix} \text{ has full rank column for } 0 \leq \omega \leq \infty$$

A similar statement could be made regarding the presence of uncontrollable modes on the imaginary axis.

2. The assumptions that

$$\begin{bmatrix} A - j\omega I & B_2 \\ C_1 & D_{12} \end{bmatrix} \text{ has full column rank for } 0 \leq \omega \leq \infty$$

and

$$\begin{bmatrix} A - j\omega I & B_1 \\ C_2 & D_{21} \end{bmatrix} \text{ has full row rank for } 0 \leq \omega \leq \infty$$

are necessary for the existence of stabilizing solutions for the associated Riccati equations in controller synthesis problems. This point is further elaborated in the next chapter.

9.5 Reduction to the Standard Problem

Many $\mathcal{H}_\infty$ optimization problems could be embedded in the generalized format resulting in what is known as the standard problem. In simple terms, the standard problem could be described as follows.

Given the plant transfer function $\boldsymbol{G}$, find a real rational matrix function $\boldsymbol{K}$ such that it minimizes the $\mathcal{H}_\infty$ norm of the transfer matrix. $\boldsymbol{T}_{zw}$ (see Figure 9.1) under the constraint that $\boldsymbol{K}$ internally stabilizes $\boldsymbol{G}$.

The procedure of reducing the given problem to standard form is now illustrated through a few examples

EXAMPLE 9.1
Consider a weighted sensitivity minimization problem $\boldsymbol{X} = \boldsymbol{W}_1(\boldsymbol{I} - \boldsymbol{GK})^{-1}\boldsymbol{W}_2$. This may be converted to a standard problem as follows

$$\begin{aligned} \boldsymbol{X} &= \boldsymbol{W}_1[\boldsymbol{I} + \boldsymbol{GK}(\boldsymbol{I} - \boldsymbol{GK})^{-1}]\boldsymbol{W}_2 \\ &= \boldsymbol{W}_1\boldsymbol{W}_2 + \boldsymbol{W}_1\boldsymbol{GK}(\boldsymbol{I} - \boldsymbol{GK})^{-1}\boldsymbol{W}_2 \end{aligned} \tag{9.49}$$

We know that for the generalized plant P

$$\mathcal{F}_l(\boldsymbol{P}, \boldsymbol{K}) = \boldsymbol{P}_{11} + \boldsymbol{P}_{12}K(I - \boldsymbol{P}_{22}K)^{-1}\boldsymbol{P}_{21} \tag{9.50}$$

Comparing (9.69) and (9.50) we have

$$\boldsymbol{P}_{11} = \boldsymbol{W}_1\boldsymbol{W}_2; \boldsymbol{P}_{12} = \boldsymbol{W}_1\boldsymbol{G}; \boldsymbol{P}_{22} = \boldsymbol{G}; \boldsymbol{P}_{21} = \boldsymbol{W}_2$$

Hence

$$\boldsymbol{P} = \begin{bmatrix} \boldsymbol{W}_1\boldsymbol{W}_2 & \boldsymbol{W}_1\boldsymbol{G} \\ \boldsymbol{W}_2 & \boldsymbol{G} \end{bmatrix}$$

and

$$\mathcal{F}_l(\boldsymbol{P}, \boldsymbol{K}) = \boldsymbol{X}$$

EXAMPLE 9.2
We are seeking a stabilizing controller $\boldsymbol{K}$ which ensures that

$$\left\| \begin{bmatrix} \boldsymbol{W}_1(\boldsymbol{I} - \boldsymbol{GK})^{-1} \\ \boldsymbol{W}_2\boldsymbol{K}(\boldsymbol{I} - \boldsymbol{GK})^{-1} \end{bmatrix} \right\|_\infty < 1$$

It is required to derive the generalized plant for this problem. Note that $\boldsymbol{W}_1$ and $\boldsymbol{W}_2$ are frequency dependent weighting matrices. In this problem we are demanding that there be good disturbance rejection properties and that the stability of $\boldsymbol{G}$ be maintained in the presence of unstructured additive uncertainty. We have

$$\boldsymbol{W}_1(\boldsymbol{I} - \boldsymbol{GK})^{-1} = \boldsymbol{W}_1[\boldsymbol{I} + \boldsymbol{GK}(\boldsymbol{I} - \boldsymbol{GK})^{-1}]$$

$$\boldsymbol{W}_2\boldsymbol{K}(\boldsymbol{I} - \boldsymbol{GK})^{-1} = 0 + \boldsymbol{W}_2\boldsymbol{K}(\boldsymbol{I} - \boldsymbol{GK})^{-1}$$

$$\begin{bmatrix} \boldsymbol{W}_1(\boldsymbol{I} - \boldsymbol{GK})^{-1} \\ \boldsymbol{W}_2\boldsymbol{K}(\boldsymbol{I} - \boldsymbol{GK})^{-1} \end{bmatrix} = \begin{bmatrix} \boldsymbol{W}_1 \\ 0 \end{bmatrix} + \begin{bmatrix} \boldsymbol{W}_1\boldsymbol{G} \\ \boldsymbol{W}_2 \end{bmatrix} \boldsymbol{K}(\boldsymbol{I} - \boldsymbol{GK})^{-1}\boldsymbol{I}$$

It is now straightforward to determine $\boldsymbol{P}$. We have

$$\boldsymbol{P} = \begin{bmatrix} \boldsymbol{W}_1 & \boldsymbol{W}_1\boldsymbol{G} \\ 0 & \boldsymbol{W}_2 \\ \boldsymbol{I} & \boldsymbol{G} \end{bmatrix} \tag{9.51}$$

Assume that $\boldsymbol{G}$, $\boldsymbol{W}_1$ and $\boldsymbol{W}_2$ have the following realizations

$$\boldsymbol{G} : (A, B, C, D); \quad \boldsymbol{W}_1 : (A_1, B_1, C_1, D_1)$$
$$\boldsymbol{W}_2 : (A_2, B_2, C_2, D_2)$$

The structure of $\boldsymbol{P}$ as given in (9.51) suggests that

$$\boldsymbol{P} = \begin{bmatrix} \boldsymbol{P}_{11} & \boldsymbol{P}_{12} \\ \boldsymbol{P}_{21} & \boldsymbol{P}_{22} \\ \boldsymbol{P}_{31} & \boldsymbol{P}_{32} \end{bmatrix} \tag{9.52}$$

Note that the realization of $\boldsymbol{W}_1\boldsymbol{G}$, given the realizations of $\boldsymbol{W}_1$ and $\boldsymbol{G}$ is

$$W_1G : \left(\begin{bmatrix} A_1 & B_1C \\ 0 & A \end{bmatrix}, \begin{bmatrix} B_1D \\ B \end{bmatrix}, [C_1 \quad D_1C], D_1D \right)$$

Hence the 'A matrix' associated with $\boldsymbol{P}$ has the structure

$$\begin{bmatrix} A_1 & B_1C & 0 \\ 0 & A & 0 \\ 0 & 0 & A_2 \end{bmatrix}$$

We have

$$\boldsymbol{P} = \begin{bmatrix} C_1(sI - A_1)^{-1}B_1 + D_1 & [C_1 \quad D_1C]\left(sI - \begin{bmatrix} A_1 & B_1C \\ 0 & A \end{bmatrix}\right)^{-1}\begin{bmatrix} B_1D \\ B \end{bmatrix} + D_1D \\ 0 & C_2(sI - A_2)^{-1}B_2 + D_2 \\ I & C(sI - A)^{-1}B + D \end{bmatrix} \tag{9.53}$$

We know from (9.52) that the 'B matrix' associated with $\boldsymbol{P}$ has two columns and the 'C matrix' associated with $\boldsymbol{P}$ has three rows. Bearing this in mind and by inspecting the structure of $\boldsymbol{P}$ as given in (9.53) we may write down a state space realization of $\boldsymbol{P}$ as follows

$$\boldsymbol{P} = \left[\begin{array}{ccc|cc} A_1 & B_1C & 0 & B_1 & B_1D \\ 0 & A & 0 & 0 & B \\ 0 & 0 & A_2 & 0 & B_2 \\ \hline C_1 & D_1C & 0 & D_1 & D_1D \\ 0 & 0 & C_2 & 0 & D_2 \\ 0 & C & 0 & I & D \end{array}\right]$$

EXAMPLE 9.3

(Adapted from Francis and Doyle 1987). This is a tracking problem where the plant $\boldsymbol{G}$ with output v is to track a reference signal r. The plant input u is generated by two controllers C_1 and C_2 as shown in Figure 9.7. The input r is not a known fixed signal, but is modelled as

$$\{r : r = \boldsymbol{W}w \text{ for } w \in \mathcal{H}_2 \text{ and } \|w\|_2 \le 1\}$$

We are adopting a two parameter controller (comprising C_1 and C_2) which is to be designed.

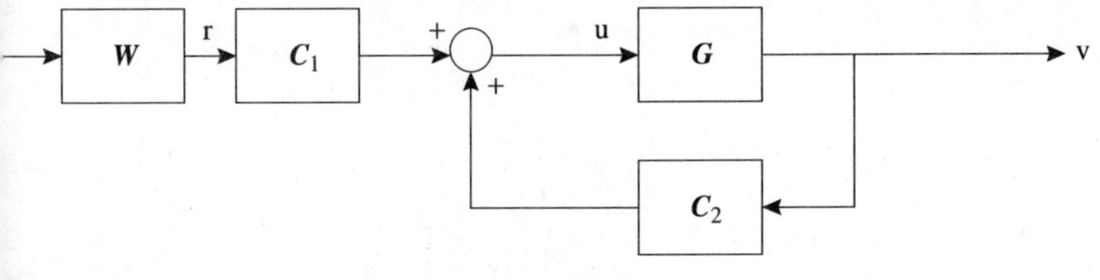

Figure 9.7 Two parameter controller for tracking.

The plant transfer function $\boldsymbol{G}$ and its weighting function $\boldsymbol{W}$ are given. The tracking error is $r - v$ and the cost to be minimized is $(\|r - v\|_2^2 + \|\rho u\|_2^2)^{\frac{1}{2}}$ where ρ is a positive scalar weighting function which places a constraint on high inputs being applied to the plant.

The problem is first recast in the generalized mould. As usual the inputs to the generalized plant are labelled w and u and the outputs z and y. From the cost function we identify z. We have

$$z = \begin{bmatrix} r - v \\ \tau u \end{bmatrix}$$

The tracking criteria is now defined as

$$\sup\{\|z\|_2 : w \in \mathcal{H}_2, \quad \|w\|_2 \leq 1\} \text{ is a minimum.}$$

From Figure 9.7 we have

$$\begin{aligned} u &= \boldsymbol{C}_1 r + \boldsymbol{C}_2 v \\ &= [\boldsymbol{C}_1 \quad \boldsymbol{C}_2] \begin{bmatrix} r \\ v \end{bmatrix} \end{aligned}$$

The plant output is hence represented as $\begin{bmatrix} r \\ v \end{bmatrix}$ and the controller $\boldsymbol{K}$ has the form $\boldsymbol{K} = [\,C_1 \quad C_2\,]$. The generalized plant has the standard form

$$\boldsymbol{P} = \begin{bmatrix} \boldsymbol{P}_{11} & \boldsymbol{P}_{12} \\ \boldsymbol{P}_{21} & \boldsymbol{P}_{22} \end{bmatrix}$$

and

$$\begin{aligned} z &= \boldsymbol{P}_{11} w + \boldsymbol{P}_{12} u \\ \begin{bmatrix} r - v \\ \rho u \end{bmatrix} &= \begin{bmatrix} \boldsymbol{W} w - \boldsymbol{G} u \\ \rho u \end{bmatrix} \\ &= \begin{bmatrix} \boldsymbol{W} \\ 0 \end{bmatrix} w + \begin{bmatrix} -\boldsymbol{G} \\ \rho \end{bmatrix} u \end{aligned}$$

Also

$$\begin{aligned} y &= \boldsymbol{P}_{21} w + \boldsymbol{P}_{22} u \\ \begin{bmatrix} r \\ v \end{bmatrix} &= \begin{bmatrix} \boldsymbol{W} w \\ \boldsymbol{G} u \end{bmatrix} \\ &= \begin{bmatrix} \boldsymbol{W} \\ 0 \end{bmatrix} w + \begin{bmatrix} 0 \\ \boldsymbol{G} \end{bmatrix} u \end{aligned}$$

Hence in standard form

$$
\boldsymbol{P} = \begin{bmatrix} \boldsymbol{W} & -\boldsymbol{G} \\ 0 & \rho \boldsymbol{I} \\ \boldsymbol{W} & 0 \\ 0 & \boldsymbol{G} \end{bmatrix}
$$

9.6 From Standard Form to Model Matching

In the generalized plant-controller problem we saw how the transfer function connecting z and w is given by (refer to (9.45))

$$
\boldsymbol{T}_{zw} = \boldsymbol{T}_{11} + \boldsymbol{T}_{12}\boldsymbol{Q}\boldsymbol{T}_{21}
$$

which corresponds to the LFT

$$
\boldsymbol{T}_{zw} = \mathcal{F}_l(\boldsymbol{T}_1\boldsymbol{Q})
$$

$$
\text{where } \boldsymbol{T} = \begin{bmatrix} \boldsymbol{T}_{11} & \boldsymbol{T}_{12} \\ \boldsymbol{T}_{21} & 0 \end{bmatrix}
$$

Regarding the sizes of matrices $\boldsymbol{T}_{12}$ and $\boldsymbol{T}_{21}$ the following possibilities may be envisaged:

1. $\boldsymbol{T}_{12}$ and $\boldsymbol{T}_{21}$ are both square. This results in $\mathcal{H}_\infty$ optimization problems of the first kind.
2. Either $\boldsymbol{T}_{12}$ has more rows than columns with $\boldsymbol{T}_{21}$ square or $\boldsymbol{T}_{21}$ have more columns than rows with $\boldsymbol{T}_{12}$ square. This results in $\mathcal{H}_\infty$ optimization problems of the second kind.
3. $\boldsymbol{T}_{12}$ has more rows than columns and $\boldsymbol{T}_{21}$ has more columns than rows. This gives rise to problems of the third kind.

Problems of the first kind lead to 1-block problems; problems of the second kind lead to 2-block problems and problems of the third kind lead to 4-block problems. These problems have already been discussed in Chapter 6.

In view of the above classification, we will now examine some of the common $\mathcal{H}_\infty$ optimization problems and ascertain to what category they belong.

Consider first, the weighted sensitivity minimization problem discussed in Example 9.1. If the plant $\boldsymbol{G}$ is assumed to be square then both $\boldsymbol{P}_{12}$ and $\boldsymbol{P}_{21}$ are square ($\boldsymbol{P}_{12} = \boldsymbol{W}_1\boldsymbol{G}$ and $\boldsymbol{P}_{21} = \boldsymbol{W}_2$). Hence the optimization problem belongs to the first kind.

In the case of the so called mixed sensitivity problem discussed in example 9.2 again assuming that G is square we have

$$P_{12} = \begin{bmatrix} W_1 G \\ W_2 \end{bmatrix}$$

which is non-square with more number of rows than columns and $P_{21} = I$ is a square matrix. The problem therefore belongs to the second kind. ■

Reverting back to the LFT defined as:

$$\mathcal{F}_l(T, Q) = T_{11} + T_{12} Q T_{21}$$

we have already obtained the state space realizations of T_{11}, T_{12} and T_{21} in (9.41), (9.42) and (9.43) respectively. It is also possible to characterize these transfer matrices in terms of known transfer functions. This is the frequency domain approach. We discuss this approach in the sequel.

Note that

$$\begin{aligned} \mathcal{F}_l(T, Q) &= \mathcal{F}_l(G, K) \\ &= G_{11} + G_{12} K (I - G_{22} K)^{-1} G_{21} \end{aligned} \tag{9.54}$$

where K is the stabilizing controller.

But we know from Theorem 9.3 that the controller which stabilizes G_{22} also stabilizes G. According to 9.22 the controller which internally stabilizes G_{22} is given by

$$K = (D_r Q - U_l)(N_r Q + V_l)^{-1}, \quad Q \in \mathcal{RH}_\infty$$

we now evaluate $\mathcal{F}_l(G, K)$ by substituting for K, the parametrized matrix given above.

Thus we have

$$(I - G_{22} K)^{-1} = (N_r Q + V_l) D_l$$

and

$$K (I - G_{22} K)^{-1} = (D_r Q - U_l) D_l$$

Substituting for $K(I - G_{22}K)^{-1}$ in (9.54) we get

$$\begin{aligned} \mathcal{F}_l(G, K) &= G_{11} + G_{12}(D_r Q - U_l) D_l G_{21} \\ &= (G_{11} - G_{12} U_l D_l G_{21}) + G_{12} D_r Q D_l G_{21} \end{aligned} \tag{9.55}$$

From (9.55) we can readily identify

$$\begin{aligned} T_{11} &= G_{11} - G_{12}U_lD_lG_{21} \\ T_{12} &= G_{12}D_r \\ T_{21} &= D_lG_{21} \end{aligned}$$

Hence

$$T = \begin{bmatrix} G_{11} - G_{12}U_lD_lG_{21} & G_{12}D_r \\ D_lG_{21} & 0 \end{bmatrix} \tag{9.56}$$

Since G could be an unstable transfer matrix there is a possibility of its submatrices also being unstable. However it is important to note that T_{11}, T_{12} and T_{21} are always stable.

In the special case when G is stable, G_{22} is stable and $N_r = N_l = G_{22}$ and $D_r = D_l = I$. Making use of the Bezout identity we can now establish that $V_r = V_l = I$ and $U_r = U_l = 0$. Hence in this special case $K = Q(I + G_{22}Q)^{-1}$.

9.7 Summary

The generalized plant and controller in Figure 9.1 can be characterized in a LFT framework. Thus we obtain

$$T_{zw} = \mathcal{F}_l(G_1, K) = G_{11} + G_{12}K(I - G_{22}K)^{-1}G_{21}$$

Internal stability for the closed loop system involves the stability of nine transfer matrices. Theorem 9.1 asserts that the feedback system is internally stable if and only if the 'A matrix' of the closed loop system is asymptotically stable. Theorem 9.2 which deals with stabilizability states that internal stability can be achieved if and only if the subsystem $G_{22}(A, B_2, C_2)$ is stabilizable and detectable. The next theorem states that a controller which stabilizes G_{22} automatically stabilizes G. The converse is also true. This theorem enables us to design a stabilizing controller for the subsystem G_{22} which will also guarantee internal stability for the composite system. The controller thus obtained may be expressed in terms of LFT as follows

$$K = \mathcal{F}_l(K_0, Q), \quad Q \in \mathcal{RH}_\infty$$

Representation of K_0 is possible both in the frequency domain (refer (9.31)) or in the time domain (refer (9.32)). The composition of two LFTs associated with G and K_0 leads to an equivalent LFT and the original problem may now be rephrased as

$$T_{zw} = \mathcal{F}_l(T, Q), \quad Q \in \mathcal{RH}_\infty$$

Here again, both the state space and frequency domain representations of T are derived in (9.40) and (9.56) respectively. It is also shown (Theorem 9.4) that if K is represented by its minimal realization, then every unobservable mode of the composite system is a zero of G_{12} and every uncontrollable mode of the composite system is a zero of G_{21}. It may be noted in passing that in the $\mathcal{H}_\infty$ optimization problem or in the associated NP interpolation problem, the set of interpolation points are precisely the right half plane zeros of T_{12} and T_{21} where the function to be optimized is in the model matching format $T_{11} - T_{12}QT_{21}$.

Given an $\mathcal{H}_\infty$ optimization problem, the path we follow is to first convert it to the standard form and thereafter reduce it to the model matching format. Three examples are given to illustrate this method. The complexity of the model matching problem depends upon whether the matrices T_{12} and T_{21} are square or rectangular. In the case where both are square inner matrices the norm preserving properties may be exploited and we may write

$$\|T_{11} - T_{12}QT_{21}\|_\infty = \|T_{12}^* T_{11} T_{21}^* - Q\|_\infty$$

which is the Hankel approximation problem. In the event that T_{12} has more rows than columns and T_{21} has more columns than rows, then the problem is complicated involving spectral and inner-outer factorization.

In the next chapter we outline a method for solving the model matching problem which marks a departure from earlier approaches and is computationally superior.

Notes and Additional References

$\mathcal{H}_\infty$ optimization studies in control engineering are of recent origin. One of the earliest books relating to design of feedback controllers using $\mathcal{H}_\infty$ optimization techniques is by Francis (1987). Francis and Doyle (1987) give a tutorial exposition of the topic and covers a lot of ground unfamiliar to design engineers at the time of their writing. An undesirable feature of controller design is that one invariably ends up with a controller whose McMillan degree is much higher than that of the plant itself. In this context the papers of Limebeer and Hung (1987) and Limebeer and Halikias (1988) are relevant. They showed that for $\mathcal{H}_\infty$ problems of the first and second kind, the associated controller requires no more than $(n - 1)$ states where n is the McMillan degree of the plant transfer function. Two papers on $\mathcal{H}_\infty$ optimal control by Hung (1989a and b) discuss in detail, the model matching problem and the synthesis of controllers.

Safonov *et al.* (1987), in a tutorial paper, discusses the application of $\mathcal{H}_\infty$ optimal control theory to the synthesis of MIMO feedback controllers. Some of these developments are adequately covered in a recent book by Green and Limebeer (1995).

Exercises

1. Two LFTs are interconnected as indicated in the figure below.

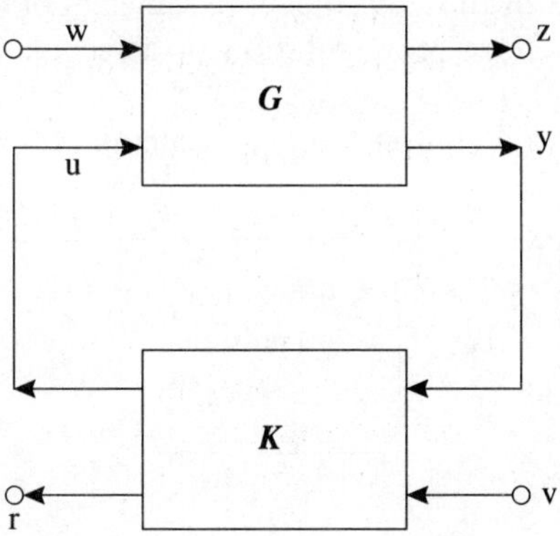

$$G = \begin{bmatrix} G_{11} & G_{12} \\ G_{21} & G_{22} \end{bmatrix} \text{ and } K = \begin{bmatrix} K_{11} & K_{12} \\ K_{21} & K_{22} \end{bmatrix}$$

Show that the composition of the two LFTs is another LFT. The composition operator $C_l(G, K)$ is a 2×2 block matrix mapping $[w' \quad v']' \rightarrow [z' \quad r']'$

2. The LFT associated with a generalized plant feedback configuration is given by

$$\mathcal{F}_l(G, K) = G_{11} + G_{12}K(I - G_{22}K)^{-1}G_{21}$$

Another form of LFT is defined as

$$\Phi(\widetilde{G}, K) = (\widetilde{G}_{11} + K\widetilde{G}_{21})^{-1}(\widetilde{G}_{12} + K\widetilde{G}_{22})$$

where:

$$G = \begin{bmatrix} G_{11} & G_{12} \\ G_{21} & G_{22} \end{bmatrix} = \left[\begin{array}{c|cc} A & B_1 & B_2 \\ \hline C_1 & D_{11} & D_{12} \\ C_2 & D_{12} & D_{22} \end{array}\right]$$

$$\tilde{G} = \begin{bmatrix} \tilde{G}_{11} & \tilde{G}_{12} \\ \tilde{G}_{21} & \tilde{G}_{22} \end{bmatrix} = \left[\begin{array}{c|cc} \tilde{A} & \tilde{B}_1 & \tilde{B}_2 \\ \hline \tilde{C}_1 & \tilde{D}_{11} & \tilde{D}_{12} \\ \tilde{C}_2 & \tilde{D}_{21} & \tilde{D}_{22} \end{array}\right]$$

Assuming that G_{12} and $\tilde{G}_{11}$ are square and invertible and that $\mathcal{F}_l(G, K) = \Phi(\tilde{G}, K)$. Obtain the transfer matrix $\tilde{G}$ in terms of the block matrix elements of G. Also derive the state model representation of $\tilde{G}$, in terms of the state model parameters of G.

3. The block diagram representation of a feedback system is given on page 223.

   The inputs and outputs of the generalized plant P are identified below.

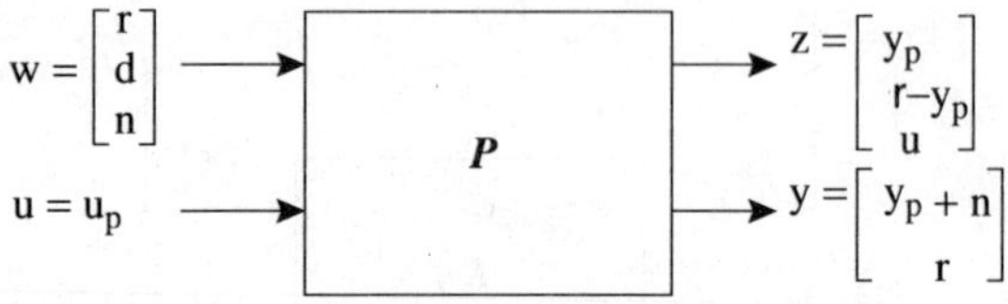

Show that

$$P = \begin{bmatrix} 0 & W_1 & 0 & P_1 \\ I & -W_1 & 0 & -P_1 \\ 0 & 0 & 0 & I \\ 0 & W_1 & I & P_1 \\ I & 0 & 0 & 0 \end{bmatrix}$$

4. This is a model matching problem where the output of the model M is matched against the output of the plant P_1. The exogenous input is r, the control input is u_p and the controlled outputs are the model matching error z_1 and the weighted actuator effort z_2. In practice a low pass weight W_1 for z_1 ensures good model match in the frequency range of interest. Similarly a low pass weight W_2 for z_2 can reflect the rate constraints of the actuator. These two weights have been omitted

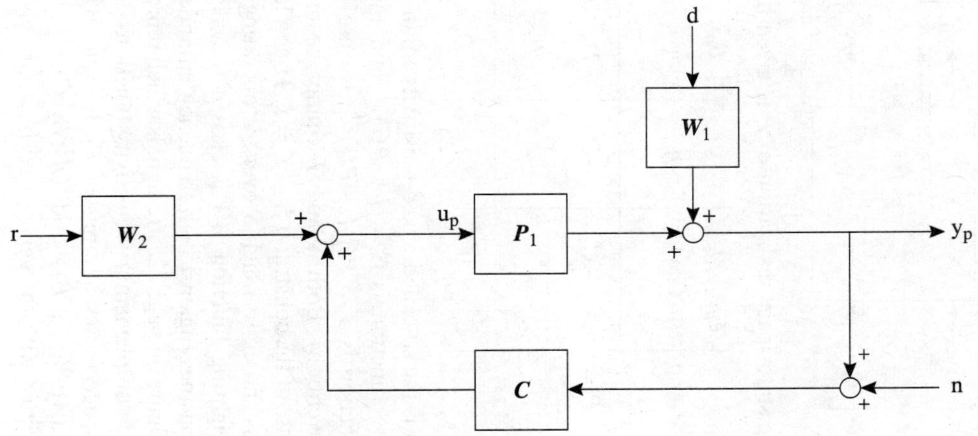

Figure pertaining to Problem 3.

to simplify the problem. The magnitude constraint on the actuator effort is reflected in ρ which is included. A schematic diagram of the feedback system is given on page 225.

The model $\boldsymbol{M}$ and plant $\boldsymbol{P}_1$ have state space realizations (A_m, B_m, C_m, D_m) and (A_p, B_p, C_p, D_p) respectively. The generalized plant is represented as follows.

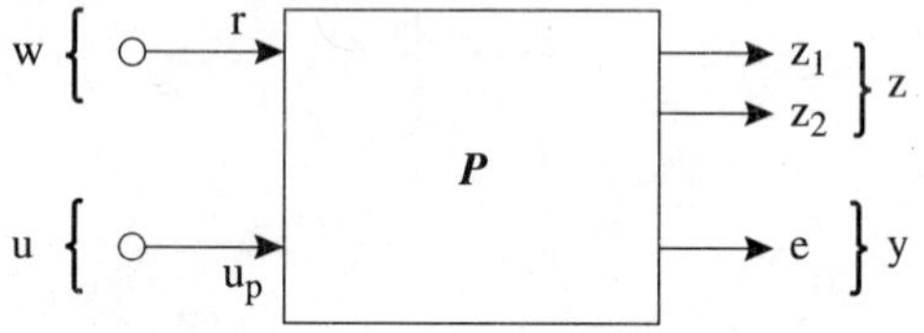

Show that the state space realization of $\boldsymbol{P}$ is given by

$$\boldsymbol{P} = \left[\begin{array}{cc|c|c} A_p & 0 & 0 & B_p \\ 0 & A_m & B_m & 0 \\ \hline C_p & -C_m & -D_m & D_p \\ 0 & 0 & 0 & \rho I \\ \hline C_p & 0 & I & -D_p \end{array}\right]$$

5. This problem is the so-called mixed sensitivity minimization problem. The sensitivity function is $\boldsymbol{S} = (I + \boldsymbol{PC})^{-1}$ and the complementary sensitivity function is $\boldsymbol{T} = \boldsymbol{PC}(I + \boldsymbol{PC})^{-1}$ (we use here the negative feedback convention). Both $\boldsymbol{S}$ and $\boldsymbol{T}$ cannot be minimized simultaneously because of the constraint $\boldsymbol{S} + \boldsymbol{T} = I$. Thus a low pass weighting function $\boldsymbol{W}_1$ is chosen to weigh $\boldsymbol{S}$ over the operating bandwidth and a high pass weighting function $\boldsymbol{W}_2$ is chosen to weigh $\boldsymbol{T}$ in order to guarantee robustness against high frequency unmodelled dynamics. z_1 and z_2 in the figure on page 225 are signals which represent the weighted sensitivity and complementary sensitivity functions.

The state space representations of P_1, W_1 and W_2 are respectively (A, B, C, D), (A_1, B_1, C_1, D_1) and (A_2, B_2, C_2, D_2). The inputs and outputs associated with the generalized plant are given below.

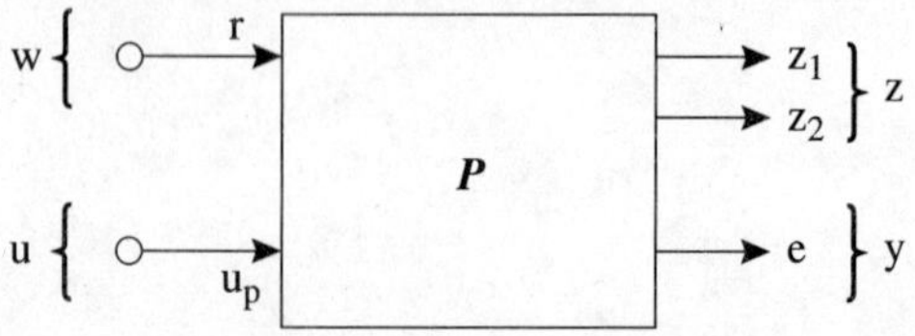

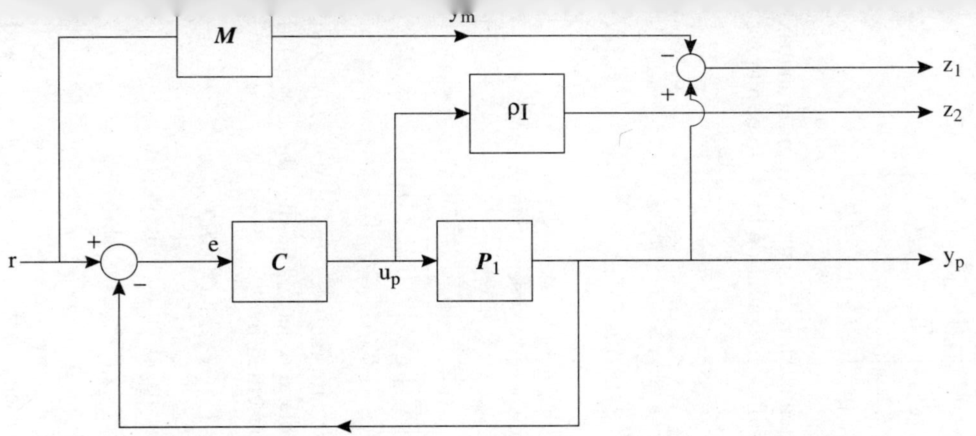

Figure pertaining to Problem 4.

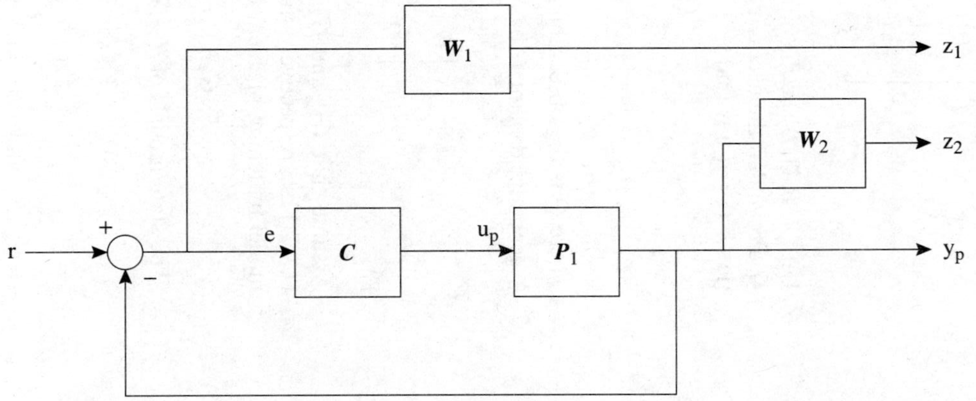

Figure pertaining to Problem 5.

Verify that the state space realization of P is given by

$$\boldsymbol{P} = \left[\begin{array}{ccc|c|c} A & 0 & 0 & 0 & B \\ -B_1C & A_1 & 0 & B_1 & -B_1D \\ B_2C & 0 & A_2 & 0 & B_2D \\ \hline -D_1C & C_1 & 0 & D_1 & -D_1D \\ D_2C & 0 & C_2 & 0 & D_2D \\ \hline -C & 0 & 0 & I & -D \end{array}\right]$$

6. Let (A, B, C, D) be the minimal realization of a square transfer matrix $\boldsymbol{G}$. Prove that the necessary and sufficient condition for $\boldsymbol{G}$ to be all pass (ie $\boldsymbol{G}^{\sim}\boldsymbol{G} = I$) is that

$$A'Q + QA + C'C = 0$$
$$D'C + B'Q = 0$$
$$D'D = I$$

where Q is the observability gramian.

7. A generalized plant $\boldsymbol{G}$ has a state space representation given by

$$\boldsymbol{G} = \begin{bmatrix} \boldsymbol{G}_{11} & \boldsymbol{G}_{12} \\ \boldsymbol{G}_{21} & \boldsymbol{G}_{22} \end{bmatrix} = \left[\begin{array}{c|cc} A & B_1 & B_2 \\ \hline C_1 & 0 & I \\ C_2 & I & 0 \end{array}\right]$$

Assume that $\boldsymbol{G}_{12}$ and $\boldsymbol{G}_{21}$ are square matrices and that (A, B_2) and (C_2, A) are respectively stabilizable and detectable. Let the associated model matching problem be represented by

$$\mathcal{F}_l(\boldsymbol{G}, \boldsymbol{K}) = \boldsymbol{T}_{11} + \boldsymbol{T}_{12}\boldsymbol{Q}\boldsymbol{T}_{21}, \quad \boldsymbol{Q} \in \mathcal{RH}_\infty$$

The state models of $\boldsymbol{T}_{12}$ and $\boldsymbol{T}_{21}$ are given as follows

$$\boldsymbol{T}_{12} : [(A + B_2F), B_2, (C_1 + F), I]$$
$$\boldsymbol{T}_{21} : [(A + HC_2), (B_1 + H), C_2, I]$$

Here F and H are so chosen that $F = -(B_2'X + C_1)$ and $H = -(YC_2' + B_1)$ where X and Y are the stabilizing solutions of the Riccati equations

$$X(A - B_2C_1) + (A - B_2C_1)'X - XB_2B_2'X = 0$$
$$(A - B_1C_2)Y + Y(A - B_1C_2)' - YC_2'C_2Y = 0$$

Show that the transfer matrices T_{12} and T_{21} are square and all pass. Hence prove that

$$\mathcal{F}_l(G, K) = \|T_{12}^{\sim} T_{11} T_{21}^{\sim} + Q\|_{\infty} = \|R + Q\|_{\infty}$$

where $Q \in \mathcal{RH}_{\infty}$ and $R \in \mathcal{RH}_{\infty}^{-}$.

8. Consider a problem similar to the one posed in problem 7 with

$$G = \left[\begin{array}{c|cc} A & B_1 & B_2 \\ \hline C_1 & 0 & D_{12} \\ C_2 & D_{21} & 0 \end{array}\right]$$

where D_{12} and D_{21} are square matrices with $D_{12}'D_{12} = I$ and $D_{21}'D_{21} = I$.

(i) obtain state models for T_{12} and T_{21}

(ii) Show that if $F = -(B_2'X + D_{12}'C_1)$ and $H = -(YC_2' + B_1D_{21}')$ where X and Y are the stabilizing solutions of associated Riccati equations, then $T_{12}^{\sim}T_{12}$ and $T_{21}^{\sim}T_{21}$ are constant matrices.

9. In the generalized plant controller configuration $\mathcal{F}_l(G, K) = G_{11} + G_{12}K(I - G_2K)^{-1}G_{21}$. Define $K(I - G_{22}K)^{-1} = Q$. Then $\mathcal{F}_l(G, K) = G_{11} + G_{12}QG_{21}$. We may now treat Q as a parameter and minimize the H_{∞} norm of $\mathcal{F}_l(G, K)$ and at the same time ensure that the closed loop system is internally stable. Compared to the Youla parameterization procedure, this appears to be more directly related to the original problem. But the snag here is that the simplification is achieved at the expense of more complex stability constraints on the parameter Q. Assuming that (A, B_1, C_1) is controllable and observable, (A, B_2) is stabilizable and (C_2, A) is detectable show that K obtained from

$$Q = K(I - G_{22}K)^{-1}$$

stabilizes P if and only if the following four transfer matrices

$$G,\ G_{12}Q,\ QG_{21} \text{ and } G_{11} - G_{12}QG_{21}$$

are asymptotically stable.

10

Controller Synthesis in State Space

10.1 Introduction

There are very many similarities between $\mathcal{H}_2$ and $\mathcal{H}_\infty$ optimal control problems. But there are also important differences. These are best brought out by treating both these problems in a unified state space frame work. However as of now they cannot be derived as special cases of a more general problem. Probably this has to wait for further advances in this area.

One strong unifying concept is that both $\mathcal{H}_2$ and $\mathcal{H}_\infty$ optimization problems are Riccati-equation based. In $\mathcal{H}_2$ optimization the separation principle plays an important role in the solution of the problem. $\mathcal{H}_\infty$ optimization also exhibits an interesting separation structure. Hence treatement of both these problems in the same setting will help bolster our understanding and provide intuitive insight. This is the central theme of this chapter.

In Section 10.2, the Full Information (FI) problem is defined and the underlying assumptions are stated. The family of stabilizing FI controllers are then delineated. In Section 10.3 Kalman filter with particular reference to the generalized regulator problem is taken up for study. This paves the way for the parametrization of all measurement feedback controllers which constitutes the subject matter of Section 10.4. The state space formulation of $\mathcal{H}_\infty$ control problems is then taken up for discussion. Here the problem is to design a suitable controller such that the closed loop system has $\mathcal{H}_\infty$ norm less than an a priori given bound $\gamma > 0$. The existence of such a controller cannot be taken for granted because of the indefinite quadratic term present in the Riccati equation. These issues are discussed in Section 10.5 culminating in the identification of all FI, $\mathcal{H}_\infty$ controllers.

In Section 10.6 the $\mathcal{H}_\infty$ filter is discussed on the same lines as was done for the $\mathcal{H}_2$ filter. However there are important differences between the two as in this case the optimal estimate of the state vector depends on the control gain matrix F unlike in the LQG problem where $\widehat{x}$ the optimal estimate does not depend on F. Finally in Section 10.7 we treat the more general output feedback problem. In the interest of brevity and also considering the scope of this book some of the important results are merely stated and explained without giving the proof. This chapter leans heavily on the definitive paper by Doyle *et al.* (1989). As frequent references are made to this paper we have used the acronym *DGKF* to indicate this source.

10.2 LQG Control – Full Information Problem

In Chapter 9 we derived the set of all controllers which internally stabilize the given plant G (9.28).

We recall that $\boldsymbol{T}_{zw} = \mathcal{F}_l(\boldsymbol{G}, \boldsymbol{K})$ where $\boldsymbol{K}$ is a stabilizing controller. Further $\boldsymbol{K} = \mathcal{F}_l(\boldsymbol{K}_0, \boldsymbol{Q})$ where $\boldsymbol{Q}$ is a free parameter belonging to $\mathcal{RH}_\infty$. We thus have

$$\begin{aligned} \boldsymbol{T}_{zw} &= \mathcal{F}_l(\boldsymbol{G}, \boldsymbol{K}) = \mathcal{F}_l(\boldsymbol{G}, \mathcal{F}_l(\boldsymbol{K}_0, \boldsymbol{Q})) \\ &= \mathcal{F}_l(\boldsymbol{T}, \boldsymbol{Q}) \end{aligned}$$

where $\boldsymbol{T}$ is obtained via the composition of the two LFTs $\boldsymbol{G}$ and K_0. Figure 10.1 shows the LFT associated with the stabilizing controller where

$$\boldsymbol{K}_0 = \left[\begin{array}{c|cc} A + B_2F + HC_2 & -H & B_2 \\ \hline F & 0 & I \\ -C_2 & I & 0 \end{array}\right]$$

While synthesizing controller K_0, we have so chosen F and H such that $A + B_2F$ and $A + HC_2$ are stable matrices. The transfer matrices from y to u

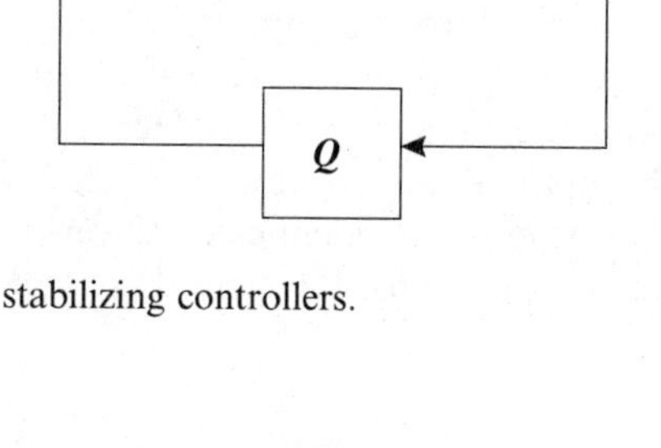

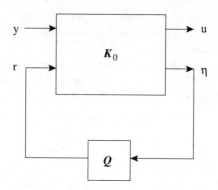

Figure 10.1 LFT for stabilizing controllers.

for $Q \in \mathcal{RH}_\infty$ give the set of all controllers K that internally stabilize the plant G. From standard LQG theory we know that for

$$G = \left[\begin{array}{c|cc} A & B_1 & B_2 \\ \hline C_1 & 0 & D_{12} \\ C_2 & D_{21} & 0 \end{array}\right] \tag{10.1}$$

if we choose $F_2 = -B_2'X_2$ and $H_2 = -Y_2C_2'$ (where X_2 and Y_2 are the solutions of the Riccati equations associated with the control and filter problems) as the control gain and filter gain matrices then, $(A + B_2F_2)$ and $(A + H_2C_2)$ are stable.

For this particular choice of F and H a unique optimal controller is obtained for the LQG problem and K_{opt} has the realization

$$K_{\text{opt}} = (A + B_2F_2 + H_2C_2; -H_2; F_2)$$

From the point of view of parametrization, this corresponds to setting $Q = 0$. For $Q \neq 0$, we obtain sub-optimal stabilizing controllers.

10.2.1 Full Information Problem

The output of the plant is normally used for feedback purpose and in this context we may envisage the following possibilities:

1. The state x and the exogenous input w are available for feedback purpose. In this case:

$$y = \begin{bmatrix} x \\ w \end{bmatrix} = \begin{bmatrix} 1 \\ 0 \end{bmatrix} x + \begin{bmatrix} 0 \\ 1 \end{bmatrix} w$$

   where we identify $C_2 = \begin{bmatrix} 1 \\ 0 \end{bmatrix}$ and $D_{21} = \begin{bmatrix} 0 \\ 1 \end{bmatrix}$. This corresponds to full information provided for feedback purpose.
2. The state x alone is available for feedback and we have the state feedback problem.
3. The output y is a linear combination of the state and exogenous variable i.e. $y = C_2x + D_{21}w$ and we have what is known as the output feedback problem.

There is a certain amount of redundancy built into the full information problem, because if w is known, then x can always be determined from state space equations knowing the initial conditions. Hence it is not surprising that in LQG problems the optimal control law for Full Information (FI) and State feedback (SF) problems are identical. However when we later on consider $\mathcal{H}_\infty$ optimization problems, the central control law for FI and SF

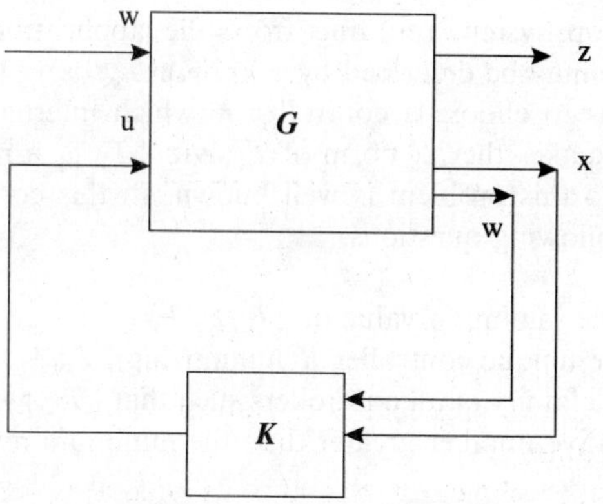

Figure 10.2 Block diagram for full information problems.

are not the same when the direct feedthrough term from disturbance to the controlled output (D_{11}) is not zero. The block diagram for full information problem is given in Figure 10.2.

While considering this problem we assume $D_{11} = 0$ and $D_{22} = 0$. This results in a major reduction of complexity of results obtained pertaining to both $\mathcal{H}_2$ and $\mathcal{H}_\infty$ control. There is not much loss of generality here because even in cases where $D_{11} \neq 0$ and $D_{22} \neq 0$, it is possible to construct an equivalent problem where both D_{11} and D_{22} are zero.

In the FI problem, the structure of $\boldsymbol{G}$ is as given in (10.1). The relevant equations are

$$\dot{x} = Ax + B_1 w + B_2 u \tag{10.2a}$$

$$z = C_1 x + D_{12} u \tag{10.2b}$$

$$y = \begin{bmatrix} I \\ 0 \end{bmatrix} x + \begin{bmatrix} 0 \\ I \end{bmatrix} w \tag{10.2c}$$

We make the following assumptions in this context:

1. (A, B_2) is stabilizable and (C_1, A) is observable on the imaginary axis i.e. $\begin{bmatrix} A - j\omega I \\ C_1 \end{bmatrix}$ has full column rank for all ω
2. $D_{12}'C_1 = 0$ and $D_{12}'D_{12} = I$

Assumption 1 is essential to ensure that the relevant Riccati equation has a stabilizing solution. Assumption 2 suppresses cross terms in the performance integral and also normalizes the coefficient of the $u'u$ term. This makes the problem formulation simpler. Assumption 2 could be dispensed with if required.

The closed loop system resulting from the application of a dynamic feedback $u = \boldsymbol{K}y$ may be described by $z = \boldsymbol{T}_{zw}w$.

The problem is to choose a controller $\boldsymbol{K}$ which internally stabilizes the system and also makes the $\mathcal{H}_2$ norm of $\boldsymbol{T}_{zw}$ viz. $\|T_{zw}\|_2$ a minimum.

The solution to this problem is well known. In this connection we seek answers to the following questions.

i) What is the minimum value of $\|\boldsymbol{T}_{zw}\|_2$?
ii) Obtain the unique controller $\boldsymbol{K}$ minimizing $\|\boldsymbol{T}_{zw}\|_2$.
iii) Obtain the family of all controllers such that $\|\boldsymbol{T}_{zw}\|_2 < \gamma$ where γ is a given positive number greater than the minimum norm of $\|\boldsymbol{T}_{zw}\|_2$.

From LQG theory we know that

$$\min \|\boldsymbol{T}_{zw}\|_2 = [\text{trace } (B_1' X_2 B_1)]^{\frac{1}{2}}$$

We will now obtain answers to the remaining two questions. Before going into detail, we explain some preliminary matters pertaining to the solution. We know from LQG theory that the optimal control law is given by $u = F_2 x$ where $F_2 = -B_2' X_2$. We now define a new control variable $\nu = u - F_2 x$ and proceed to express z in terms of w and ν. We have

$$\begin{aligned} z &= \left[\begin{array}{c|cc} A + B_2 F_2 & B_1 & B_2 \\ \hline C_1 + D_{12} F_2 & 0 & D_{12} \end{array}\right] \begin{bmatrix} w \\ \nu \end{bmatrix} \\ &= \boldsymbol{G}_c B_1 w + \boldsymbol{U}\nu \end{aligned} \tag{10.3}$$

where

$$\boldsymbol{G}_c = \left[\begin{array}{c|c} A_{F2} & I \\ \hline C_{1F2} & 0 \end{array}\right] \text{ and } \boldsymbol{U} = \left[\begin{array}{c|c} A_{F2} & B_2 \\ \hline C_{1F2} & D_{12} \end{array}\right]$$

and $A_{F2} = A + B_2 F_2$ and $C_{1F2} = C_1 + D_{12} F_2$.

In *DGKF* it is proved that $\boldsymbol{U}$ is an inner (i.e., $\boldsymbol{U}^{\sim}\boldsymbol{U} = I$) and $\boldsymbol{U}^{\sim}\boldsymbol{G}_c \in \mathcal{RH}_2^{\perp}$. We make use of both these properties of $\boldsymbol{U}$ in the sequel. We now obtain an LFT associated with G_ν as shown in Figure 10.3.

It is readily seen that

$$G_\nu = \left[\begin{array}{c|cc} A & B_1 & B_2 \\ \hline -F_2 & 0 & I \\ C_2 & D_{12} & 0 \end{array}\right]$$

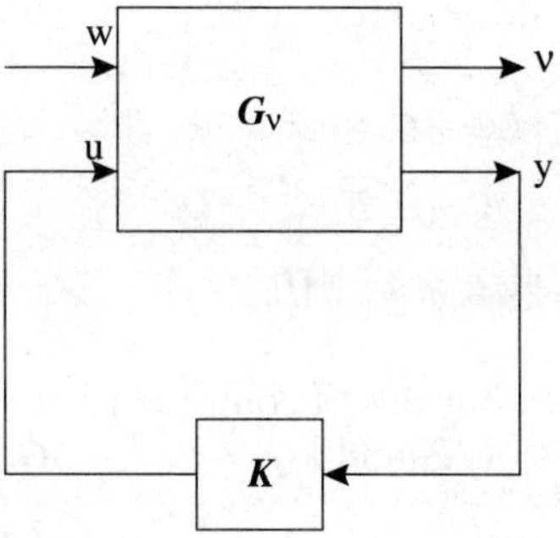

Figure 10.3 LFT associated with G_ν.

Both the composite systems – $\boldsymbol{G}$ along with $\boldsymbol{K}$ and $\boldsymbol{G}_\nu$ along with $\boldsymbol{K}$ – have the same A -matrix. Hence if $\boldsymbol{K}$ stabilizes $\boldsymbol{G}$, it stabilizes $\boldsymbol{G}_\nu$ also. Therefore $\boldsymbol{T}_{\nu w}$ belongs to $\mathcal{RH}_2$.

From Figure 10.3 we note that $\nu = T_{\nu w} w$. Substituting this in (10.3) we get

$$z = (\boldsymbol{G}_c B_1 + \boldsymbol{U}\boldsymbol{T}_{\nu w})w$$

and

$$\begin{aligned} \boldsymbol{T}_{zw} &= \boldsymbol{G}_c B_1 + \boldsymbol{U}\boldsymbol{T}_{\nu w} \\ \|\boldsymbol{T}_{zw}\|_2^2 &= \langle \boldsymbol{T}_{zw}, \boldsymbol{T}_{zw} \rangle \\ &= \langle \boldsymbol{G}_c B_1 + \boldsymbol{U}\boldsymbol{T}_{\nu w}, \boldsymbol{G}_c B_1 + \boldsymbol{U}\boldsymbol{T}_{\nu w} \rangle \\ &= \|\boldsymbol{G}_c \boldsymbol{B}_1\|_2^2 + \|\boldsymbol{T}_{\nu w}\|_2^2 \end{aligned} \tag{10.4}$$

where we made use of the properties that $\boldsymbol{U}^\sim \boldsymbol{U} = I$ and $\boldsymbol{U}^\sim \boldsymbol{G}_c \in \mathcal{RH}_2^\perp$. Because of the latter property $T_{\nu w}$ and $\boldsymbol{U}^\sim \boldsymbol{G}_c$ belong to two different spaces which are orthogonal to each other. Hence their inner product vanishes leading to the above result.

We now delineate the set of all admissible controllers which make $\|\boldsymbol{T}_{zw}\|_2 < \gamma$.

Let $\boldsymbol{K}$ be an admissible controller such that $\|T_{zw}\|_2 < \gamma$. Denote by $\boldsymbol{Q}$ the transfer matrix from w to ν. We have already shown that it belongs to $\mathcal{RH}_2$. From our definition of ν we obtain

$$u = F_2 x + \nu = F_2 x + \boldsymbol{Q}w$$

Hence

$$u = [F_2 \quad \boldsymbol{Q}] \begin{bmatrix} x \\ w \end{bmatrix}$$

Recall from (10.4) that

$$\begin{aligned}\|\boldsymbol{T}_{zw}\|_2^2 &= \|\boldsymbol{G}_c B_1\|_2^2 + \|\boldsymbol{T}_{\nu w}\|^2 \\ &= \|\boldsymbol{G}_c B_1\|_2^2 + \|\boldsymbol{Q}\|_2^2 \\ \|\boldsymbol{T}_{zw}\|_2^2 < \gamma^2 &\Rightarrow \|\boldsymbol{Q}\|_2^2 < \gamma^2 - \|\boldsymbol{G}_c B_1\|_2^2\end{aligned}$$

Hence the family of all stabilizing FI controllers with $\|\boldsymbol{T}_{zw}\|_2 < \gamma$ is given by $\boldsymbol{K} = [F_2 \quad \boldsymbol{Q}]$ where $\boldsymbol{Q} \in \mathcal{RH}_2$ and $\|\boldsymbol{Q}\|_2^2 < \gamma^2 - \|\boldsymbol{G}_c B_1\|_2^2$

10.2.2 Output Feedback – Problem Formulation

We now consider the case where the output y is a linear combination of state and the exogenous variable. The problem now separates itself into two parts one relating to optimal state control and the other relating to optimal state estimation. This is the output feedback problem with the plant description as given in (10.1).

The *Standard assumptions* made are:

1. (A, B_2) is stabilizable and (C_1, A) is observable on the imaginary axis.
2. (C_2, A) is detectable and (A, B_1) is controllable on the imaginary axis.
3. $D'_{12}C_1 = 0$ and $D'_{12}D_{12} = I$
4. $B_1 D'_{21} = 0$ and $D_{21}D'_{21} = I$

The significance of these assumptions have already been explained in the FI case. Precisely the same reasons also hold for the output feedback case.

We are now seeking a time invariant, linear, internally stabilizing controller $u = \boldsymbol{K}y$ which minimizes

$$\|\boldsymbol{T}_{zw}\|_2 = \lim_{T\to\infty} \mathcal{E}\left\{\frac{1}{T}\int_0^T z'z\mathrm{d}t\right\}^{\frac{1}{2}}$$

subject to the governing equations associated with the problem.

We have $z'z = (C_1x + D_{12}u)'(C_1x + D_{12}u)$.

In view of the assumptions that $D'_{12}D_{12} = I$ and $D'_{12}C_1 = 0$ we have

$$z'z = x'C'_1C_1x + u'u$$

and

$$\|\boldsymbol{T}_{zw}\|_2^2 = \lim_{T\to\infty} \mathcal{E}\left\{\frac{1}{T}\int_0^T (x'C'_1C_1x + u'u)\mathrm{d}t\right\}$$

We know from LQG theory that for the state feedback case (with no disturbances) the optimal u^* is given by

$$u^* = -B_2'Xx = Fx$$

where X is the positive semidefinite solution of the control Riccati equation

$$A'X + XA - XB_2B_2'X + C_1'C_1 = 0 \tag{10.5}$$

which stabilizes the A matrix of the feedback system. Hence we require that $(A - B_2B_2'X)$ be asymptotically stable. The closed loop dynamics is given by

$$\dot{x} = (A - B_2B_2'X)x + B_1w$$

If (C_1, A) has unobservable modes on the imaginary axis then for some ω

$$Ax = j\omega x \text{ and } C_1x = 0$$

Multiplying (10.5) on the left by x^* and on the right by x and making use of the above relationships we have $B_2'Xx = 0$ from which it follows that $(A - B_2B_2'X)x = jwx$ which violates our requirement that $(A - B_2B_2'X)$ be asymptotically stable. This justifies the necessity of the assumption that (C_1, A) be observable on the imaginary axis.

Presence of Cross-Coupling Terms

If we relax the condition that $D_{12}'C_1 = 0$ while retaining the assumption $D_{12}'D_{12} = I$, we have

$$\begin{aligned} z'z &= (C_1x + D_{12}u)'(C_1x + D_{12}u) \\ &= x'C_1'(I - D_{12}D_{12}')C_1x + (u + D_{12}'C_1x)'(u + D_{12}'C_1x) \end{aligned}$$

Note that since $D_{12}'D_{12} = I$ we have $(I - D_{12}D_{12}') \geq 0$.

Let

$$\tilde{u} = (u + D_{12}'C_1x)$$

and

$$\tilde{C}_1'\tilde{C}_1 = C_1'(I - D_{12}D_{12}')C_1$$

We then have

$$z'z = x'\tilde{C}_1'\tilde{C}_1x + \tilde{u}'\tilde{u}$$

Further

$$\begin{aligned} Ax + B_2u &= Ax + B_2(\tilde{u} - D_{12}'C_1x) \\ &= (A - B_2D_{12}'C_1)x + B_2\tilde{u} \end{aligned}$$

The closed loop dynamics is thus given by

$$\dot{x} = \tilde{A}x + B_2\tilde{u}$$

with

$$z'z = x'\tilde{C}_1'\tilde{C}_1x + \tilde{u}'\tilde{u}$$

where

$$\tilde{A} = A - B_2D_{12}'C_1$$

The corresponding Riccati equation is given by

$$\tilde{A}'X + X\tilde{A} - XB_2B_2'X + \tilde{C}_1'\tilde{C}_1 = 0 \tag{10.6}$$

For a stabilizing solution we require $(\tilde{A}, B_2)$ to be stabilizable and $(\tilde{C}_1, \tilde{A})$ to have no unobservable modes on the imaginary axis. Since $\tilde{A}$ can be obtained from A by state feedback, and since stabilizability is invariant under state feedback we have (A, B_2) stabilizable $\Rightarrow$ $(\tilde{A}, B_2)$ stabilizable. Further, it can be shown that the requirement that $(\tilde{C}_1, \tilde{A})$ has no unobservable modes on the imaginary axis is equivalent to the condition that $\begin{bmatrix} A - j\omega I & B_2 \\ C_1 & D_{12} \end{bmatrix}$ has full column rank for all ω.

Provided the above conditions are satisfied there is no loss of generality in assuming that no cross coupling terms are present in $z'z$. We shall assume this in all our subsequent discussions. With cross coupling present, the optimal control law is $\tilde{u}^* = -B_2'Xx$ and $u^* = -B_2'Xx - D_{12}'C_1x$. The optimal gain matrix is $F = -(D_{12}'C_1 + B_2'X)$ where X is the solution of the Riccati equation defined in (10.6). In a similar way it can be shown that for the optimal filter problem the filter gain matrix is $H = -(B_1D_{21}' + YC_2')$ where Y is the solution of the associated Riccati equation.

10.3 The Kalman Filter

In many practical situations, the output y is corrupted by plant disturbances and measurement noise and we require some method to estimate the state from output measurements. The Kalman filter provides an optimal estimate of the state either in the least square sense or in the minimum variance sense. The exogenous input now consists of two signals, w denoting the plant noise and v denoting the sensor noise. The relevant equations are

$$\dot{x} = Ax + [B_1 \quad 0]\begin{bmatrix} w \\ v \end{bmatrix} \tag{10.7}$$

$$y = C_2x + [0 \quad D_{21}]\begin{bmatrix} w \\ v \end{bmatrix} \tag{10.8}$$

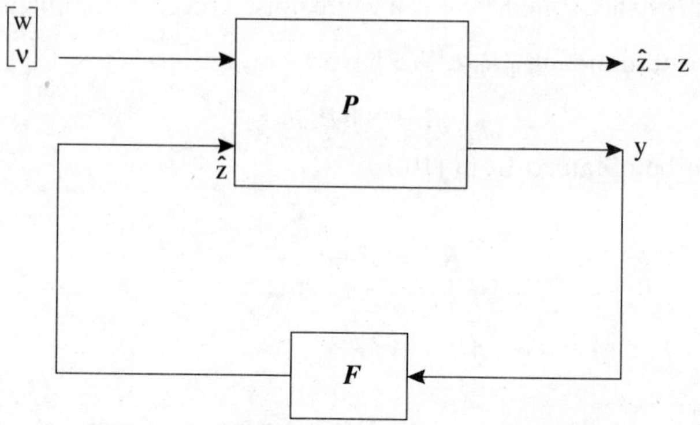

Figure 10.4 Kalman filter in LFT format.

Let $\boldsymbol{F}$ denote the optimal filter and let the signal which is to be estimated be $z = Lx$. The filter output $\hat{z} = \boldsymbol{F}y$ is the optimal estimate of z in the least square sense. Thus we seek to minimize

$$\lim_{T\to\infty} \mathcal{E}\left\{\frac{1}{T}\int_0^T (\hat{z} - Lx)'(\hat{z} - Lx)\,\mathrm{d}t\right\}^{\frac{1}{2}} \tag{10.9}$$

subject to (10.7) and (10.8).

The problem may be cast in the LFT format as in Figure 10.4. The relevant equations in the LFT format are

$$\begin{bmatrix} \dot{x} \\ \hat{z} - z \\ y \end{bmatrix} = \begin{bmatrix} A & [B_1 \;\; 0] & 0 \\ -L & [0 \;\; 0] & I \\ C_2 & [0 \;\; D_{21}] & 0 \end{bmatrix} \begin{bmatrix} x \\ \begin{bmatrix} w \\ v \end{bmatrix} \\ \hat{z} \end{bmatrix} \tag{10.10}$$

$$\hat{z} = \boldsymbol{F}y \tag{10.11}$$

Let $\boldsymbol{R}$ be the transfer matrix from $\begin{bmatrix} w \\ v \end{bmatrix}$ to $(\hat{z} - z)$

$$\boldsymbol{R} : \begin{bmatrix} w \\ v \end{bmatrix} \to (\hat{z} - z)$$

since the disturbance signal $\begin{bmatrix} w \\ v \end{bmatrix}$ is a white noise process, minimizing (10.9) is equivalent to minimizing $\|\boldsymbol{R}\|_2$. We have

$$\boldsymbol{R} = \mathcal{F}_l(\boldsymbol{P}, \boldsymbol{F})$$

where $\boldsymbol{P}$ can be obtained from (10.10).

Also

$$\boldsymbol{R}^{\sim} = \mathcal{F}_l(\boldsymbol{P}^{\sim}, \boldsymbol{F}^{\sim})$$

and

$$\|\boldsymbol{R}\|_2 = \|\boldsymbol{R}^{\sim}\|_2 \tag{10.12}$$

where $\boldsymbol{R}^{\sim}$ indicates the adjoint of $\boldsymbol{R}$. The 2-norm and ∞-norm of an operator and its adjoint are one and the same and this explains (10.12). Thus minimizing $\|\boldsymbol{R}^{\sim}\|$ is equivalent to minimizing $\|\boldsymbol{R}\|$. The adjoint of $\boldsymbol{P}$ may be obtained from the following equations.

$$\begin{bmatrix} \dfrac{\mathrm{d}}{\mathrm{d}\tau} p(\tau) \\ \tilde{z}(\tau) \\ \tilde{w}(\tau) \end{bmatrix} = \begin{bmatrix} A' & -L' & C_2' \\ \begin{bmatrix} B_1' \\ 0 \end{bmatrix} & \begin{bmatrix} 0 \\ 0 \end{bmatrix} & \begin{bmatrix} 0 \\ D_{21}' \end{bmatrix} \\ 0 & I & 0 \end{bmatrix} \begin{bmatrix} p(\tau) \\ \tilde{w}(\tau) \\ \tilde{u}(\tau) \end{bmatrix} \tag{10.13}$$

$$\tilde{u} = \boldsymbol{F}^{\sim} \tilde{w} \tag{10.14}$$

The adjoint of the Kalman filter is schematically represented in Figure 10.5.

Note that the adjoint time variable $\tau = -t$. Further, the adjoint of the filter problem is a control problem. However the controller $\boldsymbol{F}^{\sim}$ has access to only

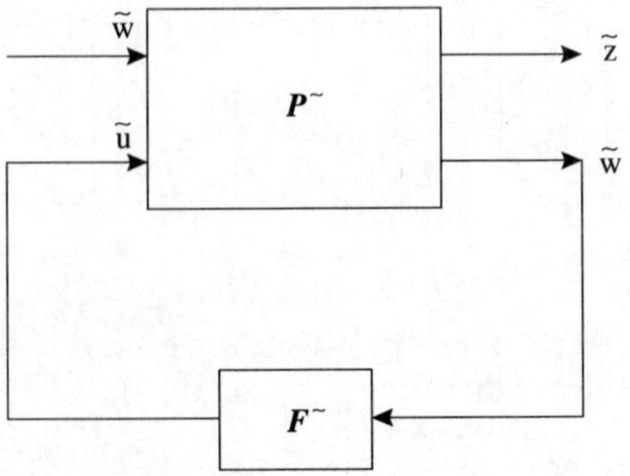

Figure 10.5 Adjoint of Kalman filter.

to $\tilde{w}$ and not to the state variable p as in the case of full information control problem. This is clear from (10.14). But this is adequate because knowledge of both $\tilde{w}$ and p are redundant. The equations relating to the control problem as obtained from (10.13) are:

$$\frac{\mathrm{d}}{\mathrm{d}\tau}p(\tau) = A'p(\tau) - L'\tilde{w}(\tau) + C_2'\tilde{u}(\tau), \qquad p(\tau)|_{\tau=0} = 0 \tag{10.15}$$

and

$$\tilde{z}(\tau) = \begin{bmatrix} B_1' \\ 0 \end{bmatrix} p(\tau) + \begin{bmatrix} 0 \\ D_{21}' \end{bmatrix} \tilde{u}(\tau) \tag{10.16}$$

We have

$$\begin{aligned}\|R^{\sim}\|_2 &= \lim_{T\to\infty} \mathcal{E}\left\{\frac{1}{T}\int_0^T \tilde{z}'(\tau)\tilde{z}(\tau)\mathrm{d}\tau\right\}^{\frac{1}{2}} \\ &= \lim_{T\to\infty} \mathcal{E}\left\{\frac{1}{T}\int_0^T (p'B_1B_1'p + \tilde{u}'\tilde{u})\mathrm{d}\tau\right\}^{\frac{1}{2}} \end{aligned} \tag{10.17}$$

The corresponding Riccati equation for the control problem is given by

$$(A')'Y + Y(A') - YC_2'C_2Y + B_1B_1' = 0 \tag{10.18}$$

The optimum controller for the adjoint problem is

$$\tilde{u}^* = -C_2Yp \tag{10.19}$$

Substituting (10.19) in (10.15) we get

$$\dot{p} = (A' - C_2'C_2Y)p - L'\tilde{w} \tag{10.20}$$

We may now obtain the optimum filter equations by taking the adjoint of the equations (10.19) and (10.20). We have

$$\dot{\widehat{x}}(t) = (A - YC_2'C_2)\widehat{x}(t) + YC_2'y(t), \quad \widehat{x}(0) = 0 \tag{10.21}$$

$$\widehat{z}(t) = L\widehat{x}(t) \tag{10.22}$$

In (10.21) let the Kalman filter gain be denoted by $H = -YC_2'$. Rearranging (10.20) we obtain

$$\dot{\widehat{x}} = A\widehat{x} - H(y - C_2\widehat{x}) \tag{10.23}$$

The signal $(y - C_2\widehat{x})$ is known as the 'innovation process'.

10.4 Parametrization of Controllers – Output Feedback

In the previous section it was shown that

$$\|\boldsymbol{T}_{zw}\|_2^2 = \|\boldsymbol{G}_c B_1\|_2^2 + \|\boldsymbol{T}_{\nu w}\|^2$$

with $\boldsymbol{T}_{\nu w} \in \mathcal{RH}_2$ mapping w to $u - u^*$ where $u^* = Fx$.

In the FI or SF case $u = u^*$ and hence $u - u^* = 0$. Therefore $\|\boldsymbol{T}_{zw}\|_2^2 = \|\boldsymbol{G}_c B_1\|_2^2$.

It can be shown that $\|\boldsymbol{G}_c B_1\|_2 = [\text{trace } (B_1' X_2 B_1)]^{\frac{1}{2}}$ where X_2 is the solution of the control Riccati equation. Hence the minimum cost in the FI case is given by $[\text{trace } (B_1' X_2 B_1)]^{\frac{1}{2}}$.

Similarly for the Kalman filter, the minimum cost is given by $[(LY_2L')]^{\frac{1}{2}}$ where Y_2 is the solution of filter Riccati equation. We are interested in the minimum cost in the output feedback case. Here the measurement feedback controller that minimizes $\|\boldsymbol{R}\|_2$ is the optimal estimator of $u^* = F_2 x$.

We have from (10.4)

$$\min \|\boldsymbol{T}_{zw}\|_2^2 = \|\boldsymbol{G}_c B_1\|_2^2 + \min \|\boldsymbol{T}_{\nu w}\|_2^2$$

It is shown in *DGKF* that $\|T_{\nu w}\|_2$ is uniquely minimized by the controller

$$\left[\begin{array}{c|c} A + B_2F_2 + H_2C_2 & -H_2 \\ \hline F_2 & 0 \end{array}\right]$$

which could immediately be identified as the optimal controller in the LQG problem. The minimal value of $\|\boldsymbol{T}_{\nu w}\|_2^2$ is therefore equal to trace $(F_2 Y_2 F_2')$ where F_2 replaces L of the previous section because we are now concerned with obtaining an optimal estimate of $F_2 x$. Therefore

$$\min \|\boldsymbol{T}_{zw}\|_2^2 = \text{trace } (B_1' X_2 B_1) + \text{trace } (F_2 Y_2 F_2') \qquad (10.24)$$

For sub-optimal controllers, $\|\boldsymbol{T}_{zw}\|_2^2$ will exceed the minimum value given in (10.24). Any bound imposed on $\|\boldsymbol{T}_{zw}\|_2$ should be greater than this minimum value. We may therefore enquire about the set of all $\boldsymbol{K}$s such that $\|\boldsymbol{T}_{zw}\|_2 < \gamma$ where $\gamma^2 >$ trace $(B_1' X_2 B_1)$ + trace $(F_2 Y_2 F_2')$. But we know that the set of all stabilizing $\boldsymbol{K}s$ is parametrized by the LFT, $\mathcal{F}_l(\boldsymbol{K}_0, \boldsymbol{Q})$ as shown in figure 10.1. With $\boldsymbol{Q} = 0$ we recover the optimal controller $\boldsymbol{K}_{\text{opt}}$.

The above result, which was derived in a heuristic manner may be summarized by the following theorem.

THEOREM 10.1 (parametrization of $\mathcal{H}_2$ controllers – output feedback case)

The family of admissible controllers for which $\|\boldsymbol{T}_{zw}\|_2 < \gamma$ is described by the set of all transfer matrices from y to u (see Figure 10.1) denoted b

K where $K = \mathcal{F}_l(M_2, Q)$. Here $Q \in \mathcal{RH}_2$ is further restricted to $\|Q\|_2^2 < \gamma^2 - [\text{trace}\,(B_1' X_2 B_1) + \text{trace}\,(F_2 Y_2 F_2')]$ and M_2 is given by

$$M_2 = \left[\begin{array}{c|cc} \hat{A}_2 & -H_2 & B_2 \\ \hline F_2 & 0 & I \\ -C_2 & I & 0 \end{array}\right]$$

with $\hat{A}_2 = A + B_2 F_2 + H_2 C_2$; $F_2 = B_2' X_2$; $H_2 = -Y_2 C_2'$. Here, X_2 and Y_2 are the solutions of the Riccati equations associated with control and estimation problems.

■

Specifically, with $Q = 0$ we obtain the optimal controller K_{opt} given by

$$K_{\text{opt}} = \left[\begin{array}{c|c} \hat{A}_2 & -H_2 \\ \hline F_2 & 0 \end{array}\right]$$

Note that M_2 is independent of γ. This completes our discussion of the $\mathcal{H}_2$ optimization problem.

10.5 State Space Formulation of $\mathcal{H}_\infty$ Control

$\mathcal{H}_\infty$ optimization was originally proposed by Zames (1981) in a input–output framework. The theory arising out of this concept was worked out in the frequency domain. However a major breakthrough in this area came with the induction of Riccati equations for solving $\mathcal{H}_\infty$ optimization problems. Since the LQG theory was already making use of Riccati equations for problem solving it was soon realized that $\mathcal{H}_2$ and $\mathcal{H}_\infty$ optimization problems follow parallel paths. This has naturally led to the formulation and solution of $\mathcal{H}_\infty$ optimization problems in the time domain using the state space framework. The LQG theory presented earlier in this chapter, provides the backdrop for developing $\mathcal{H}_\infty$ theory along the same lines.

Although minimizing $\mathcal{H}_\infty$ norm is our final goal, this does not come in easily as in the case of $\mathcal{H}_2$ norm. Therefore our immediate aim will be to obtain necessary and sufficient conditions under which we can find an internally stabilizing controller which makes the $\mathcal{H}_\infty$ norm of the closed loop system strictly less than some *a priori* given bound γ. If this is achieved then at least in principle we can obtain the infimum of the closed loop $\mathcal{H}_\infty$ norm over all internally stabilizing controllers via a search procedure.

In control theory literature, fundamental results pertaining to $\mathcal{H}_\infty$ optimization have been obtained either by using the game theoretic approach (see Green and Limebeer 1995) or the Hankel-plus – Toeplitz operator approach as given by Doyle *et al.* (1989). However, these are beyond the scope of this book, our intention being to give the reader some appreciation of the methodology involved rather than go into finer details. In this spirit, we have stated some fundamental results without proof, while some others which provide an insight into the problem have been fully stated and proved.

As in LQG theory, in this case also three types of feedback are discussed. These are the state feedback, the full information feed back and the output feedback. Similarly the questions for which we are seeking answers are:

1. Obtain the necessary and sufficient conditions for the existence of internally stabilizing controllers such that $\|\boldsymbol{T}_{zw}\|_\infty < \gamma$.
2. Obtain the family of all internally stabilizing controllers such that $\|\boldsymbol{T}_{zw}\|_\infty < \gamma$.

We begin by stating an important result pertaining to $\mathcal{H}_\infty$ control with state feedback proved by Khargonekar *et al.* (1988). This result states that the infimum of the norm of the closed loop transfer function using linear static state feedback equals the infimum of the norm of the closed loop transfer function over all stabilizing dynamic state feedback controllers including non-linear time varying controllers.

STATE FEEDBACK

To be more specific, let the system equations be

$$\dot{x} = Ax + B_1 w + B_2 u$$
$$z = C_1 x$$

where A , B_1 , B_2 , C_1 are real constant matrices of appropriate sizes and x, w, u and z have the usual significance. The class of controllers are of the form $u = \boldsymbol{K}x$ where $\boldsymbol{K}$ may be a static controller in which case it will be a constant matrix or $\boldsymbol{K}$ could be a dynamic controller with the following state and input-output equations

$$\dot{v} = Fv + Gx$$
$$u = Hv + Jx$$

where v is the state variable for the controller and x the state variable for the plant and F, G, H and J are real constant matrices of appropriate size.

In the static state feedback case, the closed loop transfer function $\boldsymbol{T}_s$ is given by

$$\boldsymbol{T}_s = C_1[sI - (A + B_2K)]^{-1}B_1$$

In the dynamic feedback case, the closed loop transfer function is denoted by $\boldsymbol{T}_d$. It may be verified that the A-matrix of the closed loop system is given by

$$\widehat{A} = \begin{bmatrix} A + B_2 J & B_2 H \\ G & F \end{bmatrix}$$

Now define

$$\mathcal{S} = \{K : A + B_2 K \text{ is a stable matrix}\}$$

and

$$\mathcal{D} = \{F, G, H, J : \widehat{A} \text{ is a stable matrix}\}$$

Further let

$$\gamma_s = \text{Inf}\{\|\boldsymbol{T}_s\|_\infty : K \in \mathcal{S}\}$$
$$\gamma_d = \text{Inf}\{\|\boldsymbol{T}_d\|_\infty : F, G, H, J \in \mathcal{D}\}$$

The computation of γ_d and then obtaining a controller such that $\|\boldsymbol{T}_d\|_\infty$ is as close to γ_d as desired is a standard problem in $\mathcal{H}_\infty$ optimization. Clearly $\gamma_d \leq \gamma_s$. What has been proved in Khargonevar *et al.* (1988) is that $\gamma_d = \gamma_s$ under the assumptions that the measured output is the state x and that there is no direct transmission from w and u to z. i.e. D_{11} and D_{12} are zero.

From the result given above, it follows that the optimal $\mathcal{H}_\infty$ controller has the structure of a static gain in the case of state feedback. It has been further proved that the infimum of the $\mathcal{H}_\infty$ norm of the closed loop transfer function using internally stabilizing static state feedback is no greater than the infimum of the operator norm of the closed loop transfer function using dynamic internally stabilizing non linear time varying state feedback.

10.5.1 Solvability Criteria

A fundamental result in $\mathcal{H}_\infty$ optimization theory relates to the existence of solutions. We shall first consider the full information $\mathcal{H}_\infty$ controller. The plant description is

$$\boldsymbol{G} = \boldsymbol{G}_{FI} = \left[\begin{array}{c|cc} A & B_1 & B_2 \\ \hline C_1 & 0 & D_{12} \\ \begin{bmatrix} I \\ 0 \end{bmatrix} & \begin{bmatrix} 0 \\ I \end{bmatrix} & \begin{bmatrix} 0 \\ 0 \end{bmatrix} \end{array}\right] \tag{10.25}$$

The corresponding equations are

$$\dot{x} = Ax + B_1 w + B_2 u$$
$$z = C_1 x + D_{12} u$$
$$y = \begin{bmatrix} I \\ 0 \end{bmatrix} x + \begin{bmatrix} 0 \\ I \end{bmatrix} w = \begin{bmatrix} x \\ w \end{bmatrix}$$

The standard assumptions for FI control problems are:

1. $(A_1 B_2)$ is stabilizable and (C_1, A) is observable on the imaginary axis.
2. $D'_{12} D_{12} = I; \quad D'_{12} C_1 = 0$

Consider a plant with its structure as given in (10.25) and satisfying the standard assumptions:
Let $\boldsymbol{T}_{zw} = \mathcal{F}_l(\boldsymbol{G}_{FI}, \boldsymbol{K})$ where $\boldsymbol{K}$ is the stabilizing controller.

THEOREM 10.2 (Existence theorem for $\mathcal{H}_\infty$ controller)
There exists an admissible controller $\boldsymbol{K}$ such that $\|\boldsymbol{T}_{zw}\|_\infty < \gamma$ if and only if

$$H_\infty = \begin{bmatrix} A & \gamma^{-2} B_1 B'_1 - B_2 B'_2 \\ -C'_1 C_1 & -A' \end{bmatrix} \in \text{dom (Ric)} \qquad (10.26)$$

and

$$X_\infty = Ric(H_\infty) \geq 0$$

Proof: omitted (Refer Green and Limebeer (1995) and Doyle *et al.* (1989)).

REMARKS

1. The associated Riccati equation for the Hamiltonian matrix H_∞ is given by

$$A' X_\infty + X_\infty A + X_\infty(\gamma^{-2} B_1 B'_1 - B_2 B'_2) X_\infty + C'_1 C_1 = 0 \qquad (10.27)$$

2. The Riccati equation (10.27) has an indefinite quadratic term. Riccati equations of this type are met with in differential game theory. They have different properties from the Riccati equations occuring in LQG theory. The Riccati equation has a positive semi definite stabilizing solution only if the quadratic term is sign definite.
3. If $H_\infty \in$ dom (Ric) and $X_\infty \geq 0$ then $A + (\gamma^{-2} B_1 B'_1 - B_2 B'_2) X_\infty$ is asymptotically stable.

10.5.2 Inner Linear Fractional Transformation

We next turn our attention to the parametrization of all full information stabilizing controllers such that $\|T_{zw}\|_\infty < \gamma$. The method described here is adapted from Zhou (1992b) where the author employs a combination of change of variables technique used by *DGKF* and Youla Parametrization theory. The method also uses a very elegant result proved by Redheffer (1960) who showed that contraction and internal stability are preserved under an inner linear fractional transformation. This result is stated below.

LEMMA 10.1
Consider a feedback system shown in Figure 10.6, where

$$P = \begin{bmatrix} P_{11} & P_{12} \\ P_{21} & P_{22} \end{bmatrix} \in \mathcal{RH}_\infty$$

Suppose that $P^{\sim}P = I$ and $P_{21}^{-1} \in \mathcal{RH}_\infty$. Then the following statements are equivalent:

a) The system is internally stable, well posed and $\|T_{zw}\| < 1$
b) $Q \in \mathcal{RH}_\infty$ and $\|Q\|_\infty < 1$

Proof: $(b) \Rightarrow (a)$
To prove well-posedness, we have to show that
$\det(I - P_{22}(\infty)\ Q(\infty)) \neq 0$.
Since $P^{\sim}P = I$, $\|P_{22}\|_\infty \leq 1$, also $\|Q\|_\infty < 1$

Hence $\|P_{22}Q\|_\infty \geq 1$. If $I = P_{22}(\infty)Q(\infty)$ then $\|P_{22}Q\|_\infty \geq 1$ which is a contradiction. Internal stability follows from the fact that both P_{22} and

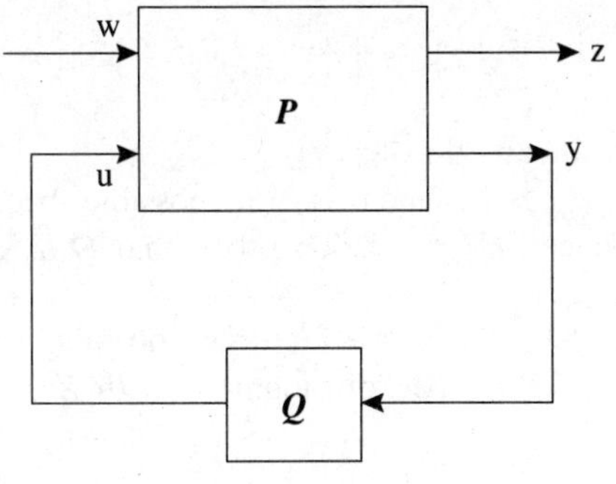

Figure 10.6 Diagram relating to lemma 10.1.

$\boldsymbol{Q}$ are stable and $\|\boldsymbol{P}_{22}\boldsymbol{Q}\|_\infty < 1$. Applying small gain argument to the loop the result follows.

Proof of $\|\boldsymbol{T}_{zw}\|_\infty < 1$

Consider any frequency $s = j\omega$ with the signals fixed as complex constant vectors. Let $\|\boldsymbol{Q}\|_\infty = \epsilon < 1$.

We have

$$\boldsymbol{T}_{wy} = \boldsymbol{P}_{21}^{-1}(I - \boldsymbol{P}_{22}\boldsymbol{Q}) \in \mathcal{RH}_\infty$$

Also let

$$\|\boldsymbol{T}_{wy}\|_\infty = k. \text{ Then } \|w\| \le k\|y\|$$

P being inner implies $\|z\|^2 + \|y\|^2 = \|w\|^2 + \|u\|^2$.

Hence:

$$\begin{aligned}\|z\|^2 &\le \|w\|^2 - (1-\epsilon^2)\|y\|^2 \\ &\le \|w\|^2 - \frac{(1-\epsilon^2)}{k^2}\|w\|^2 \\ &= \left[1 - \frac{1-\epsilon^2}{k^2}\right]\|w\|^2\end{aligned}$$

since:

$$1 - \frac{1-\epsilon^2}{k^2} < 1 \text{ we have}$$

$$\frac{\|z\|^2}{\|w\|^2} < 1 \text{ and hence } \|\boldsymbol{T}_{zw}\|_\infty < 1$$

$(a) \Rightarrow (b)$.

Proof is by contradiction. Suppose there exists frequency ω (finite or infinite) and a constant non zero vector y such that $\|\boldsymbol{Q}y\| \ge \|u\|$. Then $\|u\|^2 \ge \|y\|^2$. Since $\boldsymbol{P}$ is inner we have

$$\|z\|^2 + \|y\|^2 = \|w\|^2 + \|u\|^2$$

and $\|u\|^2 \ge \|y\|^2$ means that $\|z\|^2 \ge \|w\|^2$.

That is $\|\boldsymbol{T}_{zw}\|_\infty > 1$ which is not possible because by hypothesis $\|\boldsymbol{T}_{zw}\|_\infty < 1$. Hence $\|\boldsymbol{Q}\|_\infty < 1$. To prove that $Q \in \mathcal{RH}_\infty$, we proceed as follows.

Let $\boldsymbol{Q} = \boldsymbol{N}\boldsymbol{M}^{-1}$ be a coprime factorization of $\boldsymbol{Q}$ with $\boldsymbol{N}, \boldsymbol{M} \in \mathcal{RH}_\infty$.

Hence $\boldsymbol{X}\boldsymbol{N} + \boldsymbol{Y}\boldsymbol{M} = I$ (Bezout identity) with $\boldsymbol{X}$, $\boldsymbol{Y} \in \mathcal{RH}_\infty$.

We have

$$w = \boldsymbol{P}_{21}^{-1}(I - \boldsymbol{P}_{22}\boldsymbol{Q})y$$

Hence

$$y = (I - P_{22}Q)^{-1}P_{21}w$$
$$u = Qy = Q(I - P_{22}Q)^{-1}P_{21}w$$

Hence

$$T_{uw} = Q(I - P_{22}Q)^{-1}P_{21}$$

Substituting NM^{-1} for Q we have

$$T_{uw}P_{21}^{-1} = NM^{-1}(I - P_{22}NM^{-1})^{-1}$$
$$= N(M - P_{22}N)^{-1}$$

Since the left hand side belongs to $\mathcal{RH}_\infty$ we have $N(M - P_{22}N)^{-1} \in \mathcal{RH}_\infty$ Further, N and $(M - P_{22}N)^{-1}$ are coprime factors because

$$(X + YP_{22})N + Y(M - P_{22}N) = I$$

Hence $(M - P_{22}N)^{-1} \in \mathcal{RH}_\infty$ according to Theorem 4.7. Since $(M - P_{22}N)$ also belongs to $\mathcal{RH}_\infty$ we note that the function has neither poles nor zeros in the closed right half plane. It follows that det $(M - P_{22}N)$ has neither poles nor zeros in the closed right half plane. Hence by applying the Nyquist stability criteria the number of encirclements of the Nyquist plot relating to $\det(M - P_{22}N)$ about the origin is zero. We have

$$\det(M - P_{22}N) = \det[(I - P_{22}NM^{-1})M]$$
$$= \det M \cdot \det(I - P_{22}Q)$$

Both $\det M$ and $\det(I - P_{22}Q)$ are non zero for all $s = j\omega$, since $\det(M - P_{22}N) \neq 0$ there. Note also that $\det(I - P_{22}Q) \neq 0$ for all $s = j\omega$ because $\|P_{22}Q\|_\infty < 1$. We therefore have $\det(M - \alpha P_{22}N) \neq 0$ for all $s = j\omega$ and all $\alpha \in [0, 1]$.

Since the winding number of $\det(M - P_{22}N)$ is zero, by a continuity argument we assert that $\det M$ has also a winding number zero.

Hence $Q \in \mathcal{RH}_\infty$. This completes the proof.

■

We shall now make use of the Lemma 10.1 to parametrize all full information controllers satisfying $\|T_{zw}\|_\infty < \gamma$.

10.5.3 Parametrization of FI Controllers

THEOREM 10.3 (Parametrization of $\mathcal{H}_\infty$ controllers – FI case)
Assuming that there exists full information controllers K such that $\|T_{zw}\|_\infty < \gamma$, then the set of all admissible controllers for which

$\|\boldsymbol{T}_{zw}\|_\infty < \gamma$ is given by $\boldsymbol{K} = \mathcal{F}_l(\boldsymbol{M}_{FI}, \boldsymbol{Q})$ where

$$\boldsymbol{M}_{FI} = \left[\begin{array}{c|cc} A + B_2F_\infty & [0 \quad B_1] & B_2 \\ \hline 0 & [F_\infty \quad 0] & I \\ \begin{bmatrix} -I \\ 0 \end{bmatrix} & \begin{bmatrix} I & 0 \\ -\gamma^{-2}B_1'X_\infty & I \end{bmatrix} & 0 \end{array}\right] \tag{10.28}$$

$$F_\infty = -B_2'X_\infty, \boldsymbol{Q} = [\,\boldsymbol{Q}_1 \quad \boldsymbol{Q}_2\,] \in \mathcal{RH}_\infty \text{ and } \|\boldsymbol{Q}_2\|_\infty < \gamma$$

The controller structure is given in Figure 10.7.

Proof: Introduce two new variables v and r such that

$$v = u + B_2'X_\infty x; \quad r = w - \gamma^{-2}B_1'X_\infty x$$

From the system equations as described in (10.25) we may now obtain

$$\begin{bmatrix} \dot{x} \\ z \\ \gamma r \end{bmatrix} = \begin{bmatrix} A_{F\infty} & \gamma^{-1}B_1 & B_2 \\ C_{1F\infty} & 0 & D_{12} \\ \gamma^{-1}B_1'X_\infty & I & 0 \end{bmatrix} \begin{bmatrix} x \\ \gamma w \\ v \end{bmatrix} \tag{10.29}$$

and

$$\begin{bmatrix} \dot{x} \\ v \\ y \end{bmatrix} = \begin{bmatrix} A_{\mathrm{tmp}} & B_1 & B_2 \\ -F_\infty & 0 & I \\ \begin{bmatrix} I \\ \gamma^{-2}B_1'X_\infty \end{bmatrix} & \begin{bmatrix} 0 \\ I \end{bmatrix} & 0 \end{bmatrix} \begin{bmatrix} x \\ r \\ u \end{bmatrix} \tag{10.30}$$

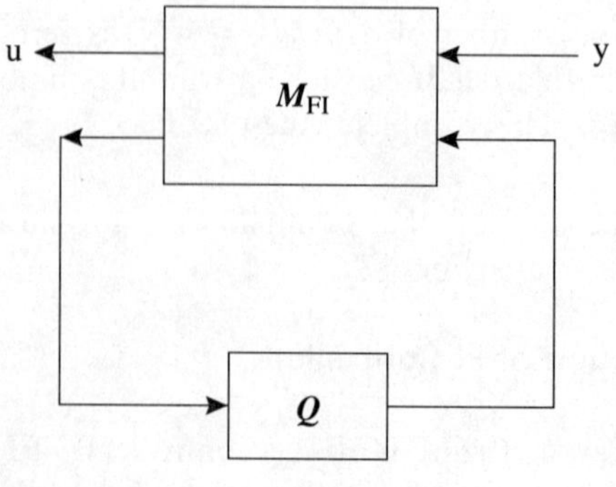

Figure 10.7 Full information $\mathcal{H}_\infty$ controller structure.

where

$$A_{F\infty} = A + B_2F_\infty; \quad C_{1F\infty} = C_1 + D_{12}F_\infty$$

and

$$A_{\text{tmp}} = A + \gamma^{-2}B_1B_1'X_\infty$$

$$\text{Let } \boldsymbol{P} = \left[\begin{array}{c|cc} A_{F\infty} & \gamma^{-1}B_1 & B_2 \\ \hline C_{1F\infty} & 0 & D_{12} \\ -\gamma^{-1}B_1'X_\infty & I & 0 \end{array}\right]$$

and

$$\widehat{\boldsymbol{G}}_{FI} = \left[\begin{array}{c|cc} A_{\text{tmp}} & B_1 & B_2 \\ \hline -F_\infty & 0 & I \\ \begin{bmatrix} I \\ \gamma^{-2}B_1'X_\infty \end{bmatrix} & I & 0 \end{array}\right]$$

A schematic block diagram giving the connections is shown in Figure 10.8.

In *DGKF* it is proved that $\boldsymbol{P}$ represents an inner LFT transformation, i.e or $\boldsymbol{P}^{\sim}\boldsymbol{P} = I$ and $\boldsymbol{P}_{21}^{-1} \in \mathcal{RH}_\infty$. Note that $\boldsymbol{G}_{FI}$ defined in (10.2) is expressed as the composition of two LFTs namely $\boldsymbol{P}$ and $\widehat{\boldsymbol{G}}_{FI}$. Because of the properties of $\boldsymbol{P}$ stated in Lemma 10.1, $\boldsymbol{K}$ is an admissible controller of $\boldsymbol{G}_{FI}$ and $\|T_{zw}\|_\infty < \gamma$, if and only if $\boldsymbol{K}$ is an admissible controller of $\widehat{\boldsymbol{G}}_{FI}$ and $\|\boldsymbol{T}_{vr}\|_\infty < \gamma$. We will now parametrize $\boldsymbol{K}$ as in Figure 10.1 and obtain a $\boldsymbol{K}_0$ which in the present context will be denoted by $\boldsymbol{M}$.

For this purpose we must find an F and H such that $A_{\text{temp}} + H\begin{bmatrix} I \\ \gamma^{-2}B_1'X_\infty \end{bmatrix}$ is stable and further $A_{\text{temp}} + B_2F$ is also stable.

If $H = [\,B_2F_\infty \quad -B_1\,]$ then the first matrix reduces to $A + B_2F_\infty$ which can be proved to be stable. Similarly the second matrix namely $A_{\text{temp}} + B_2F$ is stable if F is chosen as $F_\infty = -B_2'X_\infty$ where X_∞ is the solution of (10.27).

This is so because

$$A_{\text{temp}} + B_2F_\infty = A + \gamma^{-2}B_1B_1'X_\infty - B_2B_2'X_\infty$$

which is a stable matrix (see remark 3 under Theorem 10.2).

Hence using the Youla parametrization procedure, we have

$$\boldsymbol{K} = \mathcal{F}_l(\boldsymbol{M}, \boldsymbol{\Phi})$$

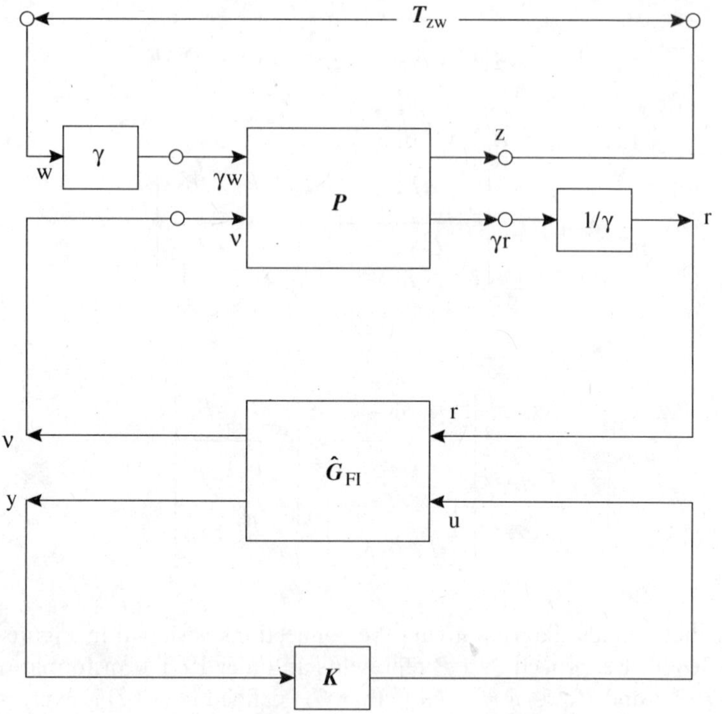

Figure 10.8 Parametrization of FI controller.

where the matrix M is given by

$$M = \left[\begin{array}{c|cc} A_{\text{tmp}} + B_2F_\infty + H\begin{bmatrix} I \\ \gamma^{-2}B_1'X_\infty \end{bmatrix} & -H & B_2 \\ \hline F_\infty & 0 & I \\ -\begin{bmatrix} I \\ \gamma^{-2}B_1'X_\infty \end{bmatrix} & I & 0 \end{array}\right] \tag{10.31}$$

with this parametrization of all stabilizing controllers the transfer matrix from r to v can be written as

$$\begin{aligned} T_{vr} &= \mathcal{F}_l(\widehat{G}_{FI}, K) = \mathcal{F}_l(\widehat{G}_{FI}, \mathcal{F}_l(M, \Phi)) \\ &= \mathcal{F}_l(N, \Phi) \end{aligned}$$

It may be verified by actual calculation that

$$N = \begin{bmatrix} 0 & I \\ \begin{bmatrix} 0 \\ I \end{bmatrix} & 0 \end{bmatrix} \tag{10.32}$$

and

$$T_{vr} = \mathcal{F}_l(N, \Phi) = \Phi_2 \text{ where } \Phi = [\Phi_1 \quad \Phi_2]$$

Hence $\|T_{vr}\|_\infty < \gamma$ if and only if $\|\Phi_2\|_\infty < \gamma$.

This implies that FI, $\mathcal{H}_\infty$ controllers may be parametrized as

$$K = \mathcal{F}_l(M, \Phi) \text{ with } [\Phi_1 \quad \Phi_2] \in \mathcal{RH}_\infty \text{ and } \|\Phi_2\|_\infty < \gamma$$

with M having the following structure

$$M = \left[\begin{array}{c|cc} A + 2B_2F_\infty & -[B_2F_\infty \quad -B_1] & B_2 \\ \hline F_\infty & 0 & I \\ -\begin{bmatrix} I \\ \gamma^{-2}B_1'X_\infty \end{bmatrix} & I & 0 \end{array}\right] \tag{10.33}$$

The structure of M may be further simplified by introducing two new free parameters Q_1 and Q_2. They are related to Φ_1 and Φ_2 as follows

$$\Phi_1 = F_\infty - \gamma^{-2}Q_2B_1'X_\infty + Q_1; \quad \Phi_2 = Q_2$$

with this change of free parameters from Φ_1 and Φ_2 to Q_1 and Q_2 the M matrix is transformed to M_{FI}. It may be verified that

$$\mathcal{F}_l(M, \Phi) = \mathcal{F}_l(M_{FI}, Q)$$

where M_{FI} is as defined in (10.28).

REMARKS

1. The LFT, $K = \mathcal{F}_l(M_{FI}, Q)$ with $Q = [Q_1 \quad Q_2]$, Q_1 and $Q_2 \in \mathcal{RH}_\infty$ and $\|Q_2\| < \gamma$ generates the set of all admissible FI Controllers which satisfy $\|T_{zw}\|_\infty < \gamma$.
2. With $Q = 0$, $K = \mathcal{F}_l(M_{FI}, 0) = [F_\infty \quad 0]$ and we obtain one of the controllers, called the central controller for which $\|T_{zw}\|_\infty < \gamma$.
3. It was pointed out by Zhou (1992b) and Mita *et al.* (1993) that the parametrization proposed in Doyle *et al.* (1989) characterizes only a subset of FI, $\mathcal{H}_\infty$ controllers. This can be seen as follows. Set $Q_1 = 0$ and obtain:

$$K = \mathcal{F}_l(M_{FI}, [0, \quad Q_2])$$

It is easily verified that $\boldsymbol{K} = [\boldsymbol{F}_\infty - \gamma^{-2}\boldsymbol{Q}_2 B_1' X_\infty \quad \boldsymbol{Q}_2]$ which is the parametrization give in *DGKF*.

Hence *DGKF* parametrization corresponds to assuming $\boldsymbol{Q}_1 = 0$ and therefore generates only a subset of controllers for which $\|\boldsymbol{T}_{zw}\|_\infty < \gamma$.

10.6 The $\mathcal{H}_\infty$ Filter

This is the counterpart of the Kalman filter in $\mathcal{H}_2$ optimization theory. However the $\mathcal{H}_\infty$ filtering problem and the Kalman filtering problem differ from each other in certain important respects. To appreciate these differences we shall briefly recapitulate the underlying premises in the two problems. In Kalman filtering there are two disturbances – one produced by a white noise process of known statistical properties driving the state space system and the other a white noise process of known statistical properties corrupting the observed output. The aim of the filter is to minimize the average RMS value of the estimation error.

In $\mathcal{H}_\infty$ filtering, white noise disturbances are replaced by deterministic disturbances of finite energy. The aim of the filter is to ensure that the energy gain from the disturbance to the estimation error is less than a prespecified level γ^2.

PROBLEM FORMULATION AND SOLUTION

The system equations are

$$\dot{x} = Ax + B_1 w \tag{10.34a}$$

$$y = C_2 x + D_{21} v \tag{10.34b}$$

Where the process disturbance w and the measurement distubance v are signals belonging to $\mathcal{L}_2[0, \infty]$

Let $d = \begin{bmatrix} w \\ v \end{bmatrix}$ denote the combined disturbance signal.

The aim is to find an estimate of $z = Lx$. This estimate is denoted by $\widehat{z}$.

The controller $\boldsymbol{K}$ is an estimator which accepts the measured signals y and generates the estimated value of z namely $\widehat{z}$. Hence

$$\widehat{z} = \boldsymbol{K}y \tag{10.35}$$

The objective is to obtain a linear time invariant causal filter such that the estimator $\boldsymbol{K}$ is stable and the mapping

$$\boldsymbol{R} : d \to (\widehat{z} - Lx)$$

is stable and satisfies

$$\|\boldsymbol{R}\|_\infty < \gamma$$

The standard assumptions are:

1. (A, C_2) is detectable.
2. (A, B_1) has no uncontrollable modes on the imaginary axis.

To solve the problem we first express the $\mathcal{H}_\infty$ filter in the LFT format. We then consider the dual of this problem which is a full information $\mathcal{H}_\infty$ control problem for which the solution is already known from the previous section. Now once again take the dual of the control problem and thus obtain the solution of the estimation problem. This is the strategy adopted in the sequel.

The LFT formulation of the filtering problem is give below

$$\begin{bmatrix} \dot{x} \\ \hat{z} - z \\ y \end{bmatrix} = \begin{bmatrix} A & [B_1 \quad 0] & 0 \\ -L & [0 \quad 0] & I \\ C_2 & [0 \quad D_{21}] & 0 \end{bmatrix} \begin{bmatrix} x \\ \begin{bmatrix} w \\ v \end{bmatrix} \\ \hat{z} \end{bmatrix} \tag{10.36}$$

$$\hat{z} = \boldsymbol{K} y \tag{10.37}$$

The LFT representation of the $\mathcal{H}_\infty$ filter in shown in Figure 10.9.

■

From figure 10.9 $\boldsymbol{R} : d \to (\hat{z} - z) = \mathcal{F}_l(\boldsymbol{P}, \boldsymbol{K})$. To obtain the equivalent control problem take the adjoint of $\mathcal{F}_l(\boldsymbol{P}, K)$ which is given by $\mathcal{F}_l(\boldsymbol{P}^\sim, \boldsymbol{K}^\sim)$. Hence

$$\boldsymbol{R}^\sim = \mathcal{F}_l(\boldsymbol{P}^\sim, \boldsymbol{K}^\sim)$$

Since the infinity norm of a system and its adjoint are equal $\|\boldsymbol{R}\|_\infty = \|\boldsymbol{R}^\sim\|_\infty$ and $\|\boldsymbol{R}\|_\infty < \gamma$ implies $\|\boldsymbol{R}^\sim\|_\infty < \gamma$.

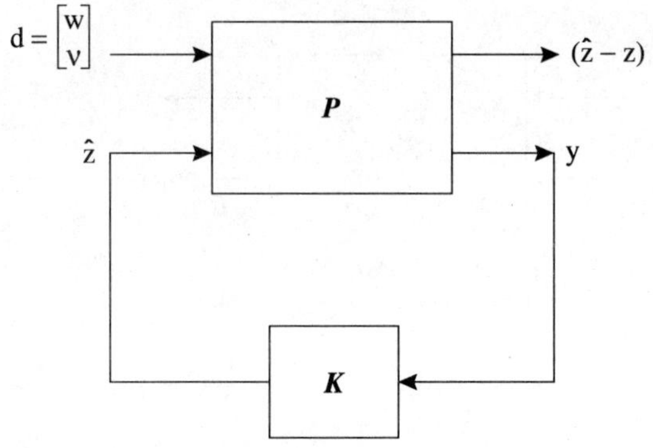

Figure 10.9 $\mathcal{H}_\infty$ filter in LFT format.

The adjoint LFT is given by

$$\begin{bmatrix} \frac{d}{d\tau}p(\tau) \\ \tilde{z} \\ \tilde{w} \end{bmatrix} = \begin{bmatrix} A' & -L' & C_2' \\ \begin{bmatrix} B_1' \\ 0 \end{bmatrix} & \begin{bmatrix} 0 \\ 0 \end{bmatrix} & \begin{bmatrix} 0 \\ D_{21}' \end{bmatrix} \\ 0 & I & 0 \end{bmatrix} \begin{bmatrix} p(\tau) \\ \tilde{w} \\ \tilde{u} \end{bmatrix} \tag{10.38}$$

$$\tilde{u} = \boldsymbol{K}^{\sim}\tilde{w} \tag{10.39}$$

The LFT representation of the adjoint of the $\mathcal{H}_\infty$ filter is given in Figure 10.10.

Note that the controller associated with the adjoint system has access only to the exogenous signal $\tilde{w}$ instead of both p and $\tilde{w}$ as in the case of FI problem. But this is alright as $p(\tau)$ can always be generated from the equation

$$\frac{d}{d\tau}p(\tau) = A'p(\tau) - L'\tilde{w}(\tau) + C_2'\tilde{u}(\tau), \quad p(\tau)|_{\tau=0} = 0$$

Hence we can apply the theory of FI Controller discussed in the earlier section to obtain a controller for the adjoint problem.

Accordingly an admissible controller for the adjoint problem $\|\boldsymbol{T}_{\tilde{z}\tilde{w}}\|_\infty < \gamma$ exists if and only if the Hamiltonian

$$J_\infty = \begin{bmatrix} A' & \gamma^{-2}L'L - C_2'C_2 \\ -B_1B_1' & -A \end{bmatrix} \in \text{Dom}(Ric) \tag{10.40}$$

and $Y_\infty = \text{Ric}\,(J_\infty) \geq 0$

In this case the controller is given by $\tilde{u}^*(\tau) = -C_2Y_\infty p(\tau)$.

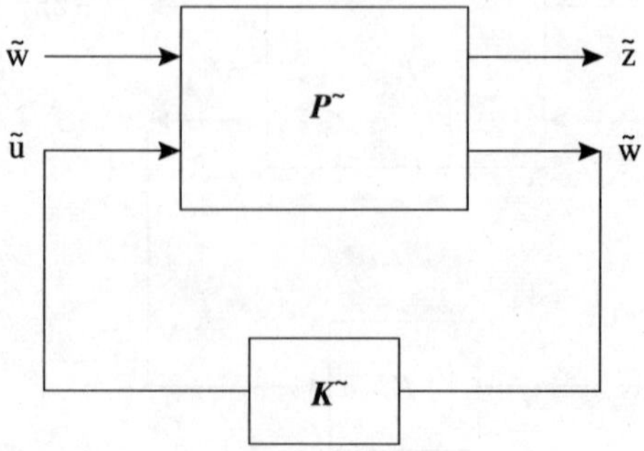

Figure 10.10 Adjoint of $\mathcal{H}_\infty$ filter in LFT format.

Where Y_∞ satisfies the Riccati equation

$$AY_\infty + Y_\infty A' + Y_\infty(\gamma^{-2}L'L - C_2'C_2)Y_\infty + B_1B_1' = 0 \tag{10.41}$$

The adjoint of the $\mathcal{H}_\infty$ filter is described by the equations

$$\dot{p} = (A' - C_2'C_2Y_\infty)p - L'\tilde{w} \tag{10.42a}$$

$$\tilde{u} = -C_2Y_\infty p \tag{10.42b}$$

Now reverting back to the original estimation problem once again we take the adjoint of the system (10.42). We thus have

$$\begin{aligned}\dot{\hat{x}} &= (A - Y_\infty C_2'C_2)\hat{x} + Y_\infty C_2'y \\ &= A\hat{x} + Y_\infty C_2'(y - C_2\hat{x})\end{aligned} \tag{10.43}$$

and

$$\hat{z} = L\hat{x} \tag{10.44}$$

The above equations of $\mathcal{H}_\infty$ filter suggest a structure reminiscent of the Kalman filter for the LQG problem. Schematic diagram of the filter is given in Figure 10.11.

From the above discussions the following points may be noted:

1. The existence of the $\mathcal{H}_\infty$ filter cannot be taken for granted. It depends upon whether the Riccati equation associated with Y_∞ has a positive semi definite stabilizing solution or not. Another way of expressing the same idea is to state that for the given value of γ, the Hamiltonian matrix J_∞ should belong to the dom(Ric) and Ric $(J_\infty) \geq 0$, for the $\mathcal{H}_\infty$ filter to exist.
2. Just as in the Kalman filter here also we have $\hat{z} = L\hat{x}$. However in this case $\hat{x}$ depends on L because the quadratic term in the Riccati equation (10.41) depends upon L. Thus Y_∞ is a function of L which

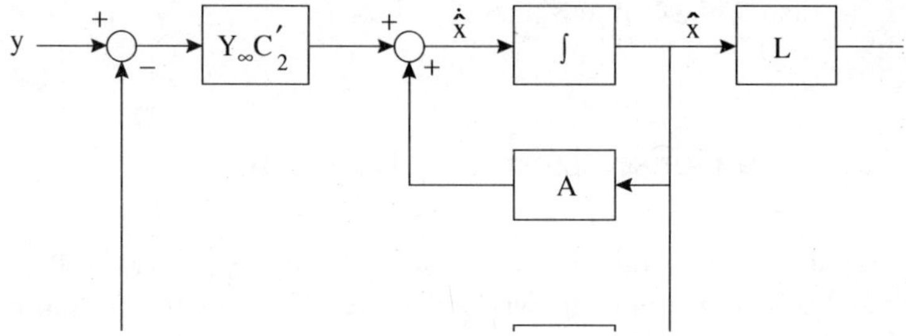

Figure 10.11 $\mathcal{H}_\infty$ filter.

implies that $\widehat{x}$ depends on L. It is relevant to point out here that in the Kalman filter $\widehat{x}$ is independent of L.

PARAMETRIZATION OF A FAMILY OF FILTERS

In the previous section we parametrized the set of all admissible controllers for the FI problem which make $\|\boldsymbol{T}_{zw}\|_\infty < \gamma$. By considering the dual of the filter problem which is an FI problem, it is possible to parametrize the set of all $\mathcal{H}_\infty$ filters which satisfy $\|\boldsymbol{R}\|_\infty < \gamma$. We are giving here only the final result and leave the problem as an exercise for the reader.

THEOREM 10.3 (parametrization of all $\mathcal{H}_\infty$ filters)
Suppose $J_\infty \in$ dom (Ric) and $Y_\infty =$ Ric $(J_\infty) \geq 0$.

Then the family of admissible controllers K for which the norm bound $\|\boldsymbol{R}\|_\infty < \gamma$ is satisfied (where $\boldsymbol{R} : d \rightarrow (\widehat{z} - Lx)$) is given by

$$\boldsymbol{K} = \mathcal{F}_l(\boldsymbol{K}_a, \boldsymbol{Q})$$

where

$$\boldsymbol{K}_a = \left[\begin{array}{c|cc} A - Y_\infty C_2' C_2 & Y_\infty C_2' & -\gamma^{-2} Y_\infty L' \\ \hline L & 0 & I \\ -C_2 & I & 0 \end{array}\right] \tag{10.45}$$

and $\boldsymbol{Q} \in R\mathcal{H}_\infty$ with $\|\boldsymbol{Q}\|_\infty < \gamma$

REMARKS

By setting $Q = 0$, we obtain the $\mathcal{H}_\infty$ filter given by

$$\boldsymbol{K}_{\text{sub}} = \left[\begin{array}{c|c} A - Y_\infty C_2' C_2 & Y_\infty C_2' \\ \hline L & 0 \end{array}\right]$$

This is only one among the family of filters for which $\|\boldsymbol{R}\|_\infty < \gamma$. This particular filter is kown as the central $\mathcal{H}_\infty$ filter and has a structure similar to that of the Kalman filter.

10.7 The Generalized $\mathcal{H}_\infty$ Regulator

In the earlier section we saw how to synthesize a controller for the FI problem and also how to synthesize a $\mathcal{H}_\infty$ filter for the estimation problem. In LQG theory the solution of these two problems leads directly to the synthesis of the regulator problem. We recall that in LQG measurement feedback problem, the filtering problem was to find an optimal estimate of the optimal FI control law $u = Fx$. The Kalman filter provided such an estimate, and combining this result with the FI control law the regulator problem was uniquely solved. A special feature of LQG theory is that the control and estimation problems are completely decoupled. Thus we need not know the control objective in order to solve the estimation problem nor need we know the statistical properties associated with the disturbances before solving the FI problem. The decoupling arises out of the fact that the optimal estimate of Fx is $F\widehat{x}$ where the evaluation of $\widehat{x}$ does not depend on F. As already pointed out in the earlier section, this is not so in the case of $\mathcal{H}_\infty$ filter where estimation of $\widehat{x}$ depends on F. Thus the $\mathcal{H}_\infty$ filter cannot be designed independently of the control objective and hence the control law. Another important point of difference is that the quadratic term in the Riccati equation associated with the FI control problem is not sign definite. This is because of the presence of the term $\gamma^{-2}B_1B_1'$ which indicates that the $\mathcal{H}_\infty$ controller depends on the disturbance through B_1 whereas the $\mathcal{H}_2$ controller does not.

Hence it may so turn out that the controller gain matrix F obtained by solving the control Riccati equation, when used in conjunction with the estimation problem may give rise to a filter Riccati equation which has no admissible solution. Hence it is intuitively clear that for the existence of admissible solutions X_∞ and Y_∞ for the two Riccati equations some sort of coupling between the two solutions should exist. We shall presently see that this is indeed the case.

PROBLEM STATEMENT

The state space description of the generalized plant $\boldsymbol{P}$ is given by

$$\dot{x} = Ax + B_1w + B_2u \tag{10.46}$$

$$z = C_1x + D_{12}u \tag{10.47}$$

$$y = C_2x + D_{21}w \tag{10.48}$$

We assume that $D_{12}'D_{12} = I$ and $D_{21}D_{21}' = I$ for convenience. This is not necessary except for the fact that D_{12} should have maximum column rank and D_{21} should have maximum row rank. Further, the assumptions in earlier chapters that $D_{12}'C_1 = 0$ and $B_1D_{21}' = 0$ are now dispensed with. This will now give rise to cross coupling terms which could easily by taken care of by methods explained earlier in this chapter.

The generalized $\mathcal{H}_\infty$ regulator problem may now be posed as follows.

For the generalized plant described by equations (10.46) to (10.48) obtain a time invariant, causal linear controller $u = \boldsymbol{K}y$ such that the

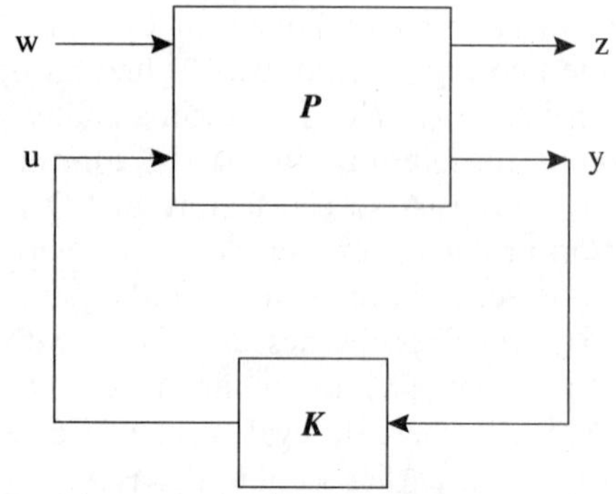

Figure 10.12 Generalized plant–controller configuration.

closed loop system $\boldsymbol{T}_{zw} = \mathcal{F}_l(\boldsymbol{P}, \boldsymbol{K})$ satisfies $\|\boldsymbol{T}_{zw}\|_\infty < \gamma$. The plant–controller configuration is shown in Figure 10.12.
The standard assumptions are as follows:

1. (A, B_2) is stabilizable and (C_2, A) is detectable.
2. $D_{12}' D_{12} = I$ and $D_{21} D_{21}' = I$.
3. $\begin{bmatrix} A - j\omega I & B_2 \\ C_1 & D_{12} \end{bmatrix}$ has full column rank for all real ω
4. $\begin{bmatrix} A - j\omega I & B_1 \\ C_2 & D_{21} \end{bmatrix}$ has full row rank for all real ω

Implications of the above assumptions may be explained as follows.

Assumption 1 is necessary and sufficient for the existence of admissible controllers. This is proved in Chapter 9 under stabilizability criteria. Assumption 2 is merely required for normalization purpose and does not have any other significance. Assumptions 3 and 4 are required to ensure that the Riccati equations have stabilizing solutions. Earlier in this chapter it was shown that the assumptions $D_{12}' C_1 = 0$, and (C_1, A) having no unobservable modes on the imaginary axis are necessary for the existence of a stabilizing solution for the control Riccati equation. With $D_{12}' C_1 \neq 0$, cross coupling terms are present and we require the more general condition stated above under assumption 3 to be satisfied. Similar explanation holds good for assumption 4 also. With these preliminaries out of the way, we now state the principal result obtained by *DGKF*.

NECESSARY AND SUFFICIENT CONDITIONS

THEOREM 10.4 (Measurement feedback problem)
For the plant described in (10.46) to (10.48) there exists a linear time invariant measurement feedback controller which ensures that the objective function $\|\boldsymbol{T}_{zw}\|_\infty < \gamma$ if and only if

1. The control Riccati equation given below has a stabilizing solution $X_\infty \geq 0$

$$X_\infty \tilde{A} + \tilde{A}' X_\infty - X_\infty (B_2 B_2' - \gamma^{-2} B_1 B_1') X_\infty + \tilde{C}' \tilde{C} = 0 \qquad (10.49)$$

where

$$\tilde{A} = A - B_2 D_{12}' C_1$$
$$\tilde{C}' \tilde{C} = C_1' (I - D_{12} D_{12}') C_1$$

2. The filter Riccati equation given below has a stabilizing solution $Y_\infty \geq 0$

$$\tilde{A} Y_\infty + Y_\infty \tilde{A}' - Y_\infty (C_2' C_2 - \gamma^{-2} C_1' C_1) Y_\infty + \tilde{B} \tilde{B}' = 0 \qquad (10.50)$$

where

$$\tilde{A} = A - B_1 D_{21}' C_2$$
$$\tilde{B} \tilde{B}' = B_1 (I - D_{21}' D_{21}) B_1'$$

3. The spectral radius of the product of the two solutions X_∞ and Y_∞ is less than γ^2, i.e.

$$\rho(X_\infty Y_\infty) < \gamma^2$$

Proof: The proof is rather lengthy and has been omitted. Interested readers may refer to either Doyle *et al*. (1989) or Green and Limebeer (1995) for a complete proof.

REMARKS

1. That Conditions 1 and 2 in the above theorem are necessary may be seen from the following argument. Let us suppose that there exists an admissible measurement feedback controller which makes $\|T_{zw}\| < \gamma$. We note that a measurement feedback controller is

also a full information controller because $u = \boldsymbol{K}y$ and $y = C_2x + D_{21}w$ implies that $u = [\,\boldsymbol{K}C_2 \quad \boldsymbol{K}D_{21}\,]\begin{bmatrix} x \\ w \end{bmatrix}$.

Hence a full information controller exists in this case. For such a controller to exist, in the FI case we require condition 1 in the above theorem to be satisfied. Condition 2 follows from a dual argument.

2. Unlike in the LQG case, the two Riccati equations (10.49) and (10.50) associated with $\mathcal{H}_\infty$ control need not always have admissible solutions. Their solutions depend on the sign definiteness of the quadratic terms in the respective equations.
3. The existence of positive semi-definite stabilizing solutions for (10.49) and (10.50) are necessary conditions but not sufficient for the $\mathcal{H}_\infty$ regulator problem to have a solution.
4. The solution of the $\mathcal{H}_\infty$ regulator problem is expressed in terms of a $\mathcal{H}_\infty$ filter that estimates the full information $\mathcal{H}_\infty$ control law, $u^* = F_\infty x$. The solution of the second Riccati equation thus depends on the solution of the first Riccati equation. The procedure is to solve the $\mathcal{H}_\infty$ full information problem first and then obtain an $\mathcal{H}_\infty$ estimator for full information control. This is the separation structure associated with $\mathcal{H}_\infty$ regulator problem.
5. Note that the Riccati equation (10.50) is associated with $\mathcal{H}_\infty$ estimation of C_1x, given the output y of the plant.
6. Since the theorem gives necessary and sufficient conditions for the existence of an admissible controller such that $||\boldsymbol{T}_{zw}||_\infty < \gamma$ we may obtain a γ_0 which is the infimum of all such solutions which satisfy Conditions 1 to 3. γ_0 can be calculated as closely as desired by a search technique.
7. As $\gamma \to \infty$, $H_\infty \to H_2$ and $X_\infty \to X_2$ and the central controller $\boldsymbol{K}_{\text{sub}} \to \boldsymbol{K}_2$ where H_2, X_2 and K_2 are respectively the Hamiltonian matrix, the control Riccati equation solution and the optimal controller pertaining to the LQG problem. The above statement can be easily verified.
8. It would be interesting to speculate which of conditions 1 to 3 pertaining to the theorem above would fail first. It has been suggested in *DGKF* that Condition 3 is most likely to fail first.

THE FAMILY OF ADMISSIBLE OUTPUT FEEDBACK CONTROLLERS

THEOREM 10.5 (Doyle *et al.* (1989))
If Conditions 1 – 3 in Theorem 10.4 are satisfied, the set of admissible controllers such that $||\boldsymbol{T}_{zw}||_\infty < \gamma$ equals the set of all transfer matrices from y to u as given in Figure 10.13.

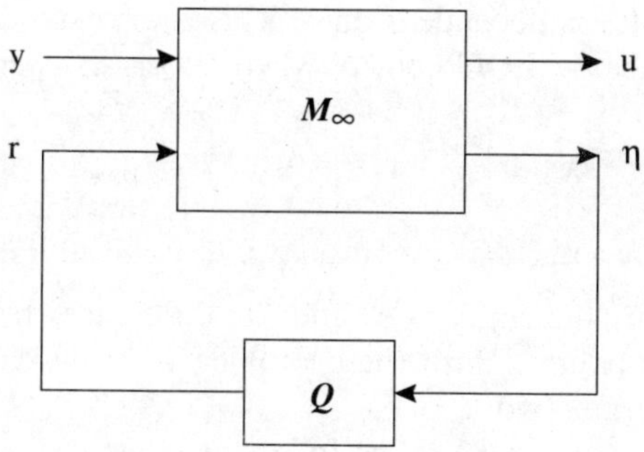

Figure 10.13 Family of output feedback controllers.

Where we have $\boldsymbol{K} = \mathcal{F}_l(\boldsymbol{M}_\infty, \boldsymbol{Q})$ and

$$\boldsymbol{M}_\infty = \left[\begin{array}{c|cc} \hat{A}_\infty & -Z_\infty L_\infty & Z_\infty B_2 \\ \hline F_\infty & 0 & I \\ -C_2 & I & 0 \end{array}\right]$$

with $\boldsymbol{Q} \in \mathcal{RH}_\infty, \quad ||\boldsymbol{Q}||_\infty < \gamma$

$$\widehat{A}_\infty = A + \gamma^{-2} B_1 B_1' X_\infty + B_2 F_\infty + Z_\infty L_\infty C_2$$
$$F_\infty = -B_2' X_\infty, \quad L_\infty = -Y_\infty C_2'; \quad Z_\infty = (1 - \gamma^{-2} Y_\infty X_\infty)^{-1}$$

Proof: For proof refer to Doyle *et al*. (1989).

Remarks

1. The above conditions have been derived under the simplifying assumptions that $D_{12}' C_1 = 0$, $B_1 D_{21}' = 0$, $D_{12}' D_{12} = I$ and $D_{21} D_{21}' = I$.
2. If $\boldsymbol{Q} = 0$ we recover the *central controller*

$$\boldsymbol{K}_{\text{sub}} = \left[\begin{array}{c|c} A_\infty & -Z_\infty L_\infty \\ \hline F_\infty & 0 \end{array}\right]$$

   This is one among the many admissible controllers for which $||\boldsymbol{T}_{zw}||_\infty < \gamma$.
3. The *central controller* has an observer based architecture and its dynamical order is the same as that of the plant.

4. Note that $\boldsymbol{M}_\infty$ is dependent on γ. This is in marked contrast to a similar problem in LQG control where $\boldsymbol{M}_2$ is independent of γ.

10.8 Summary

This chapter highlights the similarities and the differences between the $\mathcal{H}_2$ and $\mathcal{H}_\infty$ regulator problems. Both these problems are developed within a common state space frame work.

For the $\mathcal{H}_2$ optimization problem the following questions were posed and answered with respect to both full information and measurement feedback controllers.

(i) The minimum of $||\boldsymbol{T}_{zw}||_2$
(ii) The unique controller minimizing $||\boldsymbol{T}_{zw}||_2$
(iii) The family of all controllers such that $||\boldsymbol{T}_{zw}||_2 < \gamma$ where γ is greater than the minimum of $||\boldsymbol{T}_{zw}||_2$

In the $\mathcal{H}_\infty$ optimization problem, the minimum of $||\boldsymbol{T}_{zw}||_\infty$ cannot be determined in a finite number of steps. Hence the questions which are more relevant are:

(i) Necessary sufficiant conditions for the existence of a controller such that $||\boldsymbol{T}_{zw}||_\infty < \gamma$
(ii) The family of all admissible controllers such that $||\boldsymbol{T}_{zw}||_\infty < \gamma$

The major contribution of the *DGKF* paper is that it provides an answer to the first question and then proceeds to delineate the family of all controllers for which $||\boldsymbol{T}_{zw}||_\infty < \gamma$.

As in the $\mathcal{H}_2$ problem, the full information controller plays an important role in the delineation of all controllers in the general case when output feedback is considered. Hence the FI problem is taken up for detailed investigation. A very useful property of inner linear fractional transformation, which preserves contraction and internal stability is exploited to considerably simplify controller prameterization for the FI problem.

The $\mathcal{H}_\infty$ filter which is considered next is derived making use of the fact that the estimation problem is a dual of the FI controller problem. A combination of FI problem and the $\mathcal{H}_\infty$ filter yields the solution for the measurement feedback problem. Apart from the question regarding the existence of solutions to the associated Riccati equations, it turns out that the solution of the control Riccati equation enters as a parameter in the filter Riccati equation, thus coupling the two solutions in a complex manner. Finally the *DGKF* result which forms the backbone of the solution to the generalized regulator problem in state space is fully stated and explained.

Notes and Additional References

Standard references on the topic discussed in this chapter are the paper by Doyle *et al.* (1989) and the book by Green and Limebeer (1995). The latter contains an extensive bibliography on the subject. The major assumptions in the solutions of the generalized regulator problem are

(i) Certain subsystems connecting control input to the controlled output and disturbances to measurement should not have invariant zeros on the imaginary axis.
(ii) The matrices D_{12} and D_{21} should be injective and surjective respectively.

When these conditions are not met with, then we have the so called *singular systems.* The *DGKF* theory cannot solve stabilization problems arising in singular systems. However many stabilization problems end up in singular systems and their solutions merit further investigation. The work of Stoorvogel (1992) is of interest in this connection.

Exercises

1. Let

$$\dot{x} = Ax + B_1 w + B_2 u$$

$$z = \begin{bmatrix} C_1 x \\ D_{12} u \end{bmatrix}$$

with $D_{12}' D_{12} = I$ and (A, B_2) stabilizable.
If $u = -Kx$ and $(A - B_2 K)$ is asymptotically stable.
Show that $\|\boldsymbol{T}_{zw}\|_2^2 = (B_1' Q B_1)$ where Q is the unique non-negative definite solution to the equation

$$(A - B_2 K)' Q + Q(A - B_2 K) + C_1' C_1 + K' K = 0$$

2. Consider a signal generator governed by the following equations

$$\dot{x} = Ax + Bw + B_2 u$$
$$y = Cx + v$$

where u is a known control signal and w and v are disturbance signals.
Consider the filter equation in which the control signal is included i.e.

$$\dot{\hat{x}} = A\hat{x} + YC'(y - C\hat{x}) + B_2 u$$

where Y is the stabilizing solution of filter Riccati equation. Show that $(\hat{x} - x)$ is independent of u and hence conclude that $\hat{x}$ generated by the above equation is the optimal estimate of x.

3. Consider a system

$$\dot{x} = Ax + Bw \qquad x(0) = 0$$
$$y = Cx + Dv$$

where $d = \begin{bmatrix} w \\ v \end{bmatrix}$ is a unit intensity white noise process.

The innovations process in the Kalman filter is defined as $\eta = y - C\hat{x}$. Prove that η is a white noise process with unit intensity.

4. Consider a full information $\mathcal{H}_\infty$ problem where $H_\infty =$ dom (Ric) and Ric $(H_\infty) \geq 0$ for a specified γ. The central controller is given by

$$K = [F_\infty \quad 0] \text{ where } F_\infty = -B_2' X_\infty$$

The transfer function of $\boldsymbol{T}_{zw} = \left[\begin{array}{c|c} A_{F_\infty} & B_1 \\ \hline C_{1F_\infty} & 0 \end{array}\right]$

where $A_{F_\infty} = A + B_2 F_\infty$ and $C_{1F_\infty} = C_1 + D_{12} F_\infty$

(a) Verify that $\|\boldsymbol{T}_{zw}\|_\infty < \gamma$.
(b) If $H_\infty \in$ dom (Ric) then prove that A_{F_∞} is stable if and only if $X_\infty \geq 0$.

5. It is proved in this chapter that by change of variables $\nu = u - F_2 x$, we have

$$z = \left[\begin{array}{c|cc} A_{F_2} & B_1 & B_2 \\ \hline C_{1F_2} & 0 & D_{12} \end{array}\right] \begin{bmatrix} w \\ \nu \end{bmatrix} = \boldsymbol{G}_c B_1 w + \boldsymbol{U}\nu$$

Where:

$$A_{F_2} = A + B_2 F_2; \quad F_2 = -B_2' X_2$$

$C_{1F_2} = C_1 + D_{12} F_2$ and X_2 is the solution of the control Riccati equation. Show that:

(a) $\boldsymbol{U}^\sim \boldsymbol{U} = I$.
(b) $\boldsymbol{U}^\sim \boldsymbol{G}_c$ belongs to $\mathcal{RH}_2^\perp$.

6. Consider a generalized 1-block regulator problem. The off diagonal blocks of the generalized plant are square and without loss of generality we may assume that they are unit matrices. The generalized

plant has the form:

$$\begin{bmatrix} \dot{x} \\ z \\ y \end{bmatrix} = \begin{bmatrix} A & B_1 & B_2 \\ C_1 & 0 & I \\ C_2 & I & 0 \end{bmatrix} \begin{bmatrix} x \\ w \\ u \end{bmatrix}$$

We assume that the standard assumptions hold.
The standard assumptions regarding full rank on the imaginary axis now reduce to $\tilde{A} = (A - B_2C_1)$ and $\overline{A} = (A - B_1C_2)$ having no eigen values on the imaginary axis.
(a) Show that the associated LQG problem has the Riccati equations

$$X\tilde{A} + \tilde{A}'X - XB_2B_2'X = 0$$

$$\overline{A}Y + Y\overline{A}' - YC_2'C_2Y = 0$$

(b) Show that for the $\mathcal{H}_\infty$ version of the regulator problem the Riccati equations are

$$X_\infty\tilde{A} + \tilde{A}'X_\infty - X_\infty(B_2B_2' - \gamma^{-2}B_1B_1')X_\infty = 0$$

$$\overline{A}Y_\infty + Y_\infty\overline{A}' - Y_\infty(C_2'C_2 - \gamma^{-2}C_1'C_1)Y_\infty = 0$$

7. Assume that $H_\infty \in \text{dom (Ric)}$ and $\text{Ric}\ (H_\infty) \geq 0$.
Let P be a transfer matrix given by.

$$P = \begin{bmatrix} P_{11} & P_{12} \\ P_{21} & P_{22} \end{bmatrix} = \left[\begin{array}{c|cc} A_{F_\infty} & \gamma^{-1}B_1 & B_2 \\ \hline C_{1F_\infty} & 0 & D_{12} \\ -\gamma^{-1}B_1'X_\infty & I & 0 \end{array}\right]$$

where $A_{F_\infty} = A + B_2F_\infty$; $C_{1F_\infty} = C_1 + D_{12}F_\infty$; $F_\infty = -B_2'X_\infty$ and X_∞ is the solution of the control Riccati equation

$$A'X_\infty + X_\infty A + X_\infty(\gamma^{-2}B_1B_1' - B_2B_2')X_\infty + C_1'C_1 = 0$$

Prove the following:
i) $\boldsymbol{P} \in \mathcal{RH}_\infty$
ii) $\boldsymbol{P}^{\sim}\boldsymbol{P} = I$
iii) $\boldsymbol{P}_{21}^{-1} \in \mathcal{RH}_\infty$

11

Kharitonov Theory and Related Approaches

11.1 Introduction

Our main emphasis so far has been on unstructured uncertainty. To suddenly change track and take up structured uncertainty in this chapter calls for some justification (which is provided in this chapter).

Stability robustness has had a long and chequered history and one can trace it to the early works of Chebyshev and Markov. It is a classical concept and is of fundamental importance in control and system theory. In 1978, the Russian mathematician Kharitonov published a remarkable theorem regarding robust stability of polynomials. Kharitonov theorem simply stated asserts that the robust stability in the open left half complex plane of an interval polynomial whose coefficients vary independently inside a hyper rectangle (rectangular box of n dimensions) can be determined by testing for stability of just four polynomials in the real coefficient case. The implication of this theorem goes far beyond the result embodied in it and has provided the 'spark' for the activities of a large body of researchers all over the world. Such an important result even though it deals primarily with structured uncertainty cannot go unnoticed in a book on robust control. For sheer beauty and elegance if not for anything else it deserves to be known widely among the control community. This provides justification for including the topic in a book which otherwise concentrates entirely on unstructured type of uncertainty.

The organization of this chapter is as follows. Section 11.2 introduces the so called Kharitonov polynomials which form the basis for the celebrated theorem under that name. The proof of the theorem using elementary concepts is given in Section 11.3. A simplified version of the theorem for lower order polynomials is also given in this section. Kharitonov theorem has some basic limitations from the practical viewpoint. This is explained in

Section 11.4. This section goes on to explain and state a major theorem developed during the post-Kharitonov era known as the edge theorem. The conditions under which this theorem can be applied are also explained. There are two fundamental concepts, namely the value set and the zero exclusion principle which are repeatedly used while analysing robust stability of uncertain polynomials. These concepts are explained in Sections 11.5 and 11.6. In many practical problems, the coefficients of the polynomial are multi-linearly dependent on the parameters. An approximation approach to determine robust polynomial stability under such circumstances is explained in Section 11.7. The final section gives a summary of the chapter.

11.2 The Kharitonov Polynomials

Kharitonov's (1978) original proof of the theorem is somewhat cryptic and difficult to understand. After this theorem surfaced in western literature, a number of alternative proofs have appeared. But by far the easiest and conceptually elegant proof is the one provided by Minnichelli *et al.* (1989). The proof given below is an adaptation of their original proof.

Before proving the theorem we shall go through some preliminaries. Kharitonov's result pertains to a class of polynomials, whose coefficients are permitted to vary independantly over a specified but arbitrary interval. We say that *a polynomial is Hurwitz* if, and only if, all its zeros are located in the open right half of the complex plane. Polynomials whose coefficients vary independently over an interval are called *interval polynomials.* Kharitonov's result states that the whole family of interval polynomials with real coefficients is Hurwitz if and only if four specially chosen well defined polynomials known as Kharitonov polynomials are Hurwitz. Thus, by merely testing for stability of just four polynomials (eight in the case of polynomials with complex coefficients), we can ensure the stability of a whole class of polynomials.

Let the perturbation parameters be $q_0, q_1, \ldots q_{n-1}$ with known *a priori* bounds

$$\underline{q}_i < q_i < \overline{q}_i \quad i = 0, 1, \ldots, n-1.$$

Next consider the interval polynomial family $\mathcal{P}$ described by the set of nth degree monic polynomials (i.e. polynomials with unity as coefficient for the highest degree term)

$$p(s, \boldsymbol{q}) = q_0 + q_1 s + \ldots + q_{n-1} s^{n-1} + s^n$$

where the q_is take on values in the allowed range $[\underline{q}_i, \overline{q}_i]$.

The perturbation vector $\boldsymbol{q}$ is contained in a closed and bounded set Q which in the above case is a box-like structure known as hyper-rectangle of n dimensions.

We now define a value set which plays an important role in the proof of Kharitonov's theorem.

DEFINITION 11.1 (Value set)
Given a family of polynomials $\mathcal{P}$ such that $\mathcal{P} = \{p(.,\boldsymbol{q}) : \boldsymbol{q} \in Q\}$ the value set at frequency $\omega \in \mathbb{R}$ is given by

$$\boldsymbol{H}(\omega) = p(j\omega, Q) = \{p(j\omega, \boldsymbol{q}) : \boldsymbol{q} \in Q\} \tag{11.1}$$

We shall presently show that for every value of ω, $\boldsymbol{H}(\omega)$ is a rectangle in the complex plane with its sides parallel to the x and y axis respectively in the case where the bounding set Q is a hyper-rectangle. As ω changes, $\boldsymbol{H}(\omega)$ continues to be a rectangle but changes size and moves to different parts of the complex plane in a smooth and continuous manner.

In shorthand notation, we shall represent $p(s, \boldsymbol{q})$ by $p(s)$ in the analysis that follows. Thus the nth degree monic polynomial may be expressed as the sum of an even degree polynomial $g(s)$ and an odd degree polynomial $h(s)$. Therefore

$$p(s) = g(s) + h(s)$$

Now define two even polynomials

$$g_1(s) = \underline{q}_0 + \overline{q}_2 s^2 + \underline{q}_4 s^4 + \ldots$$
$$g_2(s) = \overline{q}_0 + \underline{q}_2 s^2 + \overline{q}_4 s^4 + \ldots$$

Similarly define two odd polynomials

$$h_1(s) = \underline{q}_1 s + \overline{q}_3 s^3 + \underline{q}_5 s^5 + \ldots$$
$$h_2(s) = \overline{q}_1 s + \underline{q}_3 s^3 + \overline{q}_5 s^5 + \ldots$$

Using the above polynomials we generate four new polynomials known as Kharitonov polynomials. They are:

$$k_{11}(s) = g_1(s) + h_1(s) \tag{11.2}$$
$$k_{21}(s) = g_2(s) + h_1(s) \tag{11.3}$$
$$k_{12}(s) = g_1(s) + h_2(s) \tag{11.4}$$
$$k_{22}(s) = g_2(s) + h_2(s) \tag{11.5}$$

Now substitute $s = j\omega$ in $g_1(s)$, $g_2(s)$, $h_1(s)$ and $h_2(s)$. With $\omega > 0$ a real number, we note the following facts:

1. $g_1(j\omega)$ and $g_2(j\omega)$ are always pure real.
2. $h_1(j\omega)$ and $h_2(j\omega)$ are always pure imaginary.
3. If $p(j\omega) = p_r(j\omega) + p_i(j\omega)$ where p_r and p_i denote real and imaginary parts of $p(\cdot)$,. We have:
4. $g_1(j\omega) \leq p_r(j\omega) \leq g_2(j\omega)$
5. $h_1(j\omega) \leq p_i(j\omega) \leq h_2(j\omega)$

Figure 11.1 is a typical plot of $H(\omega)$ for some $\omega > 0$. Note that $p_r(j\omega)$ can assume values only between $g_1(j\omega)$ and $g_2(j\omega)$ and $p_i(j\omega)$ can assume values only between $h_1(j\omega)$ and $h_2(j\omega)$ as q ranges over all Q. Hence the value set in this case is a rectangle with its sides parallel to the real and imaginary axis. Since $k_{11}(j\omega) = g_1(j\omega) + h_1(j\omega)$ by (11.2), the left bottom corner of $\boldsymbol{H}(\omega)$ is represented by the polynomial $k_{11}(s)$ evaluated at $j\omega$. Similarly the remaining corners are represented by the Kharitonov polynomials $k_{21}(s)$, $k_{22}(s)$ and $k_{21}(s)$ respectively. Note that every point $\boldsymbol{q}$ in the bounded set $Q \subset \mathbb{R}^n$ has a one to one correspondence with a polynomial $p(s, \boldsymbol{q})$ whose constant coefficients are precisely the coordinates of the vector $\boldsymbol{q}$. Thus every point in $\boldsymbol{H}(\omega)$ is obtained by evaluating the associated polynomial at $s = j\omega$. It is in this context that the four corners of $H(\omega)$ are identified by the four Kharitonov polynomials.

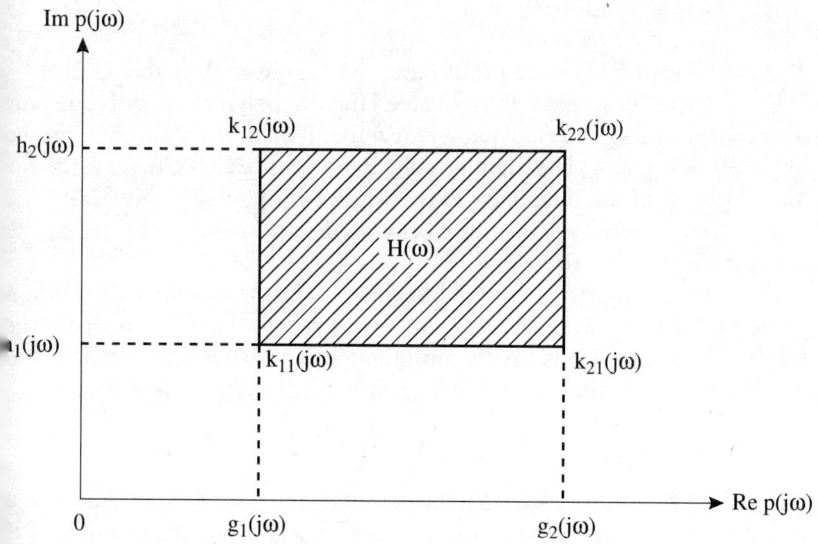

'igure 11.1 Rectangular image of $\mathcal{P}$ at $s = j\omega$.

Reverting back to the n dimensional hyper-rectangle Q, we note that its corners are points whose coordinates are permutations of maximum and minimum values of the parameters. For an 'n vector' each of whose elements can assume only two possible values (maximum or minimum) the maximum number of distinct vectors that can be generated is 2^n. Thus the hyper rectangle Q has 2^n corners and each corner is associated with one polynomial. Of these corner polynomials only four among them are associated with Kharitonov polynomials.

11.3 Proof of Kharitonov's Theorem

The Kharitonov theorem is proved in three stages. We first establish certain properties of Hurwitz polynomials. Then we prove a lemma concerning Kharitonov polynomials. Finally we make use of these properties to prove the theorem.

PROPERTY 1
If a monic polynomial $p(\cdot)$ is Hurwitz, then all its coefficients are positive. The proof is straightforward and is omitted.

PROPERTY 2
If $p(\cdot)$ is Hurwitz with degree $n \geq 1$ then $\arg[p(j\omega)]$ is a continuous and strictly increasing function of ω.

Proof: We give a heuristic proof based on Figure 11.2(a) and 11.2(b).

Consider for example a third degree Hurwitz polynomial with one pair of complex conjugate zeros and a real zero, all of which are located in the open left half plane. The angular contributions of each of these zeros for an ω located on the imaginary axis is shown in the figure. Note that the angles are measured in the anti clockwise direction with reference to the positive direction of the x axis.

As ω moves upward along the imaginary axis, the angular contribution made by the zeros with respect to $p(j\omega)$ increases. This is clear from the figure. The $\arg[p(j\omega)]$ is thus a continuously increasing function of ω and further note that because $p(s)$ is Hurwitz, $p(j\omega) \neq 0 \forall \omega$. Now define

$$\arg_{\text{net}}[p] = \lim_{\omega \to \infty} [\arg p(j\omega) - \arg p(0)]$$

where the left hand side denotes the net total angle, which is nothing but a measure of the number of encirclements of $p(j\omega)$ about the origin as ω is varied from 0 to ∞. It is clear that each zero in the open left half plane

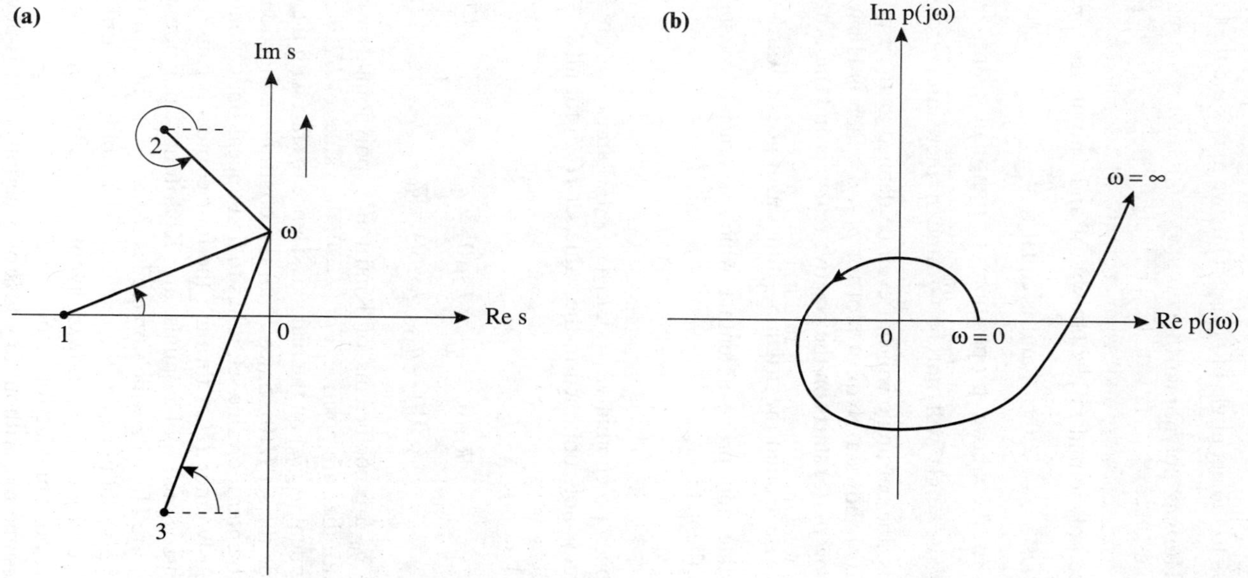

Figure 11.2 Properties of Hurwitz polynomials. (a) Angle of $p(j\omega)$ when $p(\cdot)$ is Hurwitz. (b) Plot of $p(j\omega)$ for $n = 5$.

contributes $\frac{\pi}{2}$ whereas each zero, if it is located in the open right half plane contributes $\frac{-\pi}{2}$ to the net angle of $p(j\omega)$. Thus we may state the following property for Hurwitz polynomials.

PROPERTY 3
An nth degree polynomial $p(\cdot)$ is Hurwitz if and only if $\arg_{\text{net}}[p(j\omega)]$ is well defined (i.e. $p(j\omega) \neq 0 \forall \omega$) and is equal to $\frac{n\pi}{2}$.

According to the above proposition, if right-half plane zeros are present, the net angle will not be $\frac{n\pi}{2}$ and if there are zeros on the imaginary axis, the net angle will not even be defined at these points.

Figure 11.2(b) shows a Nyquist plot for $p(j\omega)$ which is Hurwitz with $n = 5$. Note how $p(j\omega)$ starts on the positive real axis and smoothly circles counter clockwise around the origin $5\frac{\pi}{2}$ times as it goes towards infinity.

We next state and prove a lemma which is crucial in establishing Kharitonov's theorem.

LEMMA 11.1
If the Kharitonov polynomials $k_{11}(s)$, $k_{12}(s)$, $k_{21}(s)$ and $k_{22}(s)$ are Hurwitz then the origin is not contained in the value set $\boldsymbol{H}(\omega)$ for all $\omega \in \mathbb{R}$.

Proof: We have:

$$k_{11}(0) = \underline{q}_0; \quad k_{21}(0) = \overline{q}_0$$
$$k_{12}(0) = \underline{q}_0; \quad k_{22}(0) = \overline{q}_0$$

Both $\underline{q}_0$ and $\overline{q}_0$ are coefficients of Kharitonov polynomials which are Hurwitz. Hence they are positive. For $\omega = 0$, the value set $\boldsymbol{H}(\omega)$ collapses into a line segment $[\underline{q}_0, \overline{q}_0]$ on the real axis. Hence to start with the origin is located outside the set $\boldsymbol{H}(0)$. Since $H(\omega)$ is a continuous function of ω, if the origin is to enter the value set, it has to first appear on its boundary. It cannot obviously enter $\boldsymbol{H}(\omega)$ through the four corners because, at these points, the associated polynomials are Kharitonov polynomials and therefore they are Hurwitz which precludes their assuming zero value for any ω. Hence zero should enter the rectangle $\boldsymbol{H}(\omega)$ only by cutting across its parallel edges. Figure 11.3 shows a hypothetical situation where zero is located on the bottom edge for some $\omega = \overline{\omega}$.

We may assume this without any loss of generality. if ω is now varied from $\overline{\omega}$ to $\overline{\omega} + \delta\omega$ then $k_{11}j(\overline{\omega} + \delta\omega)$ moves down to the third quadrant and $k_{21}j(\overline{\omega} + \delta\omega)$ moves up to the first quadrant. This is because the angles o

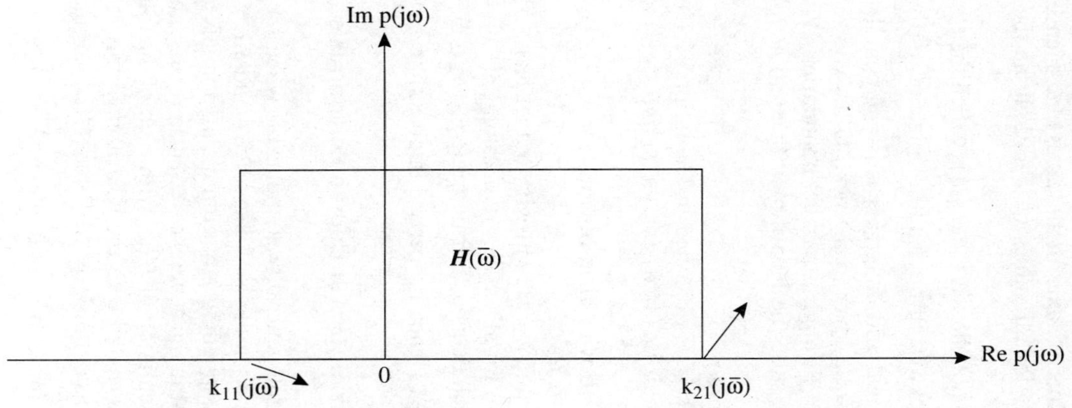

Figure 11.3 Plot of $H(\bar{\omega})$ with zero on the boundary.

Hurwitz polynomials increase continuously as ω is increased. But now we get into an anomolous situation where the imaginary part of $k_{11}j(\overline{\omega}+\delta\omega)$ is negative while the imaginary part of $k_{21}j(\overline{\omega}+\delta\omega)$ is positive. But this cannot happen because the edges of the rectangle $\boldsymbol{H}(\omega)$ should always remain parallel to the respective axis for all values of ω. Hence zero cannot enter the $\boldsymbol{H}(\omega)$ rectangle for all values of ω so long as the Kharitonov polynomials are Hurwitz.

An important point to note here is that as ω is varied the entire rectangle $\boldsymbol{H}(\omega)$ moves counter clockwise through a total angle of $\frac{n\pi}{2}$ always keeping its edges parallel to the real and imaginary axis and completely entering one quadrant before crossing over to the next. Otherwise zero will enter the inside region of $\boldsymbol{H}(\omega)$ through it parallel boundaries. With the above technicalities cleared, we state the Kharitonov theorem as follows.

THEOREM 11.1 (Kharitonov)
The class of polynomials $\mathcal{P}$ as described in Definition 11.1 is Hurwitz if and only if the polynomials $k_{11}(s)$, $k_{12}(s)$, $k_{21}(s)$ and $k_{22}(s)$ are Hurwitz.

Proof: 'only if': If $\mathcal{P}$ is Hurwitz, then obviously k_{11}, k_{12}, k_{21} and k_{22} are Hurwitz as they belong to $\mathcal{P}$.
'if': Assume k_{11}, k_{12}, k_{21} and k_{22} are Hurwitz and let $p(\cdot) \in \mathcal{P}$. According to Lemma 11.1, $0 \notin \boldsymbol{H}(\omega)$ for all ω. This implies that $p(j\omega) \neq 0$ for all ω. Hence $\arg_{\text{net}}[p(j\omega)]$ is well defined. Furthermore $p(j\omega) \in H(\omega)$ for all ω. Hence $\arg_{\text{net}}[p(j\omega)] = \frac{n\pi}{2}$. Hence Property 3 implies that $p(s)$ is Hurwitz.

11.3.1 Simplified Version for Lower Degree Polynomials

Kharitonov theorem in its full generality is applicable when the polynomial degree is six or above. For polynomials of degrees 5, 4 and 3 Kharitonov test can be simplified. The corresponding Kharitonov polynomials are only 3, 2 and 1 in number as against 4 in the general case. This property can be readily proved by a pictorial representation of $\boldsymbol{H}(\omega)$ with each of the four corners of the level rectangle being associated with the Kharitonov polynomials. What these polynomials really accomplish is to force $\boldsymbol{H}(\omega)$ to move counter clockwise about the origin through a total angle of $\frac{n\pi}{2}$ always maintaining parallelism of edges with respect to the real and imaginary axes and preventing the origin from entering the rectangle.

We first note that all monic polynomials of degree 'n' whether they are Hurwitz or not behave as $(j\omega)^n$ as $\omega \to \infty$. Hence for any choice of $\underline{q}_k$ and $\overline{q}_k$ $k = 0, 1, 2, \ldots n-1$, the rectangle $H(\omega)$ will go to infinity at an asymptotic angle of $\frac{n\pi}{2}$ (mod 2π) radians. Each encirclement of the origin corresponds to 2π radians and hence provided the Kharitonov polynomials are Hurwitz, the rectangular value set $\boldsymbol{H}(\omega)$ make $\frac{n}{4}$ counter clockwise encirclements about the origin without ever entering it as ω is varied from zero to infinity.

In the light of the above explanation, we have the following proposition for $n = 3$.

PROPOSITION 1

A polynomial $p(s, \mathbf{q}) \in \mathcal{P}$ of degree 3 is Hurwitz if and only if $k_{21}(s)$ is Hurwitz.

Proof: The 'only if' part is immediate as $k_{21}(s) \in \mathcal{P}$.

To prove the 'if' part, we assume that $k_{21}(s)$ is Hurwitz. Referring to Figure 11.4 we note that $k_{21}(s)$ occupies the lower right hand corner of $\boldsymbol{H}(\omega)$. To begin with $\boldsymbol{H}(0)$ lies in the positive real axis (because $\underline{q}_0 > 0$). The role of $k_{21}(j\omega)$ is to literally steer the rectangle $\boldsymbol{H}(\omega)$ into the open first quadrant and then move it completely into the second quadrant before it enters the third quadrant and goes to infinity at an asymptotic angle of $\frac{3\pi}{2}$. Thus $k_{21}(s)$ being Hurwitz is sufficient to ensure that the value set avoids the origin and every $p(s, \boldsymbol{q}) \in \mathcal{P}$ is therefore Hurwitz.

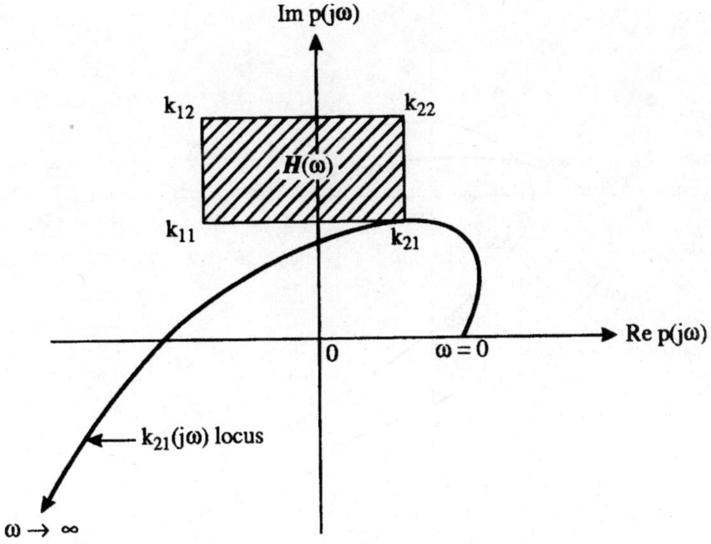

Figure 11.4 Loci of Kharitonov polynomials for $n = 3$.

Similar arguments can be used in the case of $n = 4$ where we require two Kharitonov Hurwitz polynomials to ensure stability of the members of the class $\mathcal{P}$.

PROPOSITION 2

A polynomial $p(s, \boldsymbol{q}) \in \mathcal{P}$ of degree 4 is Hurwitz if and only if $k_{21}(s)$ and $k_{22}(s)$ are Hurwitz.

Proof: As before the 'only if' part proof is trivial. To prove the 'if' part, we refer to Figure 11.5. We now assume that $k_{21}(s)$ and $k_{22}(s)$ are Hurwitz. As in the case of $n = 3$, $k_{21}(j\omega)$ pushes $H(\omega)$ off the positive real axis into the open first quadrant then completely into the second quadrant before $\boldsymbol{H}(\omega)$ can cross into the lower half plane. Once $\boldsymbol{H}(\omega)$ is in the second quadrant $k_{22}(j\omega)$ pushes $\boldsymbol{H}(\omega)$ completely into the third quadrant before $\boldsymbol{H}(\omega)$ can cross over to the right half plane. Once $k_{22}(j\omega)$ enters the open third quadrant, it can never cross the real axis again and hence $\boldsymbol{H}(\omega)$ must remain in the open lower half plane as it moves to infinity at an asymptotic angle of 2π (or zero). Thus all $p(s, \boldsymbol{q}) \in \mathcal{P}$ must be Hurwitz provided $k_{21}(s)$ and $k_{22}(s)$ are Hurwitz.

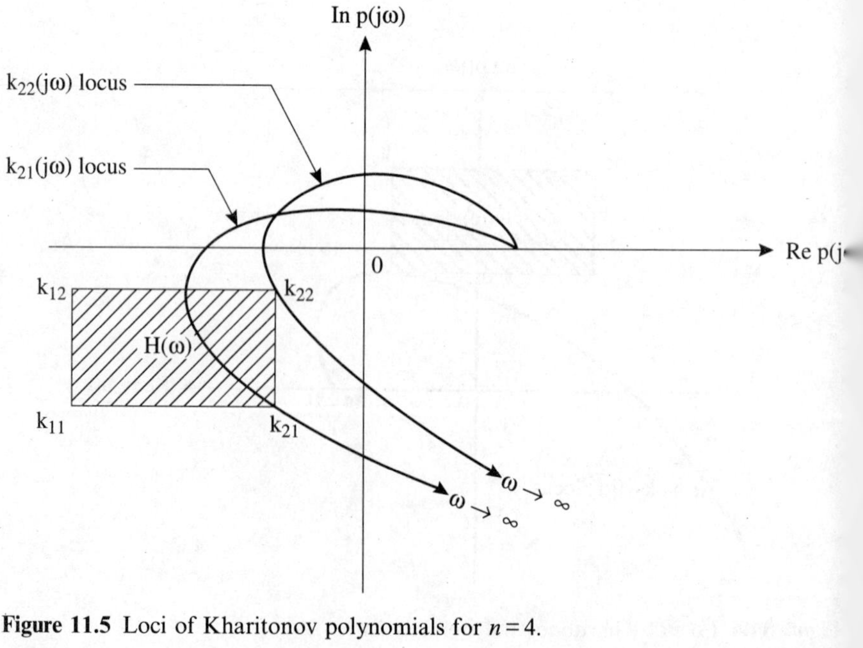

Figure 11.5 Loci of Kharitonov polynomials for $n = 4$.

PROPOSITION 3
A polynomial $p(s, \boldsymbol{q}) \in \mathcal{P}$ of degree 5 is Hurwitz if and only if $k_{21}(s)$, $k_{22}(s)$ and $k_{12}(s)$ are Hurwitz.

Proof: May be constructed on similar lines as in the case of $n = 3$ and $n = 4$.

■

A generalization of Kharitonov theorem when the polynomial coefficients are complex is possible by letting the real and imaginary parts of each coefficient vary independently in arbitrary intervals. Such a class of polynomials is Hurwitz if and only if eight special well defined polynomials are Hurwitz. For proof refer Minnichelli *et al.* (1989)

11.4 Limitations of Kharitonov's Theorem

The Kharitonov theorem suffers from two basic limitations. These are:

1. From an engineering point of view an important limitation is that in Kharitonov theorem, the polynomial coefficients are assumed to vary independently. In other words no q_i enters into more than one coefficient. This is too restrictive an assumption in physical problems. In the general case we may write

$$p(s, \boldsymbol{q}) = \sum_{i=0}^{n} a_i(\boldsymbol{q}) s^i \tag{11.6}$$

where the prescribed bounding set is $Q \subset \mathbb{R}^m$. However the coefficients a_is are now functions of $(q_1, q_2 \ldots q_m)$. For example in network theory the coefficients may be functions of capacitors resistors and inductors. It is clear that as $\boldsymbol{q}$ varies over Q, we obtain a family of polynomials $\mathcal{P}$. More precisely

$$\mathcal{P} = \{p(\cdot, \boldsymbol{q}) : \boldsymbol{q} \in Q\} \tag{11.7}$$

Kharitonov theorem cannot be applied to the situation described above, where the coefficients are linear or non-linear functions of $\boldsymbol{q}$.
2. Another important limitation is that Kharitonov theorem can only be applied to problems where the stability region corresponds to the open left half plane. Thus it cannot be applied to solve stability problems arising in discrete time systems where the stability region is the open unit disk. In this case the stability of four Kharitonov polynomials is not sufficient to ensure stability of the entire family.

Considerable research effort has gone into removing these two limitations. The most significant result in this connection is the one obtained by Bartlett *et al.* (1988). Their result, known as the edge theorem, is described in the next section.

11.4.1 The Edge Theorem

We will restrict our discussion to the case where the bounding set Q is a box and $Q \subset \mathbb{R}^m$. As already explained this hyper-rectangle has $2^m = l$ extreme points or corners. These extreme points are designated $q^1, q^2, \ldots q^l$. The ith extreme point denoted by q^i is a vector with coordinates $\boldsymbol{q}^i = [q_1^i q_2^i \cdots q_m^i]^T$ Note that $q_j^i = \underline{q}_j$ or $\overline{q}_j$ for $j = 1, 2, \ldots m$.

Associated with each of these extreme points is a polynomial. For example consider the extreme point $\boldsymbol{q}^i$. The associated polynomial is

$$p_i(s) = \sum_{j=0}^{n} a_j(\boldsymbol{q}^i)s^j \tag{11.8}$$

There are l such polynomials namely $\{p_1(s), p_2(s) \ldots p_l(s)\}$ corresponding to the l corners of the box. We now define a polytope of polynomials which is a convex hull of $p_1, p_2, \ldots, p_l$ known as the generating polynomials. We thus have

$$p(s) = \text{conv}\ \{p_1(s), p_2(s), \ldots, p_l(s)\}$$

We consider the case where the coefficient $a_j(\boldsymbol{q}^i)$ of the polynomial

$$p_i(s) = \sum_{j=0}^{n} a_j(\boldsymbol{q}^i)s^j$$

is related to $\boldsymbol{q}^i$ in an affine linear manner. That is we have

$$a_j(\boldsymbol{q}^i) = \alpha_j^T \boldsymbol{q}^i + \beta_j \quad j = 1, 2, \ldots, n \tag{11.9}$$

where α_j is a constant vector and β_j is a scalar constant. In simple words, this means that the coefficients of the generating polynomials are linear combinations of parameter values at the respective extreme points of the bounding rectangular box Q. We may now visualize two spaces. One is the parameter space which is $\boldsymbol{m}$ dimensional and contains the bounding set Q. The other is the coefficient space which is $\boldsymbol{n}$ dimensional where every point in this space has a one to one correspondence with an nth degree monic polynomial. This is the space occupied by the polytope of polynomials. It turns out that if the coefficients of the uncertain polynomial are related to the independently varying parameters inside the closed rectangular box Q in an affine linear manner, then there is a one to one correspondence between Q and the convex hull described by conv $\{p_1, p_2, \ldots, p_l\}$.

This is one of the basic assumptions made while proving the edge theorem. The other assumption relates to the stability region considered. Recall that in Kharitonov theorem, this region is the open left half plane and the boundary of this region is the imaginary axis. In the more general case the stability region $\mathcal{D}$ is an open simply connected set in the complex plane. A set $\mathcal{D}$ is called *simply connected* if every closed curve in $\mathcal{D}$ can be continuously shrunk to any point in $\mathcal{D}$ without leaving $\mathcal{D}$. For example the open left half plane and the unit disk are simply connected while an annulus defined by $\{s : 1 < |s| < 2\}$ is not simply connected.

In this context we define $\mathcal{D}$ – stability of a family of polynomials as follows.

DEFINITION 11.2 (Robust $\mathcal{D}$ stability)
A family of polynomials $\mathcal{P} = \{p(\cdot, \boldsymbol{q}) : \boldsymbol{q} \in Q\}$ is said to be robustly $\mathcal{D}$-stable if for all $\boldsymbol{q} \in Q, p(s, \boldsymbol{q})$ is $\mathcal{D}$-stable, i. e. all the roots of $p(s, Q)$ lie in $\mathcal{D}$.

A simplified version of the edge theorem of Bartlett *et al*. (1988) follows.

THEOREM 11.2 (Edge theorem)
Let $\mathcal{P} = \{p(\cdot, \boldsymbol{q}) : \boldsymbol{q} \in Q\}$ be a polytope of polynomials with invariant degree where the bounding set Q is an m dimensional hyper rectangle and let $\mathcal{D}$ be an open simply connected set in the complex plane. Then $\mathcal{P}$ is $\mathcal{D}$-stable if and only if the set of all exposed edges of $\mathcal{P}$ are $\mathcal{D}$-stable.

REMARKS

1. The bounding set Q has $l = 2^m$ extreme points. The edge theorem states that robust $\mathcal{D}$ stability of the uncertain polynomial $p(s, \boldsymbol{q})$ for $\boldsymbol{q} \in Q$ is assured if for each pair $i, j \in \{1, 2, \ldots l\}$ associates with the exposed edges, the polynomial $p_{ij}(s, \lambda) = \lambda p_i(s) + (1 - \lambda) p_j(s)$ is $\mathcal{D}$ stable for all $\lambda \in [0, 1]$ where $p_i(s)$ and $p_j(s)$ are as defined in (11.8) with their coefficients affine linearly related to the parameters in accordance with (11.9).
2. The $\mathcal{D}$ stability of the edges may be tested by a one parameter sweep method. For example we may consider the stability of the polynomial $p_i(s) + kp_j(s)$ $\left(\text{where } k = \dfrac{1-\lambda}{\lambda}\right)$ by treating k as a variable parameter and generating a root locus diagram. We have to simply check whether this root locus lies entirely inside the region $\mathcal{D}$.
3. As the number of parameters increases the $\mathcal{D}$ stability test involving pair wise combinations results in enormous computational effort.

This is what Barmish refers to as a 'combinatorial explosion'. For example if $m = 8$ (corresponding to 8 parameters) $\mathcal{P}$ has 2^8 or 256 generating polynomials leading to 32,640 pair-wise combinations of $p_{ij}(s, \lambda)$ to be considered. When m is increased from 8 to 11, the number of pair-wise combinations to be tested rises to 20, 96, 128.

4. The edge theorem is applicable to not only continuous time systems but discrete time systems as well because both the open left half plane and the unit disk are simply connected regions in the complex plane.
5. One of the hallmarks of Kharitonov's theorem is that stability of a class of polynomials may be established in a specified region by testing stability of polynomials at four vertices. The region specified is the open left half plane. The question which naturally arises is whether there are other important regions besides the left half plane for which Kharitonov-like theorem holds? It has been subsequently proved in literature that any regular $\mathcal{D}$ region belonging to the complex plane is a Kharitonov region (in the sense that stability can be established by testing for stability at a finite number of vertices) provided both $\mathcal{D}$ and its reciprocal $\frac{1}{\mathcal{D}}$ are convex. Note that in this context the unit disk is not one such region and that is why Kharitonov theorem is not applicable in this case.
6. By invariant degree we mean that all polynomials under consideration should have the same degree. This means that the coefficient of the highest degree term should not vanish for any value of the parameter.

11.5 The Value Set Concept

This is an important concept which helps us to understand robust polynomial theory irrespective of the nature of the coefficient dependency on the parameters from a single perspective. In simple terms, the value set implies the following. Given an uncertainty bounding set Q, and a fixed frequency ω, the value set is a subset of the complex plane comprising all values which can be assumed by $p(j\omega, \boldsymbol{q})$ as $\boldsymbol{q}$ ranges over Q. Formally it may be defined as follows.

DEFINITION 11.3 (The value set)
Given a family of polynomials $\mathcal{P} = \{p(\cdot, \boldsymbol{q}) : \boldsymbol{q} \in Q\}$, the value set at a real frequency ω is given by

$$p(j\omega, Q) = \{p(j\omega, \boldsymbol{q}) : \boldsymbol{q} \in Q\} \qquad (11.10)$$

The power of the value set approach is derived from the fact that it is essentially a two dimensional set in the complex plane whereas the uncertain parameter set belongs to $\mathbb{R}^l$ where l is the number of uncertain parameters. Furthermore, because of the two dimensional nature of the value set irrespective of the dimension of the parameter space, it is easier to conduct simulation studies with the help of computer graphics.

Recall that earlier in our proof of Kharitonov theorem we successfully used the value set concept. There the problem under study was robust stability of interval polynomials and it turned out that the value set $\boldsymbol{H}(\omega)$ was a level rectangle. Next we turned our attention to more complicated problems where the bounding set Q is a box as before but the polynomial coefficients are affine linearly related. In this context we came across polytope of polynomials. It can be shown that the value set in the case of a polytope of polynomials is a convex polygon in the complex plane with possibly some of the edges of Q mapped into the interior of the polygon.

The value set concept is also useful when we consider the general problem of robust $\mathcal{D}$ stability. Now the stability region is an open connected region in the complex plane. The value set at any $z \in \mathbb{C}$, for a family of polynomials $\mathcal{P} = \{p(\cdot, \boldsymbol{q}) : \boldsymbol{q} \in Q\}$ is given by $p(z, Q) = \{p(z, \boldsymbol{q}) : q \in Q\}$. In the case of stability in the open left half plane we recall how ω was varied along the boundary (the imaginary axis) and the movement of the value set was observed in the complex plane to find out whether zero is a member of the value set. In a similar manner, in the case of robust $\mathcal{D}$ stability we may choose a parameter z on the boundary of the region and sweep z along the boundary to determine $p(z, Q)$ at each and every value of z. It turns out that to ensure $\mathcal{D}$ stability, zero should not be a member of any of the value sets thus obtained. This principle, known as the zero exclusion principle is explained in the next section.

11.6 The Zero Exclusion Principle

Earlier in the context of the proof of Kharitonov theorem, the zero exclusion principle was demonstrated for interval polynomials. Actually this principle holds good in much more general situations.

Let us first consider the case where the stability region is the open left half plane. However we are not imposing any condition on the dependance of coefficients of the polynomial on the parameter vector $\boldsymbol{q}$ except that this dependance be continuous. In such a situation the value set for a fixed ω is given by $p(\jmath\omega, Q)$. Note that we are also relaxing the condition that the bounding set Q be a hyper-rectangle. It need only be pathwise connected. We assume that there exists a parameter $q^0 \in Q$ for which $p(s, \boldsymbol{q}^0)$ is stable.

We now explain the zero exclusion principle based on the following heuristic arguments.

If $p(s, \boldsymbol{q}^0)$ is stable, because of continuity assumption there exists, a neighbourhood about $\boldsymbol{q}^0$ where all polynomials $p(s, \boldsymbol{q})$ are stable. However as $\boldsymbol{q}$ distances itself from $\boldsymbol{q}^0$ instability may set in and the first sign of instability is heralded by some of the stable roots of the polynomial crossing over to the right half plane penetrating the imaginary axis. But before crossing over they must appear on the boundary and the continuity condition ensures that there exist some values of $\boldsymbol{q}$ for which this happens. If the roots are located on the boundary, then for certain specific values of ω and specific values of $\boldsymbol{q}$, we have $p(\jmath\omega^*, \boldsymbol{q}^*) = 0$.

Hence if the value set does not contain zero as ω ranges from 0 to ∞ then it means that the set of polynomials $\mathcal{P} = \{p(\cdot, \boldsymbol{q}) : \boldsymbol{q} \in Q\}$ is stable. The above principle is stated in the theorem that follows.

THEOREM 11.3 (Zero exclusion condition)
Consider a family of nth degree polynomials

$$\mathcal{P} = \{p(\cdot, \boldsymbol{q}) : \boldsymbol{q} \in Q\}$$

and associated uncertainty bounding set Q which is pathwise connected. Let the coefficients of the polynomial namely $a_i(\boldsymbol{q})$ $i = 0, 1, 2, \ldots, n$ be continuous functions of the parameter $\boldsymbol{q}$. Further, let there at least be one stable polynomial $p(s, \boldsymbol{q}^0)$. Then $\mathcal{P}$ is robustly stable if and only if the origin is excluded from the value set $p(\jmath\omega, Q)$ at all frequencies $\omega \geq 0$. That is, $\mathcal{P}$ is robustly stable if and only if

$$0 \notin p(\jmath\omega, Q) \text{ for } \forall\omega \geq 0$$

■

For the more general situation explained earlier where the stability region is a connected set $\mathcal{D}$ belonging to the complex plane, the zero exclusion principle operates exactly in the same manner as already explained. The only difference now is that the boundary is the boundary of $\mathcal{D}$ and stability means that all the zeros of the polynomial set $\mathcal{P}$ are confined to the interior of $\mathcal{D}$. Here again for instability to set in the roots of the polynomial have to first appear on the boundary in which can $p(z, \boldsymbol{q}) = 0$ where z is a variable defined on the boundary of $\mathcal{D}$.

THEOREM 11.4 (Zero exclusion condition: general case)
Consider a family of nth degree polynomials $\mathcal{P} = \{p(\cdot, \boldsymbol{q}) : \boldsymbol{q} \in Q\}$ and associated uncertainty bounding set Q which is path-wise connected. Let the coefficients of the polynomial namely $a_i(\boldsymbol{q})$, $i = 0, 1, 2, \ldots, n$ be

continuous functions of the parameter $\boldsymbol{q}$. Further let there be at least be one $\mathcal{D}$ stable member $p(s, \boldsymbol{q}^0)$. Then $\mathcal{P}$ is robustly stable if and only if

$$0 \notin p(z, Q)$$

for all $z \in \partial\mathcal{D}$ where $\partial\mathcal{D}$ denotes the boundary of $\mathcal{D}$.

The value set combined with the zero exclusion principle constitute a powerful tool to determine the robust $\mathcal{D}$ stability of a family of polynomials. A straightforward (though computationally demanding) approach for generating the value set is to grid the bounding set Q and evaluate the polynomial $p(s, \boldsymbol{q})$ at each of the grid points. This has to be repeated at each frequency. This idea is practicable only if the parameters are few in number. A graphical display of the value set at different frequencies (in the case of Hurwitz stability) may be obtained using the computer and it may be verified whether zero is a member of the value set at any of the frequencies considered.

11.7 Multilinear Dependencies

The case of affine linear dependency of the coefficients of the polynomial on the parameter vector which we discussed earlier is far too restrictive in many physical applications. Hence we now pass on to more general dependencies characterized as multilinear and polynomic.

The uncertain polynomial $p(s, \boldsymbol{q}) = \sum_{i=0}^{n} a_i(\boldsymbol{q})s^i$ is said to have a *multilinear uncertainty* structure if each of the coefficient functions $a_i(q)$ is multilinear. We say $a_i(\boldsymbol{q})$ is multilinear if we fix all but one component of the vector $\boldsymbol{q}$, then $a_i(\boldsymbol{q})$ is affine linear in the remaining variable component of $\boldsymbol{q}$.

Example 11.1
Let $p(s, q) = s^3 + (3q_1q_2 - 4q_1q_2q_3 + q_2q_3)s^2 + (5q_2q_3 - 3q_1)s + (q_2 + q_3 - 4)$.

The above polynomial has a multilinear structure. It may be easily checked that if the parameters (q_1, q_2), (q_2, q_3) or (q_1, q_3) are fixed, then the coefficients are affine linear in the remaining variable.

■

We say that $p(s, q)$ has a *polynomic structure* if each of its coefficients $a_i(\boldsymbol{q})$ is a multivariable polynomial in the components of $\boldsymbol{q}$.

Example 11.2
Let $p(s, q) = s^3 + (2q_1q_2 - 4q_2q_3 + q_1^2)s^2 + (q_1q_2q_3 - 3q_2q_3 - q_2^2)s + (3q_2 - 4q_1 + q_3)$.

Because of the presence of q_1^2 and q_2^2 terms in the coefficients $a_2(\boldsymbol{q})$ and $a_1(\boldsymbol{q})$, the above polynomial has got a polynomic uncertainty structure.

■

We saw that Kharitonov theorem is applicable only when the coefficients are independent. Similarly the edge theorem is applicable only if the uncertainty structure is affine linear. For multilinear and polynomic uncertainty structures no comparable theorems are available. But certain approximations may be tried.

For example if Q is a box and the polynomical coefficients $a_i(\boldsymbol{q})$ are multilinear in $\boldsymbol{q}$ then it can be shown that even though the value set is not a polygon, the convex hull of the value set is a polygon. This could be generated by evaluating $p(j\omega, \cdot)$ to the vertices of the box. Thus the convex hull approximation of the value set at each frequently can be used to check the zero exclusion condition. Of course such an approximation leads to conservative results. Exclusion of zero from the convex hull of the value set is therefore sufficient but not necessary for stability.

In recent years a number of matrix versions of the polynomial robust stability problem have appeared in literature. For example, matrix interval stability problem has been considered where the matrix entries a_{ij} vary over the interval $[\underline{a}_{ij}, \overline{a}_{ij}]$. It is required to determine Kharitonov-like conditions under which strict stability of the matrix is assured over the entire range of variations. It is readily seen that this is a special case of a polynomial problem with multilinear $a_i(\boldsymbol{q})$. This is so because det $(sI - A)$ has coefficients which depend multilinearly on a_{ij}. Here we have the added advantage because of the special structure of det $(sI - A)$. Interval matrix stability theory has not so far produced any results comparable to that obtained in robust polynomial stability theory.

11.8 Summary

The points discussed in this chapter may be summarized as follows:

1. Kharitonov theorem may be applied to interval polynomials with an uncertainty bound represented by a level rectangular box. The stability of the set of interval polynomials may be determined by ascertaining whether four specially chosen Kharitonov polynomials are Hurwitz.
2. Kharitonov theorem is not applicable when the stability region is the unit disk or when the coefficient perturbations are not independent.

3. The edge theorem covers problems of greater generality but at the expense of more computational effort. The theorem can be applied in cases where the polynomial coefficients are affine linear in the parameters. The stability region may be any open connected set in the complex plane and this includes the open disk.
4. If the above conditions are satisfied then robust $\mathcal{D}$ stability is guaranteed if the exposed edges of a polytope of polynomials are $\mathcal{D}$ stable.
5. Two important unifying concepts used extensively while discussing the robust stability of polynomials are the value set concept and the zero exclusion principle. They can be applied irrespective of the nature of the dependency of the polynomial coefficients on the parameters, the shape and size of the bounding set Q and the stability region chosen for study.
6. In many practical problems one comes across polynomials with coefficients with multilinear or polynomic dependance in parameters. In the former case, the value set at each frequency may be approximated by the convex hull generated by the polynomials $p(j\omega, \cdot)$ associated with the vertices of the box Q. The zero exclusion principle may now be applied to the approximate value set to obtain sufficient conditions for robust stability.
7. The quest to obtain parallel results analogous to Kharitonov and edge theorems in the case of classes of perturbed matrices, whose stability is under study, has not met with much success so far.

Notes and Additional References

Barmish is credited with introducing Kharitonov theorem to Western literature. An authoritative and complete survey of the work done so far in the area of robust polynomial stability is contained in a recently published book by Barmish (1994). The book also contains an extensive bibliography which will be useful for those who want to pursue the matter in greater depth. An IEEE publication edited by Dorato and Yedavalli (1990) on recent advances in robust control contains a large number of papers on Kharitonov and related approaches. Robust stabilization against structured perturbations are discussed in a book by Bhattacharyya (1987).

Exercises

1. Consider the interval polynomial

$$p(s, q) = q_0 + q_1 s + q_2 s^2 + q_3 s^3$$

where the coefficients vary as follows

$$q_0 \in [1, 2]; \quad q_1 \in [3, 4]$$
$$q_2 \in [5, 6]; \quad q_3 \in [7, 8]$$

Form the four Kharitonov polynomials for the problem. Check for their stability using the Hurwitz criteria. Hence conclude whether the family of polynomials are robustly stable or not.

2. Consider an interval polynomial

$$p(s, q) = s^2 + q_1 s + q_2$$

where the parameters q_1 and q_2 range in the interval

$$q_1 \in [3, 4]; \quad q_2 \in [1, 2]$$

construct the value set $p(z, \boldsymbol{q})$ for the problem at $z = 2 + j1$ and show that it is a parallelogram.

3(a) The uncertain bounding set Q for a robust stability problem is a cube of unit dimension centred about the origin. It is required to ascertain the robust stability of the polynomial

$$p(s, q) = s^3 + (2q_1 - q_2 + 3q_3)s^2 + (3q_1 - 4)s + (q_1 - q_2 - q_3 + 1)$$

Obtain the generating polynomials associated with the polytope of polynomials for the above problem. For $s = j2$ obtain the value set $p(j2, Q)$ and verify that it is convex polygon in the complex plane.

(b) Apply the edge theorem to the above problem and check for robust stability in the open left half plane.

4. Consider an uncertain polynomial

$$p(s, q) = (2q_1 - q_2 + q_3 + 1)s^3 + (3q_1 - 3q_2 + q_3 + 3)s^2 + (3q_1 + q_2 + q_3 + 3)s + (q_1 - q_2 + 2q_3 + 3)$$

and the uncertain bounding set

$$|q_i| \leq 0.245 \quad i = 1, 2, 3$$

How many extreme points are associated with the value set corresponding to the mapping function $p(j1.5, Q)$ for the above problem. If they happen to be less than eight (the number of generators for the convex polytope) how can this be explained?

5. Let $p(s, q) = qs^2 - s - 1$ have a single parameter uncertainty $q \in [0, 1]$

$$\text{at } q = 0 \quad p(s, 0) = -s - 1 \text{ and}$$
$$\text{at } q = 1, \quad p(s, 1) = s^2 - s - 1$$

Thus $p(s, 0)$ is stable and $p(s, 1)$ has an unstable root. However $p(j\omega, q) = -\omega^2 - j\omega - 1$ which is not equal to zero for any value of ω. Hence the value set never crosses the origin which may lead us to the conclusion that the uncertain polynomial is robustly stable. However we know that for parameter value $q = 1$, the polynomial is unstable. Explain this contradiction.

6. An uncertain polynomial is given by

$$p(s, q) = q_1 s^2 + (2q_1 + q_2)s + (q_1 + q_2 + 1)$$

The variations of q_1 and q_2 are in the range

$$q_1 \in [0, 1/4]; \quad q_2 \in [0, 1]$$

The uncertainty bounding set Q is now a level rectangle in parameter space. The polytope of polynomials has four exposed edges. Now consider a simply connected $\mathcal{D}$ region as shown in the Figure on page 288.

a) Show that all the four exposed edges of the polytope are $\mathcal{D}$ stable.
b) The problem appears to satisfy the conditions laid down by the edge theorem. However verify that for $q_1 = \dfrac{1}{128}$ and $q_2 = \dfrac{1}{8}$, $p(s, q)$ has a zero outside the $\mathcal{D}$ region. How does this result reconcile with the consequences of the edge theorem?
[Note: This exercise is taken from Barmish (1994)]

7. This exercise relates to the case where the coefficients of the polynomial are multilinear functions of the parameters. Consider

$$p(s, \boldsymbol{q}) = s^3 + (q_1 + q_2 + 1)s^2 + (q_1 + q_2 + 3)s + (1 + r^2 + 6q_1 + 6q_2 + 2q_1 q_2)$$

where r is a variable which could be assigned specified values. Let $r = 0.5$.

$$q_1 \in [0.3, 2.5]; \quad q_2 \in [0, 1.7]$$

The $\mathcal{D}$-stability region is taken as the open left half plane.

For all $\boldsymbol{q} \in Q$, the coefficients a_i are positive and the condition for stability applying Hurwitz test is:

$$a_1 a_2 - a_0 = (q_1 - 1)^2 + (q_2 - 1)^2 - r^2 > 0$$

This condition is satisfied outside a circle of radius $r = 0.5$, centered at $q_1 = 1$ and $q_2 = 1$ as shown in the figure on page 289. For $r = 0$ for $\boldsymbol{q}$ anywhere inside the rectangle Q the polynomial is stable except at the point (1, 1) located inside Q. Ackermann *et al.* (1990) has taken this problem as a test case for comparison of several solution

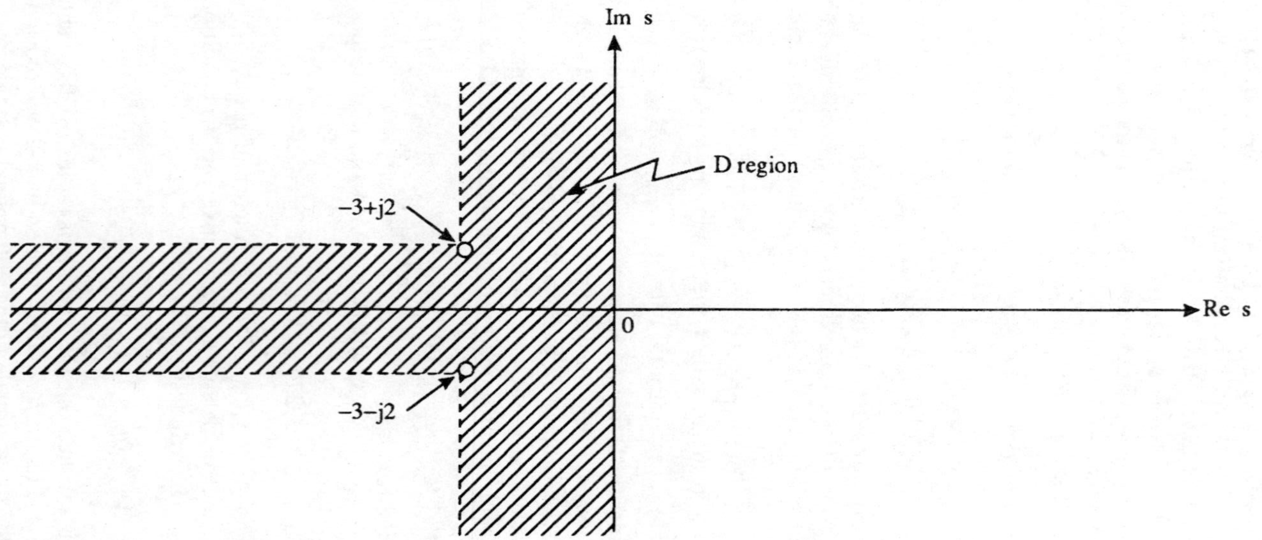

Figure pertaining to Problem 6

approaches. As the exact solution is known in this case any false conclusion could easily be detected. Test the following:

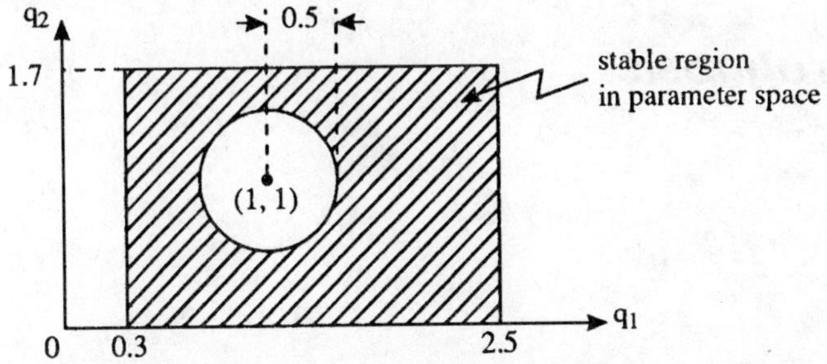

a) Omit the bilinear term $q_1 q_2$. Then the dependency is affine linear. Verify the edge theorem by checking for the stability of 4 exposed edges of the polytope.

b) With multilinear dependancy, as given in the problem, check for stability of exposed edges as well as the two diagonals. In the given problem stability test for the diagonal edges will fail. However it is possible to modify Q so that the diagonals do not pass through the unstable circular enclave. In that case the edge test (all edges including the unexposed ones) may lead us to the wrong conclusion that the polynomial is robustly stable for all $\boldsymbol{q} \in Q$. This goes to prove that for multilinear dependancy, the stability of all edges of the polytope need not guarantee robust stability.

c) When $r = 0$, there is an isolated unstable point and most robust stability tests are likely to miss this point. Even a graphical method like gridding of Q may miss such an isolated instability point

12

Prognosis

12.1 Background

During the past few years, robust multivariable control theory has come of age. Concepts such as singular values and $\mathcal{H}_\infty$ norms have provided reliable measures for specifying performance goodness in multivariable systems. A powerful $\mathcal{H}_\infty$ framework has been created within which one can now study the multiloop version of classical control theory.

While the undoubted power of these methods has come in for praise, there have been some criticisms too. One such criticism relates to the use of singular values to characterize matrix operations. It has been pointed out that singular values are gross matrix parameters and do not give proper indication of the influence of matrix structure on problem solution. This practice has also been faulted for giving results which are unduly conservative. For example, some of the perturbations which are presumed to occur in certain directions in signal space can be ruled out on purely physical grounds. But conventional singular value analysis has no built-in methodology to take care of such situations.

$\mathcal{H}_\infty$ optimization theory has also come in for its share of criticism. Unnecessary and avoidable complexity has been attributed to its implementation. Quite often – so the argument goes – the order of the controller exceeds that of the plant itself. Also when compared to its counter part namely $\mathcal{H}_2$ optimization, the computational burden appears to be excessive.

In spite of recent advances in VLSI, technology which has made it possible to implement sophisticated feedback algorithms, the cost of feedback cannot be ignored. As Horowitz (1963) one of the most articulate critics of the application of LQG theory to practical problems puts it, any feedback scheme must be related to the extent of uncertainty present and the narrowness of the performance tolerance desired. It appears to some that $\mathcal{H}_\infty$ optimization under certain circumstances amounts to an 'overkill' and that the same goal could be achieved at a much lower cost by opting for an alternative approach.

While some of the above criticisms may be valid in certain situations it cannot be denied that the $\mathcal{H}_\infty$ approach fits robust control and stability problem like a glove. No other norm has the capacity to deal with plant per turbations with equal felicity as the $\mathcal{H}_\infty$ norm. Its characteristic of prescribing a uniform bound over the entire frequency range no doubt makes the design based on $\mathcal{H}_\infty$ norm somewhat conservative but this deficiency could be remedied by proper choice of weighting functions. In this context it will be instructive to compare the performance levels attained by $\mathcal{H}_2$ and $\mathcal{H}_\infty$ approaches where such comparisons are possible. This we propose to do in the next section.

12.2 $\mathcal{H}_2$ Versus $\mathcal{H}_\infty$ Control

In control problems we use a variety of performance criteria. These include bandwidth, tracking error and robustness against plant uncertainty and disturbances. In this section the criteria specifically chosen for study are those related to $\mathcal{H}_2$ and $\mathcal{H}_\infty$ norms. Clearly the $\mathcal{H}_2$ norm identified with quadratic performance criteria is a more natural norm, better suited for system performance than the $\mathcal{H}_\infty$ norm. However, when there are uncertainties in modelling then $\mathcal{H}_2$ norm cannot serve as a suitable measure of goodness for robust control problems. On the other hand the $\mathcal{H}_\infty$ norm because of its sub-multiplicative property is an ideal tool to deal with such situations. Thus both the norms have their own strengths and limitations.

For comparing these two norms the following approach is adopted. The $\mathcal{H}_\infty$ performance of an $\mathcal{H}_2$ controller is sought to be compared with the $\mathcal{H}_\infty$ performance of an $\mathcal{H}_\infty$ controller. To spell out the problem in greater detail consider the scalar case. To be more specific the model matching problem $(T_1 - T_2Q)$ is considered for a SISO system.

Let Q_2 denote the optimal solution for the $\mathcal{H}_2$ criterion namely $\|T_1 - T_2Q\|_2$ be a minimum and let Q_∞ denote the optimal solution for the $\mathcal{H}_\infty$ criterion namely $\|T_1 - T_2Q\|_\infty$ be a minimum.

We now use the $\mathcal{H}_2$ solution to reckon $\mathcal{H}_\infty$ performance. We are therefore concerned with the value $\|T_1 - T_2Q_2\|_\infty$. Obviously, this will be greater than or equal to $\|T_1 - T_2Q_\infty\|_\infty$. Our interest is in the ratio

$$\beta = \frac{\|T_1 - T_2Q_2\|_\infty}{\|T_1 - T_2Q_\infty\|_\infty}$$

The supremum of β is considered as T_1 and T_2 range over $\mathcal{RH}_\infty$ with the proviso that T_2 has k zeros in Re $s > 0$ and no zeros on the extended imaginary axis. Francis and Doyle (1987) showed that the supremum of β is equal to $2k$ where k is the number of zeros of T_2 in the open right half plane.

In the MIMO case the problem was studied by Zhou (1992a) who made a quantitative assessment of the controller designed from $\mathcal{H}_2$ methodology and also a controller designed from $\mathcal{H}_\infty$ methodology. For a 1-block problem it was shown that the ratio of $\mathcal{H}_\infty$ performance between an $\mathcal{H}_2$ controller and $\mathcal{H}_\infty$ controller is no greater than $2k$, where k is the total number of right half plane transmission zeros of the sub system connecting disturbance and measurement and the sub system connecting control signal and the controlled output.

As already explained in Chapters 6, 9 and 10 we have

$$\begin{aligned}\|\mathcal{F}_l(G, K)\|_\alpha &= \|\boldsymbol{T}_1 + \boldsymbol{T}_2\boldsymbol{Q}\boldsymbol{T}_3\|_\alpha \\ &= \left\| \begin{bmatrix} \boldsymbol{R}_{11} - \boldsymbol{Q} & \boldsymbol{R}_{12} \\ \boldsymbol{R}_{21} & \boldsymbol{R}_{22} \end{bmatrix} \right\|_\alpha \qquad \alpha = 2, \infty\end{aligned}$$

Referring to Figure 10.1, where the $\mathcal{H}_2$ controller has been parametrized, the optimal $\mathcal{H}_2$ controller corresponds to setting $\boldsymbol{Q} = 0$. Denote the associated controller by $\boldsymbol{K}_2$. Hence

$$\|\mathcal{F}_l(\boldsymbol{G}, \boldsymbol{K}_2)\|_\infty = \left\| \begin{bmatrix} \boldsymbol{R}_{11} & \boldsymbol{R}_{12} \\ \boldsymbol{R}_{21} & \boldsymbol{R}_{22} \end{bmatrix} \right\|_\infty$$

The $\mathcal{H}_\infty$ optimal controller is a controller denoted by $\boldsymbol{K}_\infty$ which minimizes the $\mathcal{H}_\infty$ norm of $\mathcal{F}_l(\boldsymbol{G}, \boldsymbol{K})$ over all admissible $K \in \mathcal{H}_\infty$. Hence we have

$$\begin{aligned}\beta &= \frac{\|\mathcal{F}_l(\boldsymbol{G}, \boldsymbol{K}_2)\|_\infty}{\|\mathcal{F}_l(\boldsymbol{G}, \boldsymbol{K}_\infty)\|_\infty} \\ &= \frac{\left\| \begin{bmatrix} \boldsymbol{R}_{11} & \boldsymbol{R}_{12} \\ \boldsymbol{R}_{21} & \boldsymbol{R}_{22} \end{bmatrix} \right\|_\infty}{\min\limits_{Q \in \mathcal{H}_\infty} \left\| \begin{bmatrix} \boldsymbol{R}_{11} - \boldsymbol{Q} & \boldsymbol{R}_{12} \\ \boldsymbol{R}_{21} & \boldsymbol{R}_{22} \end{bmatrix} \right\|_\infty}\end{aligned}$$

In 1-block problem $\boldsymbol{R}_{12} = \boldsymbol{R}_{21} = \boldsymbol{R}_{22} = 0$ and we have

$$\beta = \frac{\|\boldsymbol{R}_{11}\|_\infty}{\min\limits_{Q \in \mathcal{H}_\infty} \|\boldsymbol{R}_{11} - \boldsymbol{Q}\|_\infty}$$

Let the realization of $\boldsymbol{G}$ be given by:

$$\boldsymbol{G} = \left[\begin{array}{c|cc} A & B_1 & B_2 \\ \hline C_1 & 0 & D_{12} \\ C_2 & D_{21} & 0 \end{array}\right]$$

and further let:

$$z_1 = \text{Number of right half plane zeros of } \left[\begin{array}{c|c} A & B_2 \\ \hline C_1 & D_{12} \end{array}\right]$$

$$z_2 = \text{Number of right half plane zeros of } \left[\begin{array}{c|c} A & B_1 \\ \hline C_2 & D_{21} \end{array}\right]$$

and let $k = z_1 + z_2$.
Then it has been proved by Zhou (1992a) that:

$$\beta \leq 2k \text{ for } k \geq 1.$$

It was further shown by Zhou that there exists a sequence of plants $\boldsymbol{G}_n$ satisfying the so called standard assumptions in $\mathcal{H}_2$ and $\mathcal{H}_\infty$ control and with fixed k for which in the limit

$$\beta = \lim_{n\to\infty} \frac{\|\mathcal{F}_l(\boldsymbol{G}_n, \boldsymbol{K}_2\|_\infty}{\|\mathcal{F}_l(\boldsymbol{G}_n, \boldsymbol{K}_\infty)\|_\infty} = 2k$$

Here D_{12} and D_{21} are assumed to be square matrices so that we are dealing with a 1-block problem. However it was pointed out that the systems which achieve a value arbitrarily close to the upper bound $2k$ are mostly pathological systems.

For $k = 0$, it can be shown that $\boldsymbol{R}_{11} = 0$ which implies that $\boldsymbol{Q} = 0$. We then have $\boldsymbol{K}_2 = \boldsymbol{K}_\infty$ and $\mathcal{F}_l(\boldsymbol{G}, \boldsymbol{K}_2) = \mathcal{F}_l(\boldsymbol{G}, \boldsymbol{K}_\infty) = 0$.

Hence the comparison is meaningful only if right half plane zeros are present, i.e. when $k \geq 1$. Normally, right half plane zeros are few in number and coupled with the fact that the upper bound of β is hardly ever achieved, we may surmise that the $\mathcal{H}_\infty$ performance is not very much different from $\mathcal{H}_2$ performance in the 1-block problem. However, the comparison we have made does not take into account other benefits accruing out of $\mathcal{H}_\infty$ control. To that extent the comparison is only of limited significance.

12.3 Summary

One of the principal contributions of modern control theory is that it has helped in developing a methodology to synthesize controllers that ensure a stable closed loop system as well as optimal system performance. In the pre-$\mathcal{H}_\infty$ era, up to the early eighties, LQG theory dominated the scene. $\mathcal{H}_\infty$ theory originated with the work of Zames (1981) who adopted frequency domain methods for $\mathcal{H}_\infty$ optimization studies. Since then, there has been a veritable explosion in research activity which has succeeded in creating a satisfactory framework for $\mathcal{H}_\infty$ optimal and sub-optimal studies. The success in this field may be attributed to the assimilation of concepts from other areas such as differential games, network theory and Riccati equations. During the past decade many new approaches have been tried and some have gained wide acceptance. The successful development of $\mathcal{H}_\infty$ theory may be traced to three factors, namely singular value analysis, stable factorization approach and Youla parametrization. Further, the Riccati equation has played a crucial part in state space synthesis of controllers. Broadly speaking, developments in $\mathcal{H}_\infty$ control have proceeded along the following lines.

i) NP interpolation approach
ii) Operator theoretic approach
iii) State space approach

From the computational view point, the state space approach has rendered the other two approaches less attractive. The work of Doyle *et al.* (1989) established the necessary and sufficient conditions for the existence of an internally stabilizing controller which makes the $\mathcal{H}_\infty$ norm of the transfer matrix less than an *a priori* given number $\gamma > 0$. These conditions are expressed in terms of the solutions of two Riccati equations. The feedback controller in the output feedback case has the structure of an observer and full information controller. Further, most importantly, they have provided a receipe to arrive at the set of all stabilizing controllers which make the $\mathcal{H}_\infty$ norm of the transfer matrix less than the prescribed positive number γ. The extra design freedom obtained through the delineation of all sub optimal controllers can be advantageously utilized to optimize other performance measures.

The first eleven chapters of this book are primarily concerned with unstructured perturbations in a dynamic system and the ways and means of containing it. In the final chapter we discuss stability polynomials in the light of the now well known Kharitonov theorem. The control-theoretic implication of this development, especially in controller synthesis is not clear at the moment. However Kharitonov's approach and its extensions are bound to influence robust control theory in the coming years.

12.4 Future Trends

Predicting the future in any fast changing technological field is a hazardous exercise. About three decades back after the triumphant emergence of LQG theory many felt that linear control theory has reached the end of its tether and that there is nothing more to be done. Yet, during the eighties linear control theory has bounced back and the challenges posed by robust control have kept the researchers busy ever since. The author's predictions of future trends are therefore merely based on extrapolation of existing activity in this field. These are:

1. Multiobjective Design

In robust control there is an increasing demand to optimize many performance criteria simultaneously. Consider a generalized plant with multiple input and output channels as shown in Figure 12.1.

Stated in its full generality, we are required to minimize the $\mathcal{H}_2$ performance of one channel (channel zero in Figure 12.1) while keeping bounds on either the $\mathcal{H}_2$ norm or the $\mathcal{H}_\infty$ norm of other channels. This is a multiobjective $\mathcal{H}_2/\mathcal{H}_\infty$ problem (Sznaier 1994, Scherer 1995, Abedar *et al.* 1995).

Thus the $\mathcal{H}_2$ norm of the transfer matrix $\boldsymbol{T}_1^0 + \boldsymbol{T}_2^0\boldsymbol{Q}\boldsymbol{T}_3^0$ is a measure of the nominal performance of the system which is to be minimized. In addition the controller should keep bounds on either the $\mathcal{H}_2$ or $\mathcal{H}_\infty$ norm of

$$\boldsymbol{T}_1^j + \boldsymbol{T}_2^j\boldsymbol{Q}\boldsymbol{T}_3^j \quad j = 1, 2, \ldots, k$$

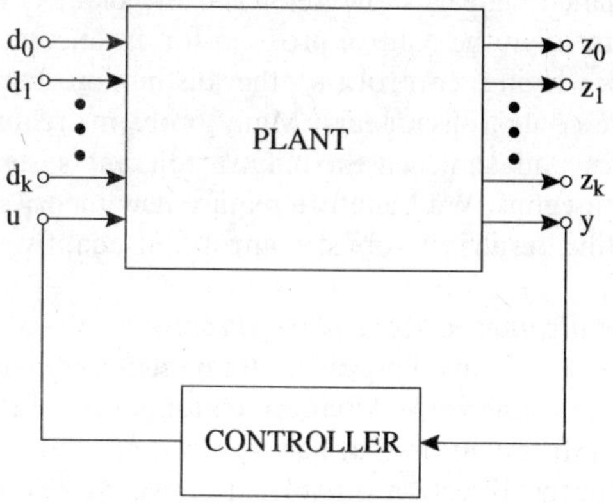

Figure 12.1 Generalized plant with multiple inputs and outputs.

Note that bounds on $\mathcal{H}_\infty$ norm keep the system robustly stable. Thus $\|\boldsymbol{T}_1^j + \boldsymbol{T}_2^j \boldsymbol{Q} \boldsymbol{T}_3^j\|_\infty < \gamma_j$ implies that the closed loop system remains internally stable under the action of uncertainty $d^j = \Delta z^j$ where Δ is an open $\mathcal{H}_\infty$ ball of radius $1/\gamma_j$. This follows from the small gain theorem There is no currently available general theory for multi objective optimization as stated in the above problem. Even problems of lesser generality such as the ones stated below are not fully solved for MIMO systems.

i) Robust stability plus nominal optimal performance.
ii) Robust stability plus robust performance. This implies the designing of a controller guaranteeing desired level of performance in the face of plant uncertainty.

2. $\mathcal{H}_\infty$ Control of Non-Linear Systems
The standard state space solution for $\mathcal{H}_\infty$ optimization problem in the case of linear time invariant systems is now well understood. This has motivated researchers to investigate the extent to which these results could be applied in the case of non-linear systems (Lu and Doyle 1994, James and Baras 1995) The time domain formulation of the $\mathcal{H}_\infty$ problem has a natural generalization to non-linear systems. Note that the $\mathcal{H}_\infty$ norm definition is not related to the concept of linearity, even though the norm has to be appropriately defined for non-linear systems. The problem then is to obtain an $\mathcal{H}_\infty$ controller parametrization when both plant and controller are non-linear and time invariant and are described by state space equations.

3. Robust Adaptive Control
In linear robust control, linear time invariant plants are controlled by fixed linear time invariant controllers. In adaptive control, the controller is non linear even when plant is linear. However when the plant is non-linear or when one attempts to solve the control problem for an uncertain plant with non-linear or time varying controllers, the distinction between robust control and adaptive control disappears. Many problems relating to sensor and actuator failures and some cases of fault tolerant systems could be brought under this category. We therefore require new theories obtained by integrating the existing results in robust control and adaptive control.

4. State Space Generalization of $\mathcal{H}_2$ and $\mathcal{H}_\infty$ Problems
At present the $\mathcal{H}_2$ and $\mathcal{H}_\infty$ control problems are treated as separate problems in a unified state space framework. A natural continuation of this work is to have a single problem formulation that has standard $\mathcal{H}_2$ and $\mathcal{H}_\infty$ theories as special cases. Work in this direction is reported in two papers by Zhou *et al*. (1994) and Doyle *et al*. (1994).

5. Directionality of Signals
It is well known that signal response in multivariable systems depends on signal direction. In many physical problems it is known that signals in certain directions are quite unlikely to occur. This property needs to be exploited. Any design based on directionality of signals will be less conservative than the one based on conventional singular value analysis.

6. MIMO Controller Design
The present procedure for designing controllers for MIMO systems is merely an extension of the procedure used for SISO systems with some marginal modifications. We require a new and radical approach for controller design in the case of MIMO systems which will take into account the robustness concept in a more natural manner.

7. Invariant Zeros
The $\mathcal{H}_\infty$ optimization problem in the model matching format is posed as

$$\operatorname*{Inf}_{Q \in \mathcal{H}_\infty} \|\boldsymbol{T}_1 - \boldsymbol{T}_2 \boldsymbol{Q} \boldsymbol{T}_3\|_\infty$$

where $\boldsymbol{T}_1, \boldsymbol{T}_2$ and $\boldsymbol{T}_3 \in \mathcal{RH}_\infty$.

The infimum in the above case is achieved only if the ranks of the two matrices $\boldsymbol{T}_2(j\omega)$ and $\boldsymbol{T}_3(j\omega)$ are constant for all $0 \le \omega \le \infty$. Following the nomenclature used in this book, the above conditions imply that the two subsystems (A, B_2, C_1, D_{12}) and (A, B_1, C_2, D_{21}) have no invariant zeros on the imaginary axis. We require a theory which will solve the $\mathcal{H}_\infty$ problem when the above conditions are relaxed.

8. Computational Aspects
Substantial computational effort is involved in solving $\mathcal{H}_\infty$ problems in the multivariable case. The computational burden is bound to increase with the introduction of multiobjective criteria. In this context we require better and more efficient algorithms to solve the various sub-problems associated with control. Even a marginal improvement in this direction will mean considerable saving of time and money.

9. Kharitonov Theory
The original work of Kharitonov has triggered a flurry of research activity and many extensions to his work have been proposed. But these results are yet to be absorbed into the main stream of controller design. Most of the results are available only in the domain of analysis. The controller synthesis problem remains largely unsolved.

10. Practical Applications
In control engineering there has always existed a wide gap between theory and practice. The gestatian period between the enunciation of a theory and its practical application has however narrowed down to some extent over the years. The role played by Kalman filter theory in aerospace as well as in many other areas of practical significance is now well known. While there have been some applications of $\mathcal{H}_\infty$ optimization in process control and aerospace industry it is yet to catch on in other areas. However, the scope for applications is vast especially because plant uncertainty is something fundamental and any system design which fails to take into consideration this aspect of the problem is totally unrealistic. Hence, we are bound to come across more and more application-oriented research in the coming years.

12.5 Conclusion

$\mathcal{H}_\infty$ optimization theory is currently an active sub-area of research in control theory and a number of different approaches are now being investigated. It is generally true that the discovery of new knowledge helps codify existing knowledge and at the same time throws up new problems for future investigation. The horizon instead of coming closer appears to recede farther from our view. One is reminded here of the indomitable Ullyssean spirit conveyed through the ringing words of Lord Tennyson:

All experience is an arch where thro'
gleams that untravell'd world,
whose margin fades
for ever and for ever when I move.

Appendix A

Introduction to Mathematical Concepts

A.1 Introduction

This appendix gives a brief introduction to some of the common mathematical concepts used in this book. The underlying idea is to collect all of them together in one place so that the reader may be spared the trouble of going through different texts to obtain a grasp of the fundamentals involved.

The topics covered include elements of abstract algebra, vector and matrix norms and functional analysis. Other topics such as Matrix Fraction Description (MED), Singular Value Decomposition (SVD), $\mathcal{H}_2$ and $\mathcal{H}_\infty$ norms are considered to be sufficiently important so as to merit separate treatment elsewhere in the book.

A.2 Sets and Equivalence Relations

The word set is used to denote a collection of objects. It may contain either finite or infinite number of elements. Sets may often be identified as subsets of a fixed universal, set U. In practical applications it is fairly obvious what a universal set is, even if it is not defined explicitly.

Given two sub-sets A and B of the universal set U we define three operations on sets.
These are:

1. Union: $A \cup B = \{x \in U: x \in A \text{ and/or } x \in B\}$
2. Intersection: $A \cap B = \{x \in U: x \in A \text{ and } x \in B\}$
3. Complementation: $A' = \{x \in U: x \notin A\}$

The set of all elements in A that are not in B is denoted by either $A \backslash B$ or $A - B$.

4. $A - B = \{x \in U : x \in A \text{ and } x \notin B\}$

The members of a set are identified by defining some characteristics which are common to all members of the set.

EXAMPLE A.1
The set $S(T)$ of signals which vanish outside a specific time interval $-T \leq t \leq T$

$$S(T) = \{x : x(t) = 0 \text{ for all } |t| > T\}$$

Note that $x \in S(T_1) \Rightarrow x \in S(T_2)$ if $T_2 \geq T_1$.

The set operations $\cap$ and $\cup$ can be used to describe the partitions of a set into a sequence of disjoint subsets i.e.

$$S = S_1 \cup S_2 \cup S_3 \ldots$$

and

$$S_i \cap S_j = \emptyset, \text{ the null set for } i \neq j$$

The motivation for partitioning sets is to create subsets which are more manageable. It is possible for example to partition a set, containing an uncountable number of elements into a finite or countable number of subsets. A partition can be generated by an equivalence relation and this is often the most convenient way to define a particular partition.

We say that two elements are equivalent denoted $x \sim y$ if the relation '$\sim$', defined for every pair of elements satisfies the following properties:

1. $x \sim x \quad \forall x$ (Reflexivity).
2. $x \sim y \Rightarrow y \sim x$ (Symmetry).
3. $x \sim y$ and $y \sim z \Rightarrow x \sim z$ (Transitivity).

It can be shown that every partition generates an equivalence relation.

EXAMPLE A.2
Equality is an equivalence relation. But the equivalence sets contains only individual elements.

Consider the partition of the set of integers

$$\{n : n = 0, \pm 1, \pm 2, \ldots\}$$

into a finite number m of equivalence sets, namely

$$S_i = \{n : n = pm + i, i = 0, 1, 2, \ldots m - 1, p \text{ an integer}\}$$

The corresponding equivalence relation

$$n_1 \sim n_2 \Rightarrow n_1 - n_2 = pm$$

We may also write $n_1 = n_2$ (mod m) which is sometimes known as modulus m congruence. Mod 2 congruence partitions the set of all integers into even and odd integers.

A.3 Mapping and Functionals

A mapping is simply a rule by which each element of a set S_1 is assigned a unique element of another set S_2. Thus

$$f : S_1 \rightarrow S_2$$

This is a compact notation for

$$y = f(x) \quad x \in S_1 \text{ and } y \in S_2$$

The element y in S_2 is called the *image* of x under mapping f. The set S_1 is the *domain* of the mapping and the set of all images of the elements of S_1 which are contained in S_2 is called the *range* of the mapping. Often this range is only a subset of S_2, known as the *codomain.* In other words, the range of a function is the image of its domain and is a proper subset of the codomain. It may be noted that the mapping f assigns a unique element $f(x)$ to each x belonging to the domain of f. It is quite possible that an element belonging to the range of f be the image of more than one element in the domain of f. Such mappings are known as *many to one* mappings. If the range of f is S_2 then f is said to be a mapping of S_1 *on to* S_2 or *surjective.* If the images of distinct elements in S_1 are distinct elements in S_2 than the mapping is *one to one* or *injective.*

If the mapping besides being one to one is also onto (i.e. both injective and surjective) then it is said to be *bijective.* A bijective mapping always possesses an inverse.

Composite mapping results from two or more consecutive mapping between sets. Thus if $f_1 : S_1 \rightarrow S_2$ and $f_2 : S_2 \rightarrow S_3$ than the composite mapping is given by $f = f_2 f_1$. When the mapping is from a set S to real or complex numbers, then the mapping is referred to as a *functional.*

Note that a mapping need not be surjective to have an inverse, but it must be one to one, i.e. the domain of f^{-1} should be limited to the range of f. A bijective function always has an inverse.

A.4 Semi-groups and Groups

We now define a single operation (e.g. addition or multiplication) between any two elements of the set. Such an operation is called a *binary operation.* *For example if $a, b \in S$*, then a binary operation between them (denoted by the dot symbol) results in

$$a \cdot b = c$$

It is not necessary that $c \in S$. However for all pairs a, b in S, if we have $a \cdot b \in S$ such an operation is said to be closed. One of the simplest algebraic structures we can think of is the semi-group.

DEFINITION A.1 (Semi-group)
A *semi-group* $(S, \cdot)$ is a non-empty set with a binary operation '·' such that

1. S is closed under '·' i.e. $a \cdot b \in S$ for all $a, b \in S$
2. '·' is associative i.e. $(a \cdot b) \cdot c = (a \cdot b) \cdot c$ for all $a, b, c \in S$

If a semi-group S has an element 'e' such that

$$a \cdot e = e \cdot a \quad \forall a \in S$$

then e is called an identity.

A group has a tighter structure than a semi-group.

DEFINITION A.2 (Group)
A *group* is a non-empty set G with a binary operation denoted by '·' such that:

1. G is closed under '·'
2. '·' is associative in G.
3. The identity element $e \in G$ satisfies:

$$a \cdot e = a = e \cdot a \quad \forall a \in G$$

4. For every $a \in G$, there exists an a^{-1} satisfying:

$$a \cdot a^{-1} = e = a^{-1} \cdot a$$

An *Abelian group* is a group $(G, \cdot)$ in which '·' operation is commutative i.e. $a \cdot b = b \cdot a \forall a, b \in G$. Depending on the binary operation chosen a group may be either additive or multiplicative.

A.5 Homomorphisms and Isomorphisms

Broadly speaking these are mappings which preserve structure. Consider two groups G_1 and G_2. Let the binary operations defined in G_1 be denoted by '$*$' and the binary operation in G_2 be denoted by '$\circ$'. These two operations are in general quite different.

In Figure A.1 if $x \to \phi(x)$ and $y \to \phi(y)$ then $\phi(x * y) \to \phi(x) \circ \phi(y)$. In general if G_1 and G_2 are groups where the above relationships are valid, then

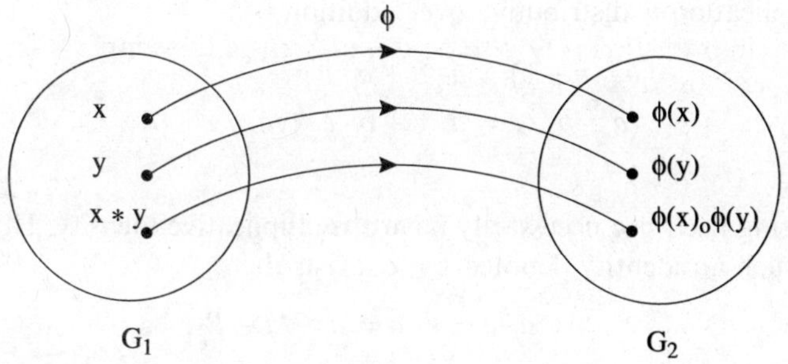

Figure A.1 Illustration of homomorphism.

G_1 is said to be *homomorphic* to G_2. In the special case when the mapping function ϕ is one to one onto, then the bijective mapping is called a (group) *isomorphism* from G_1 to G_2. The interpretation here is that element $x * y$ in G_1 correspond to element $\phi(x) \circ f(y)$ in G_2.

A.6 Rings and Fields

A ring is a non empty set R, together with two binary operations $(+)$ and $(\cdot)$, called respectively addition and multiplication, obeying the following axioms

1. $(R, +)$ is an Abelian group with respect to addition, i.e.

$$a + (b + c) = (a + b) + c \quad \forall a, b, c \in R$$
$$a + b = b + a$$

   There exists an element $0 \in R$ such that

$$a + 0 = 0 + a = a \quad \forall a \in R$$

   For every element $a \in R$, there exists '$-a$' belonging to R such that

$$a + (-a) = 0$$

2. $(R, \cdot)$ is a semi-group with respect to multiplication i.e.

$$a \cdot (b \cdot c) = (a \cdot b) \cdot c \quad \forall a, b, c \in R$$

3. Multiplication is distributive over addition i.e.

$$a \cdot (b + c) = a \cdot b + a \cdot c$$
$$(a + b) \cdot c = a \cdot c + b \cdot c \quad \forall a, b, c \in R$$

REMARKS

i) A ring need not necessarily have a multiplicative identity. However if such an identity denoted by 'e' exists then

$$a \cdot e = e \cdot a = a \quad \forall a \in R$$

ii) Even if a ring has a multiplicative identity, not all elements belonging to the ring need have an inverse in R.
The set of all such elements in R with inverse in R constitutes a unit in R and is denoted by U.

iii) A ring is said to be commutative if

$$a \cdot b = b \cdot a \quad \forall a, b \in R$$

iv) The cancellation law need not always, hold in a ring. Thus for $a, b, c \in R$ if $a \cdot b = a \cdot c$ and $a \neq 0$ it does not necessarily follow that $b = c$.

EXAMPLE A.3

Consider the non commutative ring $\mathbb{R}^{2\times2}$ with

$$a = \begin{bmatrix} 0 & 1 \\ 0 & 0 \end{bmatrix}; b = \begin{bmatrix} 1 & 1 \\ 0 & 0 \end{bmatrix}; c = \begin{bmatrix} 2 & 0 \\ 0 & 0 \end{bmatrix}$$

clearly $a \cdot b = a \cdot c = 0$. But this does not imply that $b = c$.

INTEGRAL DOMAIN

A ring R is said to be a domain or *integral domain* or entire ring if cancellation law holds, i.e. $a, b \in R$ and $a \cdot b = 0$ implies that either a or b equals zero. Thus in an integral domain the product of every pair of non zero elements, is non-zero. Some of the important rings frequently used in control and system theory are

— $\mathbb{R}[s]$, the ring of polynominals in s with coefficients in $\mathbb{R}$.
— $\mathbb{R}_p(s)$, the ring of proper rational functions in s with coefficients in $\mathbb{R}$.
— $\mathbb{R}_{p,o}(s)$, the ring of strictly proper rational functions in s with coefficients in $\mathbb{R}$.
— The subrings of $\mathbb{R}_p(s)$ and $\mathbb{R}_{p,o}(s)$ which are analytic in the closed right half plane.

In all the rings cited above, the cancellation law holds and therefore they are integral domains.

We next define a special class of rings known as Euclidean rings.

DEFINITION A.3 (Euclidean Ring)
A Euclidean ring $(R, +, \cdot, 0, 1)$ is a commutative entire ring upon which is defined a *gauge*, i.e. a function $\gamma : \boldsymbol{R} - \{0\} \to \mathbb{N}$ i.e. $a \to \gamma(a)$ such that the following axioms hold. (Here $\mathbb{N}$ denotes the set of non-negative integers).

i. $\forall a, b \in R - \{0\}, \quad \gamma(a) \le \gamma(a \cdot b)$
ii. If $a, b \in R - \{0\}$ then there exist q, r such that $a = bq + r$ where $r = 0$ or $\gamma(r) < \gamma(b)$. The Euclidean ring is thus an entire ring in which division operation is defined resulting in a quotient q and remainder r. This is known as *Euclidean Division* property. An example of a Euclidean ring is the ring of polynominals $\mathbb{R}[s]$ under pointwise addition and multiplication.

The gauge of any polynominal is its degree which is denoted by $\delta(p)$. Thus $\gamma(p) = \delta(p)$ for a polynominal. The Euclidean division theorem may be stated as follows.

Let $a, b \in \mathbb{R}[s]$ with $b \neq 0$. Then there exists a, q, r in $\mathbb{R}[s]$ such that

$$a = bq + r \text{ with either } r = 0 \text{ or } \delta(r) < \delta(b)$$

If $\delta(a) < \delta(b)$ then $q = 0$ and $r = a$. An important Euclidean ring repeatedly used in this text is the ring of all proper rational functions which are analytic in the closed right half plane. This is denoted in the text by $\mathcal{RH}_\infty$. The gauge of an element p belonging to this ring is the number of zeros of p in the closed right half plane plus the number of zeros at infinity.

We will next consider the properties of a field. As in the case of a ring, field has also two binary operations, addition (+) and multiplication $(\cdot)$. The identity element under addition is denoted by 0 and the identity element under multiplication is denoted by 1.

DEFINITION A.4 (Field)
A field $(F, +, \cdot, 0, 1)$ obeys the following axioms.

i. *Addition*

$$\begin{aligned} \text{Associative} : &\quad (a + b) + c = a \quad \forall a, b, c \in F \\ \text{Commutative} : &\quad a + b = b + a \quad \forall a, b \in F \\ \text{There exists an identity } 0 : &\quad a + 0 = 0 + a \quad \forall a \in F \\ \text{There exists an inverse} : &\quad \forall a \in F, \text{ there exists } (-a) \in F \text{ such that} \\ &\quad a + (-a) = 0. \end{aligned}$$

ii) *Multiplication*

$$\begin{aligned}\text{Associative}: &\ (a\cdot b)\cdot c = a\cdot(b\cdot c) \quad \forall a,b,c\in F\\ \text{Commutative}: &\ a\cdot b = b\cdot a \quad \forall a,b\in F\\ \text{There exists an identity } 1: &\ a\cdot 1 = 1\cdot a = a \quad \forall a\in F\\ \text{There exists an inverse}: &\ \forall a\neq 0 \text{ and } a\in F \text{ there exists } a^{-1}\in F\\ &\ \text{such that } (a)\cdot(a^{-1}) = (a^{-1})\cdot(a) = 1\end{aligned}$$

iii) *Distributive law*

$$\begin{aligned}&\mathrm{a}\cdot(b+c) = a\cdot b + a\cdot c \quad \forall a,b,c\in F\\ &(b+c)\cdot a = b\cdot a + c\cdot a \quad \forall a,b,c\in F\end{aligned}$$

Typical examples of fields are:

— The set of all real and complex numbers denoted by $\mathbb{R}$ and $\mathbb{C}$ respectively.
— The set of all rational functions $\mathbb{R}(s)$.

The set of polynominals $\mathbb{R}[s]$ does not constitute a field because the inverse of a polynominal is not a polynominal (except when it happens to be a constant) and therefore no multiplicative inverse exists within the set.

A.7 Vector Spaces and Modules

While manipulating mathematical objects such as vectors and matrices, we routinely add them or multiply them by real or complex scalars. The results of these operations once again lead to vectors or matrices as the case may be. Such a concept could be formalized by defining an entity called *Linear Vector Space.* The mathematical objects could be anything – not necessarily vectors or matrices – and the scalars used for multiplication need not be either real or complex numbers. In fact they could be members of any field. In some special cases these scalars could be chosen from a ring and the algebraic structure thus generated is called a module.

DEFINITION A.5. (Linear Vector Spaces)
A linear vector space V is a set of elements called vectors, together with two operations called addition and scalar multiplication defined in it, such that the following axioms hold.

(i) *Addition:* Commutative : $x + y = y + x \quad \forall x, y \in V$

Associative : $x + (y + z) = (x + y) + z \quad \forall x, y, z \in V$

There exists an identity

'θ' in V such that : $\forall x \in V \quad x + \theta = \theta + x = x$

Inverse : $\forall x \in V$, there exists $(-x) \in V$ such that $x + (-x) = \theta$

(ii) *Multiplication by scalars:* There is a set of elements called scalars which constitutes a field F and an operation called scalar multiplication such that for each scalar α and every vector $x \in V$, $\alpha x \in V$. Further, multiplication by scalars is

Associative : $\alpha(\beta x) = \alpha\beta x \quad \forall x \in V$ and $\alpha, \beta \in F$

$1 \cdot x = x$ and $0 \cdot x = \theta \quad \forall x \in V$

Distributive : $\alpha(x + y) = \alpha x + \alpha y \quad \forall x, y \in V$ and $\alpha \in F$

$(\alpha + \beta)x = \alpha x + \beta x \quad \forall x \in V$ and $(\alpha, \beta) \in F$

We have to distinguish between two additive identity elements in a vector space – one for vectors denoted by 'θ' and the other for scalars denoted by '0'.

If the scalars chosen belong to the field of real numbers then the vector space is called *real vector space.* Similarly the field of complex numbers gives rise to *complex vector space.*

A subset S of a linear vector space V is called a *subspace* of V if it satisfies all the axioms of the linear vector space and therefore constitutes a vector space in its own right

MODULES

Unlike in the case of vector spaces, in modules the scalars used for multiplication are chosen from a ring instead of from a field.

DEFINITION A.6 (Modules)

Let R be a ring. A *module* M over R is written as $(M, R, +, \cdot, \theta_M)$ is a set M together with a commutative ring R such that $(M, +, \theta_M)$ is an additive commutative group. Further the module is closed under multiplication by scalars where the scalars are chosen from a ring R. We have $1 \cdot m = m$ and $0 \cdot m = \theta_M$ for all $m \in M$. Here 1 and 0 are the multiplicative and additive identities of the ring R.

Addition and multiplication by scalars are related by the distributive laws.

$\forall m \in M$ and $\forall r_1, r_2 \in R \qquad (r_1 + r_2)m = r_1 m + r_2 m$

$\forall m_1, m_2 \in M$ and $\forall r \in R \qquad r(m_1 + m_2) = rm_1 + rm_2$

EXAMPLE A.4
$(\mathbb{R}^n[s], \mathbb{R}(s), +, \cdot, \theta_n)$ is not a module The reason is that $\mathbb{R}^n[s]$ is not closed under multiplication by scalars $\in \mathbb{R}(s)$. However $(\mathbb{R}^n[s], \mathbb{R}[s], +, \cdot, \theta_n)$ is a module and is embedded in the vector space

$$(\mathbb{R}^n(s), \mathbb{R}(s), +, \cdot, \theta_n)$$ ■

The polynominal matrix $M \in \mathbb{R}[s]^{n\times n}$ is the representation of a linear transformation mapping the module $(\mathbb{R}^n[s], \mathbb{R}[s])$ into the module $(\mathbb{R}^m[s], \mathbb{R}[s])$. We say that M is an $m \times n$ matrix over the ring $\mathbb{R}[s]$. A polynominal matrix $M \in \mathbb{R}[s]^{n\times n}$ is nonsingular if $\det M \neq 0$. Hence a rational inverse exists for such nonsingular matrices. However unless $\det M$ is a constant the matrix is not unimodular or invertible because for unimodular matrices, belonging to $\mathbb{R}[s]^{n\times n}$ its inverse should also belong to $\mathbb{R}[s]^{n\times n}$ To clarify this concept consider a matrix $M \in \mathbb{R}[s]^{n\times n}$. Assume further that $\det M \neq 0$ we have

$$M^{-1} = \frac{\text{Adj } M}{\det M}$$

clearly the entries of $\text{Adj}\, M$ belong to $\mathbb{R}[s]$ as they are obtained by operations permissible in the ring. If $\det M$ is not a constant $1/\det M$ is not a polynominal and every element of Adj M gets multiplied by $1/\det M$. Hence the elements of M^{-1} matrix do not belong to $\mathbb{R}[s]$. In other words M is not unimodular or invertible within the ring.

A.8 Vector and Matrix Norms

Norms measure the 'size' of vectors and matrices. We use different types of norms but there are certain properties common to all norms. They are described below.

DEFINITION A.7 (Vector norms)
Let V be a vector space over a field F ($\mathbb{R}$ or $\mathbb{C}$). A function $\|\cdot\| : V \to \mathbb{R}$ is a vector norm if $\forall x, y \in V$.

i) $\|x\| \geq 0$ (non negative).
ia) $\|x\| = 0$ if and only if $x = 0$ (positive).
ii) $\|cx\| = |c|\ \|x\|$ for all scalars $c \in F$ (Homogenity).
iii) $\|x + y\| \leq \|x\| + \|y\|$ (triangle inequality).

A function that satisfies (i), (ii) and (iii) but not (ia) is called a *semi norm*. A semi norm generalizes the notion of the norm in the sense that some vectors other than zero vectors are allowed to have zero length.

EXAMPLE A.5
Examples of vector norms are

1. The Euclidean norm (or l_2 norm) on $\mathbb{C}^n$

$$\|x\|_2 = [|x_1|^2 + |x_2|^2 + \cdots + |x_n|^2]^{1/2}$$

2. The sum norm (or l_1 norm) on $\mathbb{C}^n$.

$$\|x\|_1 = |x_1| + |x_2| \cdots + |x_n|$$

3. The maximum norm (or l_∞ norm) on $\mathbb{C}^n$

$$\|x\|_\infty = \max\{|x_1|, |x_2|, \cdots |x_n|\}$$

4. The l_p norm on $\mathbb{C}^n$

$$\|x\|_p = \left[\sum_{i=1}^{n} |x_i|^p\right]^{1/p} \qquad p \geq 1$$

It can be shown that each of the norms defined above satisfy all the axioms of the vector norm

A norm is said to be *unitarily invariant* if $\|Ux\| = \|x\| \quad \forall x \in \mathbb{C}^n$ for all unitary matrices U. It is easily verified that l_2 and l_∞ norms are unitarily invariant.

The definition of vector norm does not require that the vector space be finite dimensional. Consider for example the infinite dimensional vector space $C[a, b]$ comprising all continuous functions (real or complex valued) on the real interval $[a, b]$. We may define the following vector norms on $C[a, b]$. Thus for $f \in C[a, b]$ we have

$$\mathcal{L}_2 \text{ norm} : \|f\|_2 = \left[\int_a^b |f(t)|^2 dt\right]^{1/2}$$

$$\mathcal{L}_1 \text{ norm} : \|f\|_1 = \int_a^b |f(t)| dt$$

$$\mathcal{L}_p \text{ norm} : \|f\|_p = \left[\int_a^b |f(t)|^p dt\right]^{1/p}$$

$$\mathcal{L}_\infty \text{ norm} : \|f\|_\infty = \max\{|f(t)|; t \in [a, b]\}$$

We next consider how matrix norms are defined. Since an $n \times n$ matrix M has n^2 elements, it is natural to consider these elements as belonging to

$\mathbb{C}^{n^2}$ and use vector norm definition on $\mathbb{C}^{n^2}$. However, in such a definition we are not exploiting the natural multiplication operation associated with the matrix which is often useful in relating the 'size' of AB with the 'sizes' of A and B where A and B are compatible matrices. The following matrix norm definition with an additional axiom (axiom iv) incorporated takes care of this requirement.

DEFINITION A.8 (Matrix norm)
A function $\|\cdot\| : \mathbb{C}^{n\times n} \to \mathbb{R}$ is said to be a matrix norm if the following axioms are satisfied.

i) $\|A\| \geq 0$ (non-negative).
ia) $\|A\| = 0$ if and only if $A = 0$ (positive).
ii) $\|cA\| = |c|\ \|A\|$ for all complex scalars c (Homogenity).
iii) $\|A + B\| \leq \|A\| + \|B\|$ (Triangle inequality).
iv) $\|AB\| \leq \|A\| \cdot \|B\|$ (Sub-multiplicative).

Note that axioms (i) to (iii) are identical to those for vector norms. If axiom (iv) is not satisfied then the norm is known as *generalized matrix norm.* By deleting axiom (i.a) we obtain *matrix semi-norm.*

Associated with each vector norm $\|\cdot\|$ on $\mathbb{C}^n$, is a natural matrix norm $\|\cdot\|_i$ induced by $\|\cdot\|$ on $\mathbb{C}^{n\times n}$. This brings us to that definition of induced norms.

DEFINITION A.9 (induced norm)
Let $\|\cdot\|$ be a given norm on $\mathbb{C}^n$. Then for each matrix $A \in \mathbb{C}^{n\times n}$, we define $\|A\|_i$ as

$$\|A\|_i = \sup_{\substack{x \neq 0 \\ x \in \mathbb{C}^n}} \frac{\|Ax\|}{\|x\|} = \sup_{\|x\|=1} \|Ax\| = \sup_{\|x\|\leq 1} \|Ax\|$$

$\|A\|_i$ is known as the *induced norm* of A corresponding to the vector norm $\|\cdot\|$. Induced norm can be given a simple geometric interpretation. It is the least upper bond of the ratio $\|Ax\|/\|x\|$ as x varies over $\mathbb{C}^n$. In this sense $\|A\|_i$ can be thought of as the maximum gain of the mapping A. It can be shown that for every norm defined on $\mathbb{C}^n$, the corresponding induced norm $\|\cdot\|_i$ is a matrix norm satisfying all the postulates in Definition A.8. In fact one way to show that a certain functional on $\mathbb{C}^{n\times n}$ is a matrix norm is to show that it is induced by some vector norm. While every norm of a matrix induced by a vector norm is a matrix norm, the converse statement is not always true.

Consider for example the *Forbenius norm* defined as

$$\|A\|_2 = \left[\sum_{i,j=1}^{n} |a_{ij}|^2\right]^{1/2}$$

This norm satisfies all postulates in Definition A.8 including the submultiplicative clause and hence is a matrix norm. However no vector norm on $\mathbb{C}^n$ exists such that $\|A\|_2$ defined above is an induced norm. Thus the Forbenius norm is not an induced norm but yet it is a matrix norm. We now state without proof the induced matrix norms corresponding to l_1, l_2 and l_∞ norms on $\mathbb{C}^n$. (For proof refer to Horn and Johnson 1985.)

Norm on $\mathbb{C}^n$	*Induced norm on* $\mathbb{C}^{n\times n}$
$\|x\|_1 = \sum_i |x_i|$	$\|A\|_{i1} = \max_j \sum_i |a_{ij}|$ (Column sum)
$\|x\|_2 = \left[\sum_i |x_i|^2\right]^{1/2}$	$\|A\|_{i2} = \lambda_{\max}^{1/2}(A^*A)$ where $\lambda_{\max}$ is the maximum eigen value of A^*A
$\|x\|_\infty = \max_i |x_i|$	$\|A\|_{i\infty} = \max_i \sum_j |a_{ij}|$ (row sum)

The $\|A\|_{i2}$ norm is known as the *spectral norm.*

A.9 Background Concepts in Functional Analysis

Some of the basic concepts in functional analysis are given in the sequel. A linear vector space provides the proper setting for performing linear operations from one space to another. However in the absence of definition of distance in the respective spaces it is not possible to discuss concepts such as convergence and continuity. This is taken care of in *normed linear spaces* which are nothing but linear vector spaces with the length of the vector defined in the respective spaces.

DEFINITION A.10 (Normed linear space)
A normed linear space is represented by the ordered pair $(X, \|\cdot\|)$ where X is a linear vector space and $\|\cdot\|$ is a real valued function on X called the norm function such that.

i) $\|x\| \geq 0 \quad \forall x \in X, \|x\| = 0$ if and only if $x = \theta$ the zero vector.
ii) $\|\alpha x\| = |\alpha|\ \|x\| \quad \forall x \in X$ and scalar α.
iii) $\|x + y\| \leq \|x\| + \|y\| \quad \forall x, y \in X$.

The notion of norm enables us to characterize distance between any two vectors $x, y \in X$ as the norm of the difference vector $\|x - y\|$. Consider a linear space $(X, \|\cdot\|)$ and let $\{x_n\}_1^\infty$, be a sequence of elements in this space. The sequence is said to converge to an element $x_0 \in X$ if $\|x_n - x_o\| \to 0$ as $n \to \infty$. To put it differently, for every $\epsilon > 0$, there exists an integer $N(\epsilon)$ such that whenever $n > N(\epsilon)$ $\|x_n - x_o\| \leq \epsilon$. Very often we do not know whether the limit of the given sequence exists. The concept of *Cauchy sequence* enables us to discuss the convergence properties even when we have no knowledge about the limit of the sequence.

Definition A.11 (Cauchy Sequence)
A sequence in a normed linear space $(X, \|\cdot\|)$ is said to be a Cauchy sequence if for every $\epsilon > 0$ there exists an integer $N(\epsilon)$ such that $\|x_n - x_m\| < \epsilon$ whenever $n, m > N(\epsilon)$. In other words, for sufficiently large values of n and m the members of the sequence crowd around a limit point and as a consequence the distance between them becomes arbitrarily small. While it can be shown that every convergent sequence in a normed space is a Cauchy sequence, the converse is not always true. In other words a sequence could be Cauchy without it being convergent. This happens when the limit of the sequence though it exists lies outside the set considered and hence is not a member of the set

If in a normed linear space $(X, \|\cdot\|)$ every Cauchy sequence in X converges to an element in X, then the space is called a *Banach space* named after the Polish mathematician Stephan Banach. Important physical concepts such as angle between vectors or energy dissipated by the vectors cannot be expressed in a normed linear spaces unless we impose some additional structure on it. This is precisely what is done in inner product spaces, whose definition follows.

Definition A.12 (Inner product spaces)
An *inner product space* is a linear vector space X with an associated field F (normally belonging to $\mathbb{R}$ or $\mathbb{C}$) together with a function $\langle\cdot,\cdot\rangle$ called the inner product function which maps any two elements in X to F such that the following axioms are satisfied for $x, y, z \in X$ and $\lambda \in \mathbb{C}$ or $\mathbb{R}$.

i) $\langle x, y\rangle = \overline{\langle y, x\rangle}$ where 'over-bar' indicates complex conjugate.
ii) $\langle \lambda x, y\rangle = \lambda\langle x, y\rangle$.
iii) $\langle x + y, z\rangle = \langle x, z\rangle + \langle y, z\rangle$.
iv) $\langle x, x\rangle > 0$ whenever $x \neq \theta$, the identity element in the vector space.

The inner products in $\mathbb{C}^n$ and l_2 are defined as follows

$$\text{For } x, y \in \mathbb{C}^n \quad \langle x, y \rangle = \sum_{i=1}^{n} x_i \bar{y}_i$$

$$\text{For } x, y \in l_2 \quad \langle x, y \rangle = \sum_{i=1}^{\infty} x_i \bar{y}_i$$

The inner product in the complex vector space $C[0, 1]$ of all continuous complex valued functions on [0,1] with point wise addition and scalar multiplication is given by

$$\langle f, g \rangle = \int_0^1 f(t)\overline{g(t)}dt$$

for $f, g \in C[0, 1]$.

It is routine to check that all the above formulae for inner products satisfy the condition laid down in definition A.12. A useful inequality arising out of the definition of inner product is the *Cauchy-Schwarz inequality* which states that for $x, y \in V$, the inner product satisfies

$$|\langle x, y \rangle| \leq \langle x, x \rangle^{1/2} \langle y, y \rangle^{1/2}$$

(for proof refer to Horn and Johnson 1985). In the above inequality if x is proportional to y (i.e. they are related by a scalar) then equality is achieved. We than have

$$\langle x, y \rangle = \langle x, x \rangle^{1/2} \langle y, y \rangle^{1/2}$$

EXAMPLE A.6

In systems and control theory, the vector space which one often comes across is the space of rational functions analytic in the right half plane ($\mathcal{RH}_2$ space) or analytic on the imaginary axes ($\mathcal{RL}_2$ space).

Consider $\mathcal{RL}_2$ space where the boundary is the unit circle in the complex plane. Elements of this space have no poles of modulus 1.
Let the inner product in this space be defined as

$$\langle f, g \rangle = \frac{1}{2\pi} \int_{-\pi}^{\pi} f(e^{j\theta})\overline{g(e^{j\theta})}d\theta$$

where $f, g \in \mathcal{RL}_2$.
Consider a typical case where

$$f(z) = \frac{1}{z - \alpha} \text{ and } g(z) = \frac{1}{z - \beta}$$

with $|\alpha| < 1$ and $|\beta| < 1$. Note that $z = e^{i\theta}$ on the unit circle along which the integration is performed. We then have

$$\langle f, g\rangle = \frac{1}{2\pi j}\int_{\partial D} f(z)\overline{g(z)}\frac{dz}{z}$$

where ∂D denotes the unit circle. Substituting for f and g in the above formula we have

$$\begin{aligned}\langle f, g\rangle &= \frac{1}{2\pi j}\int_{\partial D} \frac{1}{z-\alpha}\cdot\frac{1}{\bar{z}-\bar{\beta}}\frac{dz}{z}\\ &= \frac{1}{2\pi j}\int_{\partial D} \frac{1}{z-\alpha}\cdot\frac{1}{1-\bar{\beta}z}dz\\ &= \frac{1}{2\pi j}\int_{\partial D} \frac{h(z)}{z-\alpha}dz.\end{aligned}$$

where $h(z) = \dfrac{1}{1-\bar{\beta}z}$ is analytic on the closed unit disk $\mathcal{D}$ since $|\beta| < 1$.

Hence applying Cauchy's integral formula yields:

$$\langle f, g\rangle = h(\alpha) = \frac{1}{1-\bar{\beta}\alpha}$$

■

Earlier we saw how the norm in certain vector spaces could be induced by defining an inner product in those spaces. An inner product space which is complete with respect to the norm induced by the inner product is called a *Hilbert Space.*

Clearly every Hilbert space is a Banach space but there are many Banach spaces which are not Hilbert spaces.

For example consider the l_∞ space of bounded sequences $x = \{x_n\}_{n=1}^{\infty}$ with component-wise addition and scalar multiplication. The norm on l_∞ is given by $\|x\|_\infty = \sup_{n\in\mathbb{N}} |x_n|$.

It can be shown that it is not possible to define an inner product on l_∞ such that $\langle x, x\rangle = \|x\|_\infty^2$ for all $x \in l_\infty$. Hence l_∞ space though a Banach space is not a Hilbert space.

All inner product spaces need not be Hilbert spaces. An example of an inner product space which is not a Hilbert space is the space $C[0, 1]$. The reason is that this space is not complete with respect to the inner product defined, because one can construct a Cauchy sequence in this space which converges outside this space. It is our insistence on continuity, which stands in the way of completeness. The right setting is provided by $\mathcal{L}_2[0, 1]$ which permits inclusion of Lebesgue measurable functions. It can be shown that this space is complete for the inner product defined and therefore qualifies to be a Hilbert space.

An important concept in Hilbert space is that of orthonormal sequences. This is a sequence $\{e_n\}$ such that $\langle e_i, e_j \rangle = 0$ if $i \neq j$ and $\langle e_i, e_j \rangle = 1$ if $i = j$.

If every vector x in the space considered could be written as $x = \sum_{n=1}^{\infty} a_n e_n$ we say that $\{e_n\}$ constitutes an orthonormal basis for the space.
In this case, we have

$$a_n = \langle x, e_n \rangle \text{ and } \quad \|x\|^2 = \sum_{n=1}^{\infty} |a_n|^2$$

We may use the orthogonality concept to split a Hilbert space into a sum of subspaces. The orthogonal complement of a subset E of a Hilbert space H is the set $\{x \in H : \langle x, y \rangle = 0 \quad \forall y \in E\}$. This space is denoted by $H - E$ or $E^{\perp}$. It can be shown $E^{\perp}$ is a closed linear subspace of H.

For any closed subspace M of Hilbert space H we have $M \cap M^{\perp} = \{0\}$. Hence $H = M \oplus M^{\perp}$.

Notes and Additional References

Linear algebra is an indispensable tool in the modern approach to engineering design and applications. The book by Strang (1976) gives a comprehensive coverage of this topic. The applications of linear algebra in control theory are highlighted in the books by Callier and Desoer (1982) and Vidyasagar (1985). A standard reference book for real and complex analysis is Rudin (1974). Matrix norms and their properties are extensively discussed in Horn and Johnson (1985). A very readable account of functional analysis with particular reference to Hilbert space is given in the book by Young (1988).

Appendix B

Review of Control Theory

B.1 Introduction

As the main body of control theory grew over the years, it has generously borrowed ideas and concepts from diverse disciplines – from mathematics, from econometrics, and from systems theory to cite a few areas. It will be difficult to locate information pertaining to all these topics in one single place. Some of the important concepts such as controllability, observability and stability have been absorbed into the mainstream of control theory and they are now commonplace in control literature. They do not bear any repetition. There are other topics like matrix Lyapunov equation, algebraic Riccati equation, balanced state space realization, singular value decomposition and Hankel norm approximation which are so important in the modern context that they need special attention. Hence they have been treated separately, elsewhere in this text. Some peripheral topics still remain and this chapter is intended to fill in this gap. Therefore, as is to be expected, the reader may find the topics rather disjointed – an amalgam of many things.

B.2 Transfer Functions and their Realizations

We deal only with rational transfer functions. A matrix transfer function $\boldsymbol{G}(s) \in \mathbb{R}(s)^{p\times m}$ has a realization given by

$$\boldsymbol{G}(s) = C(sI - A)^{-1}B + D$$

for some $A \in \mathbb{R}^{n\times n}$, $B \in \mathbb{R}^{n\times m}$, $C \in \mathbb{R}^{p\times n}$ and $D \in \mathbb{R}^{p\times m}$. We represent to this realization in two ways.
In the first method of representation we write

$$\boldsymbol{G} : (A, B, C, D)$$

In the second method of representation we write

$$G = \left[\begin{array}{c|c} A & B \\ \hline C & D \end{array}\right]$$

Both ways of representing the realization are used in this text.

A basis change T applied to a realization $\boldsymbol{G} = (A, B, C, D)$ results in a transformation $(A, B, C, D) \rightarrow (TAT^{-1}, TB, CT^{-1}, D)$. Note that under this transformation, the transfer function $\boldsymbol{G}$ remains invariant.

ADJOINT SYSTEM

Let $\boldsymbol{G} : (A, B, C, D)$.

The adjoint of the system with realization A, B, C, D is denoted by $\boldsymbol{G}^{\sim}$ and has a realization

$$\boldsymbol{G}^{\sim} : (-A', -C', B', D')$$

and

$$\boldsymbol{G}^{\sim}(s) = -B'(sI + A')^{-1}C' + D'$$

$$\text{clearly } \boldsymbol{G}^{\sim}(s) = \boldsymbol{G}'(-s)$$

INVERSE SYSTEMS

Let $\boldsymbol{G} : (A, B, C, D)$ and let D be non-singular. Then the inverse system $\boldsymbol{G}^{-1}(s)$ has a realization

$$\boldsymbol{G}^{-1} : (A - BD^{-1}C, BD^{-1}, -D^{-1}C, D^{-1})$$

If the $p \times m$ matrix D is singular with maximum row rank then a right inverse $D^{\dagger}$ of D exists and we have

$$DD^{\dagger} = I_p$$

The right inverse system $G^{\dagger}(s)$ is then given by

$$\boldsymbol{G}^{\dagger} : (A - BD^{\dagger}C, BD^{\dagger}, -D^{\dagger}C, D^{\dagger})$$

Similarly if D has maximum column rank then the left inverse of D exists and

$$D^{\dagger}D = I_m. \text{ We have}$$

$$\boldsymbol{G}^{\dagger} : (A - BD^{\dagger}C, BD^{\dagger}, -D^{\dagger}C, D^{\dagger})$$

In the above formula $\boldsymbol{G}^{\dagger}$ is the left inverse of G and $D^{\dagger}$ is the left inverse of D.

MINIMAL REALIZATION

For a realization $\boldsymbol{G}:(A,B,C,D)$

$$\text{We have } \boldsymbol{G}(s) = C(sI - s)^{-1}B + D$$
$$= \frac{C\text{Adj}(sI - A)B}{\det(sI - A)} + D$$

$C\,\text{Adj}(sI - A)B$ is a polynominal matrix and $\det(sI - A)$ is a polynominal. Det $(sI - A) = 0$ defines the eigenvalues of A. Since there is a possibility of the zeros of the polynominal matrix $C\,\text{Adj}(sI - A)B$ being cancelled by the zeros of $\det(sI - A)$, not all eigenvalues of A, appear as poles of $\boldsymbol{G}(s)$. Such cancellations occur when (A, B) is uncontrollable or (C, A) is unobservable. If the realization is both controllable and observable, then there is no cancellation and the realization is said to be *minimal.* It is a realization for which the A-matrix has minimal dimension. In fact the degree of the pole polynominal $p(s)$ as obtained from the Smith-McMillan form of $\boldsymbol{G}(s)$ gives the dimension of A associated with the minimal realization.

TRANSMISSION ZEROS, SMITH ZEROS AND SYSTEM ZEROS

The points at which the transfer matrix $\boldsymbol{G}(s)$ loses normal rank will be referred to as the *transmission zeros* of $\boldsymbol{G}(s)$.

The points where the system matrix

$$\begin{bmatrix} sI - A & -B \\ C & D \end{bmatrix}$$

loses normal rank will be referred to as the *Smith zeros* of the realization.

Rosenbrock (1973) has shown that the uncontrollable modes (also called *input decoupling zeros*) are given by the points at which the matrix $[sI - A \quad -B]$ loses normal rank. Similarly the unobservable modes (also called *output decoupling zeros)* are the points at which the matrix $\begin{bmatrix} sI - A \\ C \end{bmatrix}$ loses normal rank.

The term *system zeros* is used to refer to any zero belonging to the set

$$\{\text{transmission zeros}\} \cup \{\text{Input and/or output decoupling zeros}\}$$

Given $G(s)$ and a specific realization (A, B, C, D) we have

$$\{\text{Transmission zeros}\} \subseteq \{\text{Smith zeros}\} \subseteq \{\text{system zeros}\}$$

For a minimal realization all the three sets of zeros are identical.

B.3 System Interconnection

If $G_1 : (A_1, B_1, C_1, D_1)$ and $G_2 : (A_2, B_2, C_2, D_2)$ then it can be easily shown that G_1G_2 has a realization given by

$$G_1G_2 = \left[\begin{array}{c|c} A_1 & B_1 \\ \hline C_1 & D_1 \end{array}\right] * \left[\begin{array}{c|c} A_2 & B_2 \\ \hline C_2 & D_2 \end{array}\right]$$

$$= \left[\begin{array}{cc|c} A_1 & B_1C_2 & B_1D_2 \\ 0 & A_2 & B_2 \\ \hline C_1 & D_1C_2 & D_1D_2 \end{array}\right]$$

Even if G_1 and G_2 are associated with minimal realizations, the realization of the cascaded system G_1G_2 need not be minimal because of pole-zero cancellation.

The parallel connection of G_1 and G_2 (both having the same input and the output being that sum of the two outputs of G_1 and G_2) gives rise to a realization.

$$G_1 + G_2 = \left[\begin{array}{cc|c} A_1 & 0 & B_1 \\ 0 & A_2 & B_2 \\ \hline C_1 & C_2 & D_1 + D_2 \end{array}\right]$$

given

$$G(s) = \begin{bmatrix} G_{11} & G_{12} \\ G_{21} & G_{22} \end{bmatrix} = \left[\begin{array}{c|cc} A & B_1 & B_2 \\ \hline C_1 & D_{11} & D_{12} \\ C_2 & D_{21} & D_{22} \end{array}\right]$$

Then

$$G_{ij}(s) = C_i(sI - A)^{-1}B_j + D_{ij} \quad i,j = 1,2$$

B.4 Inners, Outers and Blashke Products

A convenient method of factorizing a function $F \in \mathcal{RH}_\infty$ is by expressing it as a product of two functions – an inner function F_i and an outer function F_0.

A matrix valued function $\boldsymbol{F} \in \mathcal{RH}_\infty$ is *inner* if $\boldsymbol{F}^\sim \boldsymbol{F} = I$ where $\boldsymbol{F}^\sim = \boldsymbol{F}'(-s)$. A matrix valued function $\boldsymbol{F} \in \mathcal{RH}_\infty$ is *co-inner* if its transpose is inner. If F is co-inner $\boldsymbol{F}\boldsymbol{F}^\sim = I$. A matrix valued function $\boldsymbol{F} \in \mathcal{RH}_\infty$ is *outer* if it has full row rank in $\mathrm{Re}\, s > 0$. If the transpose of a matrix valued function $\boldsymbol{F} \in \mathcal{RH}_\infty$ is outer, then it is termed *co-outer.*

LEMMA B.1

For every matrix $\boldsymbol{F} \in \mathcal{RH}_\infty$ there exists inner, outer, co-inner and co-outer matrices $\boldsymbol{F}_i$, $\boldsymbol{F}_o$, $\boldsymbol{F}_{ci}$ and $\boldsymbol{F}_{co}$ respectively such that

$$\boldsymbol{F} = \boldsymbol{F}_i \boldsymbol{F}_o = \boldsymbol{F}_{co} \boldsymbol{F}_{ci}.$$

If $\boldsymbol{F}$ has constant rank on the extended imaginary axis (i.e for all real values of ω including $\omega = \infty$) then $\boldsymbol{F}_o$ has a right inverse in $\mathcal{RH}_\infty$ and $\boldsymbol{F}_{co}$ has a left inverse in $\mathcal{RH}_\infty$.

REMARKS

1. Left multiplication by an inner preserves norms in both $\mathcal{H}_2$ and $\mathcal{H}_\infty$ Thus if $\boldsymbol{G} \in \mathcal{RH}_\infty$, is an inner and $\boldsymbol{F}$ and $\boldsymbol{H}$ are matrices belonging to $\mathcal{RH}_\infty$ and $\mathcal{RH}_2$ respectively then

$$\|\boldsymbol{GF}\|_\infty = \|\boldsymbol{F}\|_\infty \text{ and } \|\boldsymbol{GH}\|_2 = \|\boldsymbol{H}\|_2$$

   Similarly if $\boldsymbol{G}$ is an inner, right multiplication by $\boldsymbol{G}^\sim$ preserves norms in both $\mathcal{H}_2$ and $\mathcal{H}_\infty$. This norm preserving property is central to many of the analytical results derived in this book.
2. Suppose $\boldsymbol{F} \in \mathcal{RH}_\infty^{p \times m}$ with $p < m$ has maximum row rank, then $\boldsymbol{F} = \boldsymbol{F}_i \boldsymbol{F}_o$ where $\boldsymbol{F}_i \in \mathcal{RH}_\infty^{p \times p}$ and $\boldsymbol{F}_o \in \mathcal{RH}_\infty^{p \times m}$. Note that $\boldsymbol{F}_i$ is square and $\boldsymbol{F}_o$ is rectangular in the above case. However if $p > m$ than $\boldsymbol{F}_i$ is a $p \times m$ matrix and $\boldsymbol{F}_o$ is an $m \times m$ matrix and the roles are reversed.
3. If $\boldsymbol{F}$ is a square matrix, of full rank than $\boldsymbol{F}$ can be factored either way. Thus $\boldsymbol{F} = \boldsymbol{F}_i \boldsymbol{F}_o$ or $\boldsymbol{F} = \boldsymbol{G}_o \boldsymbol{G}_i$ where $\boldsymbol{F}_i$ and $\boldsymbol{G}_i$ are inners and $\boldsymbol{F}_o$ and $\boldsymbol{G}_o$ are outers.
4. It can be verified that if $\boldsymbol{G}$ is a square inner, then $\det \boldsymbol{G}$ is an inner function and $(\det \boldsymbol{G}) \cdot \boldsymbol{G}^{-1}$ known as Adj $\boldsymbol{G}$ is an inner matrix.

BLASHKE PRODUCTS

The concept of Blashke product is best explained through an example. Consider a non-minimum phase stable transfer function

$$\boldsymbol{T}(s) = \frac{(s-1)(s-2)(s+3)}{(s+4)(s+5)(s+6)}$$

$T(s)$ could be rewritten as

$$\begin{aligned} T(s) &= \frac{(s+1)(s+2)(s+3)}{(s+4)(s+5)(s+6)} \cdot \frac{(s-1)(s-2)}{(s+1)(s+2)} \\ &= T_1(s) \cdot T_2(s) \end{aligned}$$

Note that $T_1(s)$ is a minimum phase transfer function because it has no zeros in the right half plane All right half plane zeros of $T(s)$ are absorbed in $T_2(s)$. It is easily verified that $T_2(j\omega)$ is an all pass function of unit norm i.e. $|T_2(j\omega)| = 1\ \forall\omega$. Thus a non-minimum phase transfer function can be viewed as the product of a minimum phase transfer function and an all pass transfer function of unit norm. Further on the imaginary axis $|T(j\omega)| = |T_2(j\omega)|$. This is the underlying idea behind factoring a stable transfer function as the product of inner and outer transfer functions, the former corresponding to an all pass transfer function of unit norm.

It should be noted that the zeros if any on the imaginary axis are included in the outer function and therefore it cannot be strictly called a minimum phase function in the usually accepted sense of the term.

The inner function associated with the given transfer function may be routinely constructed by forming what is known as the *Blashke product*. Thus for a scalar transfer function with zeros $\alpha_1, \alpha_2, \ldots \alpha_q$ in the open right half plane, the Blashke product is defined as

$$b(s) = \prod_{i=1}^{q} \frac{\alpha_i - s}{\bar{\alpha}_i + s}$$

Note that $b(s) \in \mathcal{RH}_\infty$ and $|b(j\omega)| = 1\ \forall\omega$. Zeros on the imaginary axis do not find a place in $b(s)$. A Blashke product associated with the unit disk in the complex plane could be similarly defined.

Let $g(z)$ be a rational function analytic in the closed unit disk $\mathcal{D}$ and let $\alpha_1, \alpha_2, \ldots \alpha_q$ be the zeros of $g(z)$ (not necessarily distinct) such that $0 < |\alpha_i| < 1$ for $i = 1, 2, \ldots, q$.. Further, let $g(z)$ have m zeros at the origin $z = 0$. (If $g(0) \neq 0$ set $m = 0$). Then the Blashke product associated with $g(z)$ is given by

$$b(z) = z^m \prod_{i=1}^{q} \frac{|\alpha_i|}{\alpha_i} \frac{z - \alpha_i}{1 - \bar{\alpha}_i z}$$

Clearly $b(z)$ is analytic in $\mathcal{D}$ and $|be^{i\theta}| = 1\ \forall\theta \in [0, 2\pi]$.

B.4 PBH Rank Test

Consider a realization (A, B, C, D) of a linear time invariant system. We shall now describe the Popov–Belevitch–Hautus (PBH) rank test for controllability and observability. According to this test, the pair (A, B) is controllable if and only if the matrix $[A - sI - B]$ has full row rank for all $s \in \mathbb{C}$.

Similarly the pair (C, A) is observable if and only if the matrix $\begin{bmatrix} A - sI \\ C \end{bmatrix}$ has full column rank for all $s \in \mathbb{C}$.

Those points at which the rank condition fails for the respective matrices are precisely the uncontrollable and unobservable modes associated with the system. Proof for the PBH rank test may be found in Kailath (1980).

Notes and Additional References

Though published in the early sixties, the book by Zadeh and Desoer (1963) continues to be a classic text for linear system theory. Among the later publications the book by Kailath (1980) covers a wide ground in an authoritative manner. Several excellent publications have appeared in recent years, e.g. Chen (1984), Brogan (1985), Kwakernaak and Sivan (1991) and Rugh (1993).

Appendix C

Singular Value Decomposition

C.1 Introduction

When signals pass through systems, they normally suffer attenuation. In SISO systems, we know that this attenuation varies with frequency and the frequency response curve specifies precisely the nature of this variation over the entire frequency range. But in the case of MIMO systems, the phenomena is more complex. For every $\omega \in \mathbb{R}$, the transfer function is now represented by a matrix $\boldsymbol{G}(j\omega)$ whose elements belong to either $\mathbb{R}$ or $\mathbb{C}$. Further, the input to the system is now a vector which may be complex or real. It is now not possible to speak about a unique gain for each frequency. Gain is now direction specific and depends upon the direction of the input vector. However, it turns out that every constant matrix has the property that it imparts the maximum gain for a particular direction of the input vector and further there exists another direction corresponding to which the gain attained is a minimum. In between, for different directions of the input vector the gain may vary between the maximum and the minimum. We are thus led to replace our ideas regarding a single gain in the case of SISO systems by a range of gains with upper and lower bounds for MIMO systems for each frequency considered.

The next question is how exactly to determine these bounds? The underlying theory of Singular Value Decomposition (SVD) which we present later in this chapter shows how these gains may be evaluated.

C.2 Four Fundamental Subspaces

Before we explain the theory of SVD, we present certain facts about the subspaces on which the input–output map operates.

Consider a linear map $A : X \rightarrow Y$ between two complex vector spaces X and Y of finite dimensions n and m respectively. A matrix $A \in \mathbb{C}^{m \times n}$ is associated with this mapping. The domain of A in this case is $\mathbb{C}^n$ and its codomain $\mathbb{C}^m$. It is well known (For example, Zadeh and Desoer 1963) that relative to the mapping A, the two spaces $\mathbb{C}^n$ and $\mathbb{C}^m$ may be decomposed as follows

$$\mathbb{C}^n = \text{dom}\ (A) = \mathcal{R}(A^*) \overset{\perp}{\oplus} \mathcal{N}(A)$$

$$\mathbb{C}^m = \text{codom}\ (A) = \mathcal{R}(A) \overset{\perp}{\oplus} \mathcal{N}(A^*)$$

$$\text{Rank}\, A = \text{Rank}\, A^* = r$$

Where A^* is the complex conjugate transpose of A and represents the adjoint mapping of A. If we restrict the domain of the mappings A and A^* to $\mathcal{R}(A^*)$ and $\mathcal{R}(A)$ respectively, then it can be proved that the mappings are bijective (one to one on to). Further, the following statements are true.

$$\mathcal{R}(A^*) = \mathcal{R}(A^*A); \mathcal{N}(A) = \mathcal{N}(A^*A) \tag{C.1}$$

$$\mathcal{R}(A) = \mathcal{R}(AA^*); \mathcal{N}(A^*) = \mathcal{N}(AA^*) \tag{C.2}$$

$$\text{Rank}(AA^*) = \text{Rank}(A^*A) = \text{Rank}\, A = r \tag{C.3}$$

Figure C.1 shows how the input and output spaces are decomposed into four fundamental subspaces. The diagram is self explanatory.

From equation (C.1) it follows that the eigen vectors of A^*A spans the space $\mathbb{C}^n$. We further note that A^*A is hermitian positive semi definite from which it follows that the eigenvalues of A^*A are real and non-negative. Similar reasoning may be applied in the case of the transformation AA^*. Both A^*A and AA^* have identical nonzero eigenvalues. However the number of zero eigenvalues may be different for A^*A and AA^*. With these preliminaries out of the way, we now state and prove a theorem concerning SVD.

THEOREM C-1 (Singular Value Decomposition)
Let the matrix $A \in \mathbb{C}^{m\times n}$ and let its rank r be $r \le \min(m, n)$ Then there exists a unitary matrix $U \in \mathbb{C}^{m\times m}$ and a unitary matrix $V \in \mathbb{C}^{n\times n}$ such that

$$A = U\Sigma V^* \tag{C.4}$$

where:

$$\Sigma = \begin{bmatrix} \Sigma_1 & 0 \\ 0 & 0 \end{bmatrix}; \Sigma_1 = (\sigma_1, \sigma_2, \ldots \sigma_r)$$

with $\sigma_1 \ge \sigma_2 \ge \cdots \ge \sigma_r > 0$

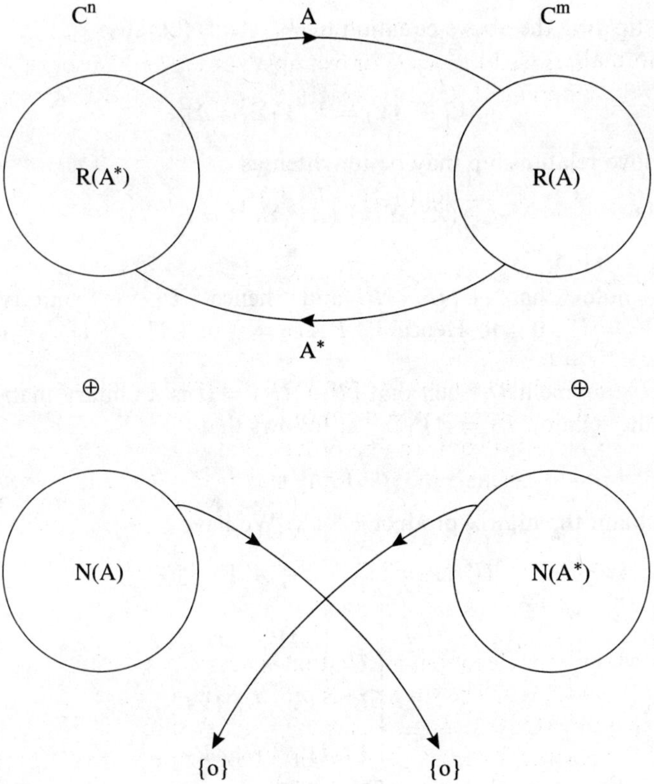

Figure C.1 Orthogonal space decomposition.

Proof: Let $\lambda_i(A^*A) = \sigma_i^2$ where $\lambda_i(A^*A)$ is the ith eigen value of A^*A. The n eigen values of A^*A (r of them positive and the remaining $(n-r)$ of them zero) can be ordered as

$$\sigma_1^2 \geq \sigma_2^2 \geq \cdots \geq \sigma_r^2 > 0 = \sigma_{r+1}^2 \ldots = \sigma_n^2$$

Let $v_1, v_2, \ldots, v_n$ be the corresponding orthonormal eigenvectors of A^*A. Further let $V_1 = [v_1 v_2 \ldots v_r]$ and $V_2 = [v_{r+1} v_{r+2} \ldots v_n]$.

We have

$$(A^*A)v_i = \sigma_i^2 v_i \quad i = 1, 2, \ldots, r$$

Hence

$$A^*AV_1 = V_1\Sigma_1^2$$

Premultiplying the above equation by V_1^* yields (because $v_1, v_2, \ldots, v_r$ are orthonormal)

$$V_1^* A^* A V_1 = V_1^* V_1 \Sigma_1^2 = \Sigma_1^2.$$

The above relationship may be rewritten as

$$(A V_1 \Sigma_1^{-1})^* (A V_1 \Sigma_1^{-1}) = I$$

Let $U_1 = A V_1 \Sigma_1^{-1}$.

We note that $U_1^* U_1 = I$ and hence U_1 is unitary, also $(A^*A) V_2 = V_2 \cdot 0 = 0$. Hence $V_2^* A^* A V_2 = 0$ or $(AV_2)^*(AV_2) = 0$ which implies $AV_2 = 0$.

Let U_2 augment U_1 such that $[\,U_1 \quad U_2\,] = U$ is a unitary matrix. From the relation $U_1 = A V_1 \Sigma_1^{-1}$ it follows that

$$A V_1 = U_1 \Sigma_1$$

Now obtain the matrix product U^*AV. We have

$$\begin{aligned} U^*AV &= [\,U_1 \quad U_2\,]^* A [\,V_1 \quad V_2\,] \\ &= \begin{bmatrix} U_1^* \\ U_2^* \end{bmatrix} [\,AV_1 \quad AV_2\,] \\ &= \begin{bmatrix} U_1^*AV_1 & U_1^*AV_2 \\ U_2^*AV_1 & U_2^*AV_2 \end{bmatrix} \\ &= \begin{bmatrix} \Sigma_1 & 0 \\ 0 & 0 \end{bmatrix} = \Sigma \end{aligned}$$

From the above equation it follows that

$$A = U \Sigma V^* \text{ which is required to be proved.}$$

The real numbers $\sigma_1, \sigma_2 \ldots \sigma_r$ along with $\sigma_{r+1} = \sigma_{r+2} \ldots = \sigma_n = 0$ are called the *singular values* of A. They are the positive square roots of the eigenvalues of A^*A which we know are real and non negative. The choice of A^*A instead of AA^* for determining eigenvalues is purely arbitrary. The latter choice would also lead to identical results.

Remarks

From the proof of theorem C.1, the following conclusions may be drawn

1. $\boldsymbol{v}_1, \boldsymbol{v}_2, \ldots, \boldsymbol{v}_r$ are the basis vectors of $\mathcal{R}(A^*)$ and $\boldsymbol{v}_{r+1}, \boldsymbol{v}_{r+2}, \ldots, \boldsymbol{v}_n$ are the basis vectors of $\mathcal{N}(A)$.
2. $\boldsymbol{u}_1, \boldsymbol{u}_2, \ldots, \boldsymbol{u}_r$ are the basis vectors of $\mathcal{R}(A)$ and $\boldsymbol{u}_{r+1}, \boldsymbol{u}_{r+2}, \ldots, \boldsymbol{u}_n$ are the basis vectors of $\mathcal{N}(A^*)$.

3. From the relation $AV_1 = U_1\Sigma$ it is clear that

$$A\boldsymbol{v}_i = \boldsymbol{u}_i\sigma_i \quad i = 1, 2, \ldots, r$$

C.3 Geometric Interpretation of SVD

Geometric interpretation helps bolster ones intuition and is therefore extremely useful. In the present context, the matrix operation on unit vector $\boldsymbol{v}_i$ in $\mathcal{R}(A^*)$ may be visualized as a combination of dilation and rotation. This is clear from the equation $A\boldsymbol{v}_i = \boldsymbol{u}_i\sigma_i$ in which the unit vector $\boldsymbol{v}_i \in \mathcal{R}(A^*)$ is mapped onto a unit vector $\boldsymbol{u}_i \in \mathcal{R}(A)$ and further dilated by a factor σ_i in the direction of $\boldsymbol{u}_i$.

The geometric viewpoint is pictorially expressed in Figure C.2. To simplify matters, it is assumed that $\mathcal{R}(A^*)$ and $\mathcal{R}(A)$ are both spanned by vectors $(\boldsymbol{v}_1, \boldsymbol{v}_2)$ and $(\boldsymbol{u}_1, \boldsymbol{u}_2)$ respectively. $\boldsymbol{v}_1$ is mapped onto $\boldsymbol{u}_1$ and gets multiplied by σ_1. Similarly $\boldsymbol{v}_2$ is mapped onto $\boldsymbol{u}_2$ and the dilation constant is σ_2.

To consider a more general case, we examine the action of the mapping function, when its domain is confined to a hyper sphere of unit radius with centre at the origin in $\mathbb{C}^n$ space. We will presently show how the unit sphere is mapped into an r-dimensional ellipsoid with the basis vectors $\boldsymbol{u}_1, \boldsymbol{u}_2 \ldots \boldsymbol{u}_r$ constituting the principal axes and $\sigma_1, \sigma_2, \ldots, \sigma_r$ representing the lengths of the semi axes.

Let the coordinates of a vector on the unit sphere with reference to the basis vectors $\boldsymbol{v}_1, \boldsymbol{v}_2, \ldots, \boldsymbol{v}_n$ be $(x_1, x_2, \ldots, x_n)$. We have $\sum_{i=1}^{n} x_i^2 = 1$ and $x = \boldsymbol{v}_1 x_1 + \boldsymbol{v}_2 x_2 + \cdots + \boldsymbol{v}_n x_n$. Hence $Ax = (A\boldsymbol{v}_1)x_1 + \cdots + (A\boldsymbol{v}_n)x_n$. But $A\boldsymbol{v}_i = \sigma_i \boldsymbol{u}_i \quad i = 1, 2, \ldots, r$ and $A\boldsymbol{v}_i = 0 \quad i = r + 1, r + 2, \ldots, n$.

Hence

$$Ax = \boldsymbol{u}_1(\sigma_1 x_1) + \boldsymbol{u}_2(\sigma_2 x_2), \ldots, \boldsymbol{u}_r(\sigma_r x_r) \tag{C.5}$$

We may also write

$$Ax = \boldsymbol{u}_1 y_1 + \cdots + \boldsymbol{u}_r y_r \tag{C.6}$$

Comparing (C.5) and (C.6) component-wise we get

$$y_i = \sigma_i x_i \tag{C.7}$$

or

$$x_i = \frac{y_i}{\sigma_i} \quad i = 1, 2, \ldots, r \tag{C.8}$$

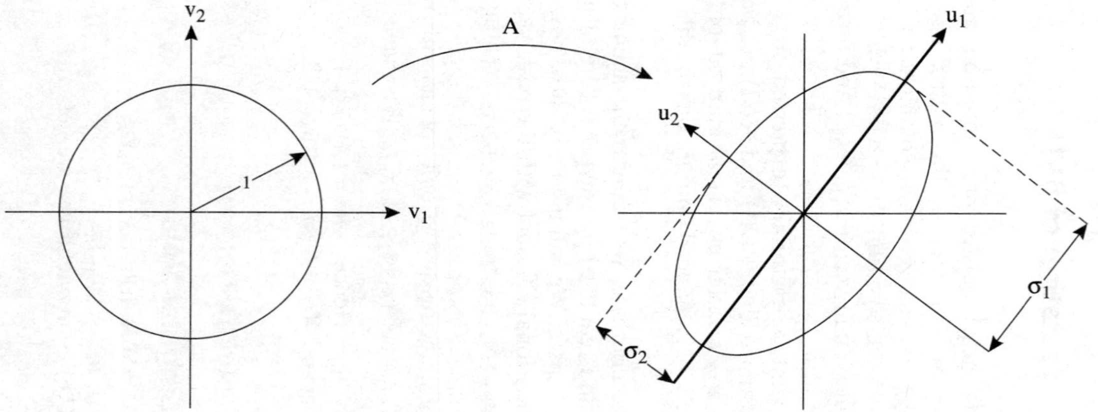

Figure C.2 Geometric interpretation of SVD.

Consider the most general case with $r < \min(m, n)$. A vector on the unit sphere in $\mathbb{C}^n$ is first projected on to the $\mathcal{R}(A^*)$ before the transformation is applied. Hence

$$x_1^2 + x_2^2 + \cdots + x_r^2 \leq 1$$

substituting for x_i from (C.8) we get

$$\frac{y_1^2}{\sigma_1^2} + \frac{y_2^2}{\sigma_2^2} + \cdots + \frac{y_r^2}{\sigma_r^2} \leq 1 \tag{C.9}$$

The unit sphere is thus mapped into an interior of an ellipsoid. In the light of the above discussion, consider the definition of matrix norm of A namely

$$\|A\| = \max\{\|Ax\|_2 : \|x\|_2 = 1\} \tag{C.10}$$

The matrix norm is thus seen to be the size of the largest vector of the image under A of the unit sphere in $\mathbb{C}^n$. From Figure C.2 it is clear that this is nothing but the largest singular value σ_1 of A. Hence

$$\|A\| = \sigma_{\max}(A) = \sigma_1(A) \tag{C.11}$$

Similarly it can be shown that

$$\sigma_{\min}(A) = \min\{\|Ax\|_2 : \|x\|_2 = 1\} \tag{C.12}$$

To recapitulate, the geometric interpretation of singular values shows how the magnitude of a vector x under transformation A is direction specific. Thus, the maximum amplification occurs when the direction of the x vector coincides with the direction of $\boldsymbol{v}_1$. Its image under A is in the $\boldsymbol{u}_1$ direction with amplification equal to σ_1. The ellipsoid in $\mathcal{R}(A)$ gives a visual picture of how the amplification continuously changes depending upon the direction of the x vector. In standard notation, $\sigma_{\max}(A)$ is denoted by $\bar{\sigma}(A)$ and $\sigma_{\min}(A)$ is denoted by $\underline{\sigma}(A)$.

C.4 Properties of Singular Values

We have already seen how the maximum singular value defines an induced norm. This has several consequences such as satisfaction of triangle inequality and sub-multiplicative property associated with induced norms. We therefore have the following inequalities.

$$\bar{\sigma}(A + B) \leq \bar{\sigma}(A) + \bar{\sigma}(B) \tag{C.13}$$
$$\bar{\sigma}(AB) \leq \bar{\sigma}(A)\bar{\sigma}(B) \tag{C.14}$$

The above inequalities repeatedly occur in the applications of singular values in MIMO systems. Further, these inequalities lead to certain additional relationships which are quite useful in applications. These are

$$|\bar{\sigma}(A) - \bar{\sigma}(B)| \leq \bar{\sigma}(A+B) \leq \bar{\sigma}(A) + \bar{\sigma}(B) \tag{C.15}$$

$$\underline{\sigma}(A)\bar{\sigma}(B) \leq \bar{\sigma}(AB) \leq \bar{\sigma}(A)\bar{\sigma}(B) \tag{C.16}$$

$$\max\{\underline{\sigma}(A) - \bar{\sigma}(B), \underline{\sigma}(B) - \bar{\sigma}(A)\} \leq \underline{\sigma}(A+B) \leq \underline{\sigma}(A) + \bar{\sigma}(B) \tag{C.17}$$

$$\underline{\sigma}(A)\underline{\sigma}(B) \leq \underline{\sigma}(AB) \leq \bar{\sigma}(A)\underline{\sigma}(B) \tag{C.18}$$

The proofs of these inequalities are left as exercises for the reader. However, we briefly indicate the lines on which the proofs may be constructed. For example, to prove the left hand inequality in (C.15), we proceed as follows.

First we note that the right hand inequality directly follows from the fact that the maximum singular value defines an induced norm (Refer to C.13). Now replace A and B in the right hand inequality in (C.15) by $(A+B)$ and $(-B)$ we then have

$$\bar{\sigma}(A) = \bar{\sigma}[A + B + (-B)] \leq \bar{\sigma}(A+B) + \bar{\sigma}(B)$$

Hence

$$|\bar{\sigma}(A) - \bar{\sigma}(B)| \leq \bar{\sigma}(A+B)$$

To take yet another example, consider the proof of left hand inequality in (C.18). We note that

$$\begin{aligned}
\underline{\sigma}(AB) &= \min_{\|u\|=1} \|ABu\| \\
&= \min_{\|u\|=1} \left[\frac{\|ABu\|}{\|Bu\|}\|Bu\|\right] \\
&\geq \min_{\|v\|=1} \|Av\| \min_{\|u\|=1} \|Bu\| \\
&= \underline{\sigma}(A)\underline{\sigma}(B)
\end{aligned}$$

Other inequalities could be similarly proved. The establishment of the following relationships are straightforward and they are listed below for easy reference

$$\sigma_i(A) = \lambda_i^{1/2}(A^*A) = \lambda^{1/2}(AA^*) \quad i = 1, 2, \ldots, r$$

$$\bar{\sigma}(A) = \max_{\|x\|_2 \neq 0} \frac{\|Ax\|_2}{\|x\|_2}$$

$$\underline{\sigma}(A) = \min_{\|x\|_2 \neq 0} \frac{\|Ax\|_2}{\|x\|_2}$$

Let A be an $n \times n$ matrix and let $\underline{\sigma}(A) > 0$. This implies that A is invertible. We then have

$$\sigma_j(A^{-1}) = \frac{1}{\sigma_l(A)} \text{ where } l = (n - j + 1)$$

Specifically we have

$$\bar{\sigma}(A^{-1}) = \frac{1}{\underline{\sigma}(A)}$$
$$\bar{\boldsymbol{u}}(A^{-1}) = \underline{\boldsymbol{v}}(A)$$
$$\underline{\boldsymbol{u}}(A^{-1}) = \bar{\boldsymbol{v}}(A)$$

where $\boldsymbol{u}$ and $\boldsymbol{v}$ are the left and right singular vectors associated with the argument. The over-bar and under-bar indicate that the singular vectors relate to the maximum and minimum singular values respectively.
Since

$$\bar{\sigma}(A) = \max_{\|x\| \neq 0} \frac{\|Ax\|_2}{\|x\|_2} \geq \frac{\|A\bar{x}\|_2}{\|\bar{x}\|_2} = |\bar{\lambda}(A)|$$

and

$$\underline{\sigma}(A) = \min_{\|x\| \neq 0} \frac{\|Ax\|_2}{\|x\|_2} \leq \frac{\|A\underline{x}\|_2}{\|\underline{x}\|_2} = |\underline{\lambda}(A)|$$

where $\bar{\lambda}$ and $\underline{\lambda}$ are the maximum and minimum eigenvalues of A and $\bar{x}$ and $\underline{x}$ the corresponding eigen vectors we have

$$\bar{\sigma}(A) \geq |\lambda_i(A)| \geq \underline{\sigma}(A) \quad i = 1, 2, \ldots, n$$

Thus, the maximum and minimum singular values of A bound its eigen values.

If A is hermitian, then

$$A^* = A \text{ and } \lambda_i(A^*A) = \lambda_i(A^2) = \lambda_i^2(A)$$

Hence

$$\sigma_i(A) = \lambda_i^{1/2}(A^*A) = \left[\lambda_i^2 \quad (A)\right]^{1/2} = \lambda_i(A)$$

Thus for hermitian matrices, its eigenvalues and singular values are one and the same.

An attractive feature of singular values is that they are not very sensitive to changes in the matrix and therefore they are computationally robust (see Stewart 1973). This is in marked contrast to eigenvalues which are known to be highly sensitive to matrix perturbatians. As a matter of fact, for rank determination of a matrix, SVD offers one of the most reliable methods. The minimum singular value of a matrix not only gives information about the rank of a matrix but also indicates how far that matrix is away from the nearest matrix of lower rank. Let $\sigma_1, \sigma_2, \ldots, \sigma_m$ be the singular values of B matrix and $\tilde{\sigma}_1, \tilde{\sigma}_2, \ldots, \tilde{\sigma}_m$ be the singular values of $B + \delta B$ matrix. Then it can be shown that $|\sigma_i - \tilde{\sigma}_i| \le \|\delta B\| \quad i = 1, 2, \ldots, m$. Ideally, if $\sigma_1 \ge \sigma_2 \ge \cdots \ge \sigma_r > 0$ and $\sigma_{r+1} = \cdots = \sigma_m = 0$, then B has a rank r. Further, for any $j \le r$, it will require a change $\|\delta B\| \ge \sigma_j$ to produce a perturbed matrix $B + \delta B$ of rank less than j. In fact $B + \delta B = U \operatorname{diag}[\sigma_1, \sigma_2, \ldots, \sigma_{j-1}, 0, \ldots, 0] V^*$ is such a matrix with $\|\delta B\| = \sigma_j$.

C.5 Applications of Singular Values

Singular values find application in a variety of problems relating to analysis and design. Firstly, as already explained the minimum singular value of a square matrix is a measure of the distance from that matrix to one that is singular. Secondly, the unitary matrices associated with SVD can be used for row and column compression as follows.

Recall that for an $m \times n$, matrix A of rank r

$$A = U \Sigma V^*$$

where U and V are unitary matrices and

$$\Sigma = \begin{bmatrix} \Sigma_r & 0 \\ 0 & 0 \end{bmatrix} \text{ where } \Sigma_r = \{\sigma_1, \sigma_2, \ldots, \sigma_r\}$$

we may use the matrices U and V to obtain row and column compression as follows. Thus

$$U^* A = \Sigma V^* = \begin{bmatrix} R \\ 0 \end{bmatrix} \text{ and}$$
$$AV = U\Sigma = [S \quad 0]$$

Clearly R has r linearly independent rows and S has r linearly independent columns. Another use to which SVD can be put is in determining the *pseudo-inverse* of a matrix A. We have $A = U \Sigma V^*$.

The pseudo-inverse of A denoted by $A^\dagger$ is given by

$$A^\dagger = V \begin{bmatrix} \Sigma_r^{-1} & 0 \\ 0 & 0 \end{bmatrix} U^*$$

It may be verified that the pseudo-inverse thus obtained satisfies the required properties namely $AA^{\dagger}A = A$ and $A^{\dagger}AA^{\dagger} = A^{\dagger}$.

Notes and Additional References

A tutorial exposition of SVD dealing with both computational aspects and applications is given in Klema and Laub (1980). The book by Stewart (1973) is a standard reference text for SVD.

References

Abedar J., Nagpal K., Khargonekar P.P. and Poolla K. (1995) Robust regulation in the presence of norm-bounded uncertainty. *IEEE Trans. Auto. contr.* **AC-40**, 147–152.

Ackermann J., Hu H.Z. and Kaesbauer. D. (1990) Robustness analysis: A case study. *IEEE Trans. Auto. contr.* **AC-35**, 352–356.

Adamjan V.M., Arov D.Z. and Krein M.G. (1978). Infinite block Hankel matrices and related extension problems. *Amer. Math. Soc. Trans.* **III**, 133–156.

Athans M.A. (1971a). The role and use of the stochastic Linear – Quadratic – Gaussian Problem. *IEEE Trans. Auto. contr.* **AC-16**, 529–552.

Athans M.A. (Guest editor) (1971b). Special issue on LQG problem. *IEEE Trans. Auto. contr.* **AC-16**, 527–869

Ball J.A. and Helton J.W. (1983) A Burling-Lax theorem for the Lie group U(m,n) which contains most classical interpolation theory. *J. Operator Theory* **Vol. 9**, 107–142.

Barmish B.R. (1994) *New Tools for Robustness of Linear Systems.* New York: Macmillan

Bartlett A.C., Hollot C.V. and Huang L. (1988) Root locations of an entire polytope of polynomials: It suffices to check the edges. *Maths. contr. Signals and Systems* **Vol. 1** 61–71

Bhattacharyya S.P. (1987) *Robust Stabilization Against Structure Perturbations.* New York: Springer-Verlag.

Bode H.W. (1945) *Network Analysis and Feedback Amplifier Design.* Princeton NJ: Van Nostramd.

Brogan W.L. (1985). *Modern Control Theory.* Englewood NJ: Prentice Hall.

Callier F.M. and Desoer C.A. (1982) *Multivariable Feed back Systems.* New York: Springer-Verlag.

Chang B.C. and Pearson J.B. (1984). Optimal disturbance reduction in linear multivariable systems. *IEEE Trans. Auto. contr.* **AC-29, 880–887.**

Chen C.T. (1984). *Linear Systems Theory and Design.* New York: Holt Rinehart and Winston

Delsarte P.H., Genin Y. and Kamp Y. (1979) The Nevanlinna – Pick problem for matrix valued functions. *SIAM J. Appl. Math* **vol. 36**. 47–61

Delsarte P.H., Genin Y. and Kamp Y. (1981). The Nevalinna – Pick problem in circuit and system theory. *Int. J. Circuit Theory and Applications* **vol. 9**. 177–187

Desoer C.A., Liu R.W., Murray J. and Saeks R. (1980) Feed back systems design: The fractional representation approach to analysis and synthesis. *IEEE Trans. Auto. contr.* **AC-25**, 339–412

Desoer C.A. and Gustafson C.L. (1984). Algebraic theory of linear multivariable feedback systems. *IEEE Trans. Auto. contr.* **AC-29**, 909–917

Dorato P. (Editor) (1987). *Robust Control. New York: IEEE Press*

Dorato P. and Yedavalli R.K. (Editors) (1990) *Recent Advances in Robust Control. New York: IEEE Press*

Doyle J.C. (1978). Guaranteed margins for LQG regulators *IEEE Trans. Auto. contr.* **AC-23**, 756–757

Doyle J.C. and Stein G. (1981) Multivariable feedback design: concepts for a classical modern synthesis. *IEEE Trans. Auto. contr.* **AC-26**, 4–16

Doyle J.C., Glover K., Khargonekar P.P. and Francis B.A. (1989) State-space solutions to standard $\mathcal{H}_2$ and $\mathcal{H}_\infty$ control problems. *IEEE Trnas. Auto. contr.* **AC-34**, 831–847

Doyle J.C., Francis B.R., and Tannenbaum A.R. (1992) *Feedback Control Theory. New York: Macmillan*

Doyle J.C., Zhou K., Glover K and Bodenheimer B. (1994). Mixed $\mathcal{H}_2$ and $\mathcal{H}_\infty$ performance objectives II: optimal control. *IEEE Trans. Auto. contr.* **AC-39**, 1575–1586

Duren P.L. (1970). *Theory of $\mathcal{H}_p$ Spaces.* New York: Academic Press

Francis B.A. and Zames G. (1984) On $\mathcal{H}_\infty$ optimal sensitivity theory for SISO systems. *IEEE Trans. Auto. contr.* **AC-29**, 9–16

Francis B.A., Helton J.W. and Zames, G. (1984) $\mathcal{H}_\infty$ optimal controller for linear multivariable systems. *IEEE Trans. Auto. contr.* **AC-29**, 888–990

Francis B.A. (1987) A Course in $\mathcal{H}_\infty$ Control. Lecture notes in control and information sciences **No 88**, Berlin: Springer-Verlag

Francis B.A. and Doyle J.C. (1987). Linear control theory with $\mathcal{H}_\infty$ optimality criteria *SIAM J. contr. and optimization* **vol. 25**, 815–841

Freudenberg J.S. and Looze D.P. (1986a) An analysis of $\mathcal{H}_\infty$ optimization design methods *IEEE Trans. Auto. contr.* **AC-31**, 194–200

Freudenberg J.S. and Looze D.P. (1986b) A relation between open - loop and closed - loop properties of multivariable feedback systems. *IEEE Trans. Auto. contr.* **AC-31** 333–340

Glover K. (1984) Optimal Hankel norm approximation of linear multivariable systems and their $\mathcal{L}_\infty$ error bounds *Int. J. Control* **vol. 39**, 1115–1193

Glover K. (1986) Robust stabilization of linear multivariable systems: relations to approximation *Int. J. Control.* **vol. 43**, 741–766

Green M. and Limebeer D.J.N. (1995) *Linear Robust Control.* Englewood cliffs, NJ: Prentice Hall

Helton J.W. (1985) Worst case analysis in the frequency domain: The $\mathcal{H}_\infty$ approach to control. *IEEE Trans. Auto. contr.* **AC-30**, 1154–1170

Horn R.A. and Johnson C.A. (1985) *Matrix Analysis* Cambridge, England: Cambridge University Press

Horowitz I.M. (1963) *Synthesis of Feedback Systems.* New York: Acadamic Press.

Hung Y.S. (1989a) $\mathcal{H}_\infty$ optimal control, Part 1 Model matching. *Int. J. Control* **vol. 49** 1291–1330

Hung Y.S. (1989b) $\mathcal{H}_\infty$ optimal Control. Part 2 Solution for controllers *Int. J. Control* **vol. 49** 1331–1359

James M.R. and Baras J.S. (1995) Robust $\mathcal{H}_\infty$ output feedback for non linear systems. *IEEE Trans. Auto. contr.* **AC-40**, 1107–1110.

Kalman R.E. (1960). Contributions to the theory of optimal control. *Boletin de la Sociedad Mathematica Mexicana* 102–119

Kalman R.E. (1963). Mathematical description of linear dynamical systems. *SIAM J. Control* **vol. 1**, 151–192

Kalman R.E. (1964). When is a linear system optimal? *ASME Trans. Series D. J. Basic Engg* **No. 86**, 51–60

Kalman R.E. (1965). Irreducible realizations and the degree of rational matrix. *SIAM J. App. Math.* **vol.** 13 520–544

Kailath, T. (1980). *Linear Systems.* Englewood Cliffs, NJ: Prentice Hall

Kazerooni H. (1988) Loop shaping design related to LQG/LTR for SISO minimum phase plants. *Int. J. Control*, 241–255

Khargonekar P.P., Petersen, I.R. and Rotea M.A. (1988). $\mathcal{H}_\infty$ optimal control with state feedback. *IEEE Trans. Auto. contr.* **AC-33**, 786–788

Kharitonov V.L. (1978) Asymptotic stability of an equilibrium position of systems of linear differential equations. *Differentsial'nye Uravneniya* **vol. 14**, 2086–2088

Kimura H. (1984) Robust stabilization of a class of transfer functions. *IEEE Trans. Auto. contr.* **AC-29**, 778–793

Klema V.C. and Laub A.J. (1980) The singular value decomposition – its computation and some applications. *IEEE Trans. Auto. contr.* **AC-25** 164–176

Kolmogrov A.N. (1941) Interpolation and extrapolation of stationary random sequences. *Bull Acad. Sci. USSR* **vol. 5** Translation RAND Corporation U.S.A. (1962)

Koosis P. (1980) *The Theory of $\mathcal{H}_p$ Spaces.* Cambridge, England: Cambridge University Press

Kucera V. (1972). A contribution to matrix quadratic equations. *IEEE Trans. Auto. contr.* **AC-17**, 344–349

Kung S.Y. and Lin D.W. (1981) Optimal Hankel norm model reductions: Multivariable systems. *IEEE Trans. Auto. contr.* **AC-26** 832–852

Kwakernaak H. and Sivan R. (1991) *Modern Signals and Systems.* Englewood Cliffs NJ: Prentice-Hall

Laub A.J., Heath M.T., Page C.C. and Ward R.C. (1987) Computation of balancing transformation and other applications of simultaneous diagonalization algorithms. *IEEE Trans. Auto. contr.* **AC-32**, 115–122

Lehtomaki N.A., Sandell N.R. and Athans M. (1981) Robustness results in Linear-Quadratic – Gaussian based multivariable control designs. *IEEE Trans. Auto. contr.* **AC-26**, 75–93

Limebeer D.J.N. and Hung Y.S. (1987) Analysis of the pole zero cancellations in $\mathcal{H}_\infty$ optimal control problems of the first kind. *SIAM J. contr. optimization* **vol. 25**. 1457–1493

Limebeer D.J.N. and Halikias G.D. (1988) Controller degree bound for $\mathcal{H}_\infty$ problems of the second kind. *SIAM J. contr. optimization* **vol. 26**, 646–677

Lu W.M. and Doyle J.C. (1994) $\mathcal{H}_\infty$ control of non linear systems via output feedback: controller parametriation. *IEEE Trans. Auto. contr.* **AC-39**, 2517–2521

MacFarlane A.G.J. (Editor) (1979). *Frequency Response Methods in Control Systems.* New York: IEEE Press

Maciejowski J.M. (1989) *Multivariable Feedback Design.* New York: Addison Wesley

Minnichelli R.J., Anagnost J.J. and Desoer C.A. (1989) An elementary proof of Kharitomov's theory with extensions. *IEEE Trans. Auto. contr.* **AC-34**, 995–998

Mita T., Liu K.Z. and Ohuchi. S. (1993) Correction of the FI result in $\mathcal{H}_\infty$ control and parametrization of $\mathcal{H}_\infty$ state feedback controllers. *IEEE Trans. Auto. contr.* **AC-38**, 343–347

Moore B.C. (1981) Principal componet analysis in linear systems: Controllability, observability and model reduction. *IEEE Trans. Auto. contr.* **AC-26**, 17–32

Morari M. and Zafiriou E. (1989) *Robust Process Control.* Englewood Cliffs NJ: Prentice Hall

Nehari Z. (1957) On bounded bilinear forms. *Ann. of Math.* **Vol.(2) 65**, 153–162

Nett C.N., Jacobson C.A. and Balas M.J. (1984) A connection between state space and doubly coprime fractional representation. *IEEE Trans. Auto. contr.* **AC-29**, 831–832

Nyquist N. (1932) Regenerative theory. *Bell Syst. Tech. J.* January issue

Partington J.R. (1988) *An Introduction to Hankel operators.* London Mathematical Society Student Texts. **Vol 13**, Cambridge Unversity Press

Pernebo L. and Silverman L.M. (1982) Model reduction by balanced state space representations. *IEEE Trans. Auto. contr.* **AC-27**, 382–387

Potter J.E. (1966) Matrix quadratic solutions *SIAM J. Appl. Math.* **Vol. 14**, 496–501

Redheffer R. (1960) On a certain linear fractional transformation. *J. Math. Physics* **Vol. 39**, 269–286

Rosenbrock H.H. (1970) *State Space and Multivariable Theory,* London: Nelson

Rosenbrock H.H. (1974). *Computer-Aided Control System Design.* New York: Academic Press

Rudin W. (1974) *Real and Complex Analysis.* New York: McGraw Hill.

Rugh W.J. (1993). *Linear System Theory* Englewood Cliffs, N.J: Prentice Hall

Safonov M.G. and Athans M. (1977). Gain and phase margins of multi loop LQG regulators. *IEEE Trans. Auto. contr.* **AC-22**, 173–178

Safonov M.G. Jonckheere E.A. and Verma M. (1987) Synthesis of positive real multivariable feedback systems. *Int. J. Control* **vol. 45**, 817–842

Sarason D. (1967). Generalized interpolatian in $\mathcal{H}_\infty$. *Trans. Amer. Math. Society* **No. 127**, 179–203

Scherer C.W. (1995) Multi objective $\mathcal{H}_2/\mathcal{H}_\infty$ control. *IEEE Trans. Auto. contr.* **AC-40**, 1054–1062

Stein G. and Athans M. (1987). The LQG/LTR procedure for multivariable feedback control design. *IEEE Trans. Auto. contr.* **AC-32**, 105–114

Stewart G.W. (1973) *Introduction to Matrix Computations.* New York: Academic Press

Stoorvogel A. (1992) *The $\mathcal{H}_\infty$ Control Problem: A State Space Approach.* London: Prentice Hall

Strang G. (1976) *Linear Algebra and its Applications.* New York: Academic Press

Sznaier M. (1994) An exact solution to general SISO mixed $\mathcal{H}_2/\mathcal{H}_\infty$ problems via convex optimization. *IEEE Trans. Auto. contr.* **AC-39** 2511–2517

Vidyasagar M. (1985). *Control System Synthesis-A Factorization Approach.* Cambridge MA: MIT Press

Vidyasagar M. and Kimura H. (1986) Robust controllers for uncertain linear multivariable systems *Automatica* **vol. 22**. 85–94

Wiener N. (1949) *Extrapolation, Interpolation and Smoothing of Stationary Time Series With Engineering Applications.* New York: Wiley

Wonham W.M. (1968) On a matrix Riccati equation of stochastic control. SIAM. J. contr. **6**, 681–698

Youla D.C., Bongiorno J.J. and Lu C.N. (1974) Single loop feedback stabilization of linear multivariable plants. *Automatica* **Vol. 10**. 159–173.

Youla D.C., Bongiorno J.J. and Lu C.N. (1976a) Modern Wiener-Hopf design of optimal controllers, Part I: The single input case. *IEEE Trans. Auto. contr.* **AC-21**. 3–14

Youla D.C., Jabar H.A. and Bongiorno J.J (1976b) Modern Wiener-Hopf design of optimal controllers Part II: The multivariable case. *IEEE Trans. Auto. contr.* **AC-21** 319–338

Young N. (1988). *An Introduction to Hilbert space.* Cambridge: Cambridge University Press

Young S, and Francis B.A. (1985) Sensitivity trade offs for multivariable plants. *IEEE Trans. Auto. contr.* **AC-30**, 625–632

Zadeh L. and Desoer C.A. (1963) *Linear Systems-State Space Approach.* New York: McGraw Hill

Zames G. (1981) Feedback and optimal sensitivity: Model reference transformations, multiplicative semi-norms and approximate inverses. *IEEE Trans. Auto. contr.* **AC-26** 301–320

Zames G. and Francis B.A. (1983) Feedback, minimax sensitivity and optimal robustness. *IEEE Trans. Auto. contr.* **AC-28** 585–601

Zhou K. (1992a) Comparison between $\mathcal{H}_2$ and $\mathcal{H}_\infty$ controllers. *IEEE Trans. Auto. contr.* **AC-37**, 1261–1265

Zhou K. (1992b) On the parameterization of $\mathcal{H}_\infty$ controllers. *IEEE Trans. Auto. contr.* **AC-37**. 1442– 1446

Zhou, K., Glover K., Bodenheimer B. and Doyle J.C. (1994) Mixed $\mathcal{H}_2$ and $\mathcal{H}_\infty$ performance objective I: Robust performance analysis. *IEEE Trans. Auto. contr.* **AC-39**, 1564–1574

Index